Extinct and Endangere

in the collections of

The Natural History Museum

Alan G. Knox
&
Michael P. Walters

with the assistance of Philip Howes, Ian James,
Richard Stephenson and Susan Tyler

Published by
THE BRITISH ORNITHOLOGISTS' CLUB
Occasional Publications. No. 1
1994

The British Ornithologists' Club
% B.O.U.
The Walter Rothschild Zoological Museum
Akeman Street, Tring, Herts HP23 6AP

Registered Charity No. 279583

ISBN 0 9522886 0 5

British Library Cataloguing-in-Publication Data
A catalogue record for this book is available from the British Library

This volume was published by the British Ornithologists' Club with the help of a loan from the Bird Exploration Fund which sponsored the compilation of the catalogue.

The Bird Exploration Fund was established by trust deed in 1950 and is a registered charity No. 313019. The Fund is a small one but has over the years given assistance to various ornithological projects both at home and abroad, usually by supporting field studies. It has also provided aid for the curation and improvement of the ornithological research collections of The Natural History Museum, which was the original aim of the BEF. Further details of the Fund may be obtained from the Secretary, BEF, % The Walter Rothschild Zoological Museum, Akeman Street, Tring, Herts HP23 6AP.

Printed in Great Britain by Henry Ling Ltd at the Dorset Press, Dorchester, Dorset

EDITOR'S FOREWORD

For some time the British Ornithologists' Club has been considering assisting the publication of significant ornithological work which individuals or their parent bodies have found themselves regrettably unable to finance. This important compilation by Dr Alan Knox and Michael Walters comes emphatically into this category and it is a pleasure for the BOC to make this the first in a series of Occasional Publications which it is hoped will prove valuable to the ornithological world, its museums and institutions. It is a faithful response to one of the objects of the Club stated at the beginning of the rules, namely, 'to facilitate the publication of scientific information connected with ornithology'. This has been done for 100 years by publishing without interruption the *Bulletin of the British Ornithologists' Club* in several parts annually, together with supplementary volumes culminating in Volume 112a, a symposium on taxonomy and systematics with papers from leading specialists in this field worldwide, at the celebration of its Centenary in 1992.

The authors started this work while both were employed at The Natural History Museum, completing it subsequently. As Dr David Snow points out in his Preface, it is hoped the book will stimulate other institutions to follow their example. The authors, and the Bird Exploration Fund, who instigated the project, are to be congratulated on this important initiative.

James Monk
December 1993

EXTINCT AND ENDANGERED BIRDS

CONTENTS

PREFACE 1

INTRODUCTION 2
Acknowledgements 2

THE INFORMATION IN THE CATALOGUE 4
THE COLLECTIONS COVERED 4
CATALOGUING PROCEDURES 4

THE FORMS CONSIDERED IN THIS WORK 6
EXTINCT BIRDS 6
ENDANGERED BIRDS 6
SEQUENCE AND NOMENCLATURE 7
LIST OF EXTINCT AND ENDANGERED BIRDS OF THE WORLD . . 8

CATALOGUE OF EXTINCT AND ENDANGERED BIRDS 25
CONVENTIONS AND ABBREVIATIONS USED IN THE CATALOGUE 25
ACCOUNTS OF THE TAXA 29
SPECIES PRESUMED EXTINCT PRIOR TO 1600 277
ADDENDA 278

REFERENCES 279

INDEX 281

PREFACE

The relationship between scientific collecting and the extinction of animals, or the threat of their extinction, has inevitably become somewhat controversial as more and more animal populations have been reduced to low levels by man's activities. To what extent collecting has contributed to the loss of bird species in the past is debatable; certainly it has not been a significant factor in nearly all cases. As far as further collecting is concerned – and this is now of the greater importance – two extreme positions may be adopted. Some conservationists argue against any collecting at all, even if it is unlikely to have an adverse effect on the population. Others, usually academic biologists, say, in effect: this species is going to become extinct in any case, so let's at least have specimens and find out as much about it as we can before it goes for good. Most people, of course, take a position somewhere in between. Strong feelings are aroused, and rightly so, as our ability to alter whole ecosystems, and even the world itself, has increased our responsibility for other organisms far beyond what it once was.

Whatever arguments one may find most convincing with regard to future collecting, there can be no questioning the need to conserve as carefully as possible and obtain the greatest possible amount of information from specimens of extinct and threatened species that have already been collected. An important first step in so doing is to publish comprehensive, annotated lists of specimens in museum collections, as I argued at the Hague in 1970 (Snow, 1972, pp. 21–2 in *Proceedings XV International Ornithological Congress*, Leiden). Such lists could themselves be an aid to conservation. If it is known that an adequate series of specimens already exists and is available for examination, arguments for further collecting become very weak. This is especially true now that recent developments in DNA analysis make it possible to establish relationships between species on the basis of very small samples, taken from museum skins or without harming a live donor.

I welcome the publication of this book and hope that similar lists, covering other major bird collections, will follow soon.

D.W. SNOW

INTRODUCTION

Largely through the actions of man, many species of birds have become extinct in the recent past and more will shortly join them. It has been estimated that over 1000 species (11 % of the world's avifauna) are currently under threat (Collar & Andrew, 1988a). Of those that have already gone, some are known only from the paintings and writings of early travellers. For others, we are fortunate to have more tangible remains, in the form of skins, bones, eggs or even whole bodies, preserved in public museums and private collections.

The study of extinct species can be of both academic and practical value. From it we may learn not only about forms that are no more, but also lessons for the future – particularly those characteristics which make species vulnerable and, perhaps, how best we may help those that are currently threatened.

Some of the birds that are now most severely endangered were once much commoner and more widespread within the recent past. Others have always been rare or localised over the same period. The practical value of the data relating to these species may be more immediate than with the extinct birds. Details of former ranges, movements and habits may yet be of assistance in the conservation of living species and their habitats.

The bird collections of The Natural History Museum, London, held at Tring, are possibly the largest in the world and are exceptionally rich in historic material, much of it dating from the days of colonial exploration and the former British Empire. As such, they contain many specimens of extinct and endangered birds and information on these forms is frequently requested by ornithologists and conservationists throughout the world. The specimens are totally irreplaceable and are not normally available for loan by the Museum. This catalogue was produced to list the Museum's holdings of the relevant species and subspecies and to make the data associated with the specimens more widely accessible.

Acknowledgements

Compilation of this catalogue was sponsored by the Bird Exploration Fund. The initial data–gathering and computer–entry was carried out by staff provided by the Manpower Services Commission under a Community Programme run by the British Trust for Ornithology (BTO). Many of the bird skins, skeletons and anatomical specimens were surveyed by Ian James and Sue Tyler from November 1986 to October 1987, and the eggs by Philip Howes and Richard Stephenson from November 1987 to October 1988. Philip Howes and Richard Stephenson also assisted with other parts of the work. The Community Programme was administered by Jim Wolf and Sara Cross of the BTO. Mark Adams, Jo Bailey and Peter Colston of the Museum staff helped to gather and check data. Entries for skeletons and anatomical specimens were carefully examined by Graham Cowles and Philip Burton respectively. The Museum library staff was most helpful when we were checking the references and the unpublished paintings. Nigel

Collar and Jon Fjeldså advised on the list of taxa to be considered. Mr I.R. Bishop, Mr D.J. Brooks, Mr N.J. Collar, Mr G.S. Cowles, Dr R. Prŷs–Jones, Dr C. Violani and Mrs F.E. Warr read and commented on part or all of the manuscript. We are grateful to these people and to the many others who helped in a variety of ways, usually concerning individual specimens or collectors.

THE INFORMATION IN THE CATALOGUE

THE COLLECTIONS COVERED

This is a catalogue of the extinct and endangered birds in the collections of The Natural History Museum (NHM; formerly British Museum (Natural History)) at Tring, in Hertfordshire, U.K.. Also listed are relevant specimens in the public galleries of the Walter Rothschild Zoological Museum at Tring (part of the NHM) and the public galleries of The Natural History Museum at South Kensington in London. Not included are the holdings of birds in the Department of Palaeontology, also at South Kensington. Several original paintings of extinct and endangered birds are held in the Zoology Library of the Department of Library Services at South Kensington and at Tring. Those that came to our attention (many of them constituting the only known 'specimens' of the relevant taxa) have been listed here but no systematic search for others was undertaken. For example, most of the original plates for Lord Rothschild's monograph on extinct birds (Rothschild, 1907) are at Tring but have not been listed in this catalogue.

In the Tring bird collections, representatives of any particular species may be found in any or all of the main sections (spirit, egg, nest, skeleton, skin; Knox & Walters, 1992). Within each of these areas there are usually a number of special collections. Although most of these were surveyed for the catalogue, it was not possible to do so for a few (including some containing old specimens). Among the eggs, most of the passerines are only roughly sorted and are difficult to use. In addition, an estimated 30,000–40,000 eggs remain unincorporated, but these are unlikely to comprise much relevant material. The NHM holds over one million bird skins, one million eggs, 13,000 spirit specimens and 8,000 skeletons. While it is hoped that this catalogue is reasonably complete, we have undoubtedly overlooked some specimens that should have been included.

CATALOGUING PROCEDURES

The initial data–gathering for this catalogue was done from the labels on the specimens themselves except for some of the eggs, for which there was an existing card index. Information was copied onto data–sheets from which it was keyed into a computer database. Two–thirds of the data were gathered by non–specialist staff: printouts of this section were checked against the original specimens by Museum curators. The remaining third of the data was extracted by Museum staff (mainly by MPW) and printouts of this were checked against the data–sheets. For the part of the egg collection for which there were card entries, data were keyed from, and checked against, the cards.

The other egg entries (concerning most of the passerines and some of the near-passerines) were done from the eggs and associated data-slips.

The registers were searched for additional information on each entry, apart from the eggs which had been card-indexed and for which this was assumed to have been done already. As the entry for each specimen was examined, adjacent sections were scanned to look for registered specimens which had not been found in the collection, but the registers were not searched line-by-line for entries relating to other missing specimens. Neither was it possible to cross-check all the specimens against the registers. In particular, tens of thousands of skins have never been registered. In other cases, specimens had been dealt with in blocks – for instance, a large part of the great Shelley collection occupies only seven lines in the register.

The monumental, 27 volume, *Catalogue of Birds in the British Museum* (Sharpe and others), which appeared between 1874 and 1898, contains details of the specimens in the collections at the time each group was revised for publication. This work was consulted for each species and subspecies, frequently revealing further information on extant specimens and indicating where others had since disappeared, been reidentified or exchanged with other museums. Suitably annotated catalogue entries have been made here for all missing specimens using the data available in the registers, the *Catalogue of Birds* or other sources. A number of eviscerated spirit specimens are listed for the collection by mistake in *Bull. Br. Mus. nat. Hist. (Zool.)* 34 (1979): 125–180: most of these[1] do not exist. Additional information about many specimens has been drawn from published works and unpublished material within the Museum and elsewhere. The search for this was by no means exhaustive.

The quality of the data associated with many of the specimens, particularly the older ones, is very poor and often contradictory. Individual cases frequently required hours of work to research their background. It was clearly impracticable to spend this amount of time on each specimen and we are intensely aware of the shortcomings of this cataloguc and somc of the potential sources of error and omission. We hope they do not detract from the general usefulness of the book.

For further information relating to the interpretation of data associated with specimens listed here see Knox & Walters (1992).

[1] *Mergus australis, Nesoclopeus poeciloptera, Porzanula palmeri, Pennula millsi, P. sandwichensis, Pareudiastes pacificus, Numenius borealis, Ectopistes migratorius, Conuropsis carolinensis, Acrocephalus familiaris, Psittirostra kona, Ciridops anna, Aplonis corvina, Heterolocha acutirostris.*

THE FORMS CONSIDERED IN THIS WORK

There are many species of birds that will never again be seen alive. Many taxa are also now threatened in their natural habitats. *For the purposes of this volume we have examined the Museum's collections only for those forms that have become extinct in recent times together with the most highly endangered extant forms*. These are given below in a combined list, as it is often difficult to establish whether or not a form is truly extinct. Several species have recently been rediscovered many decades after they had last been seen. For others, no recent surveys have been undertaken to determine their present status.

EXTINCT BIRDS

The list includes all species and subspecies believed to have become extinct since 1600. Each is known from specimens (skins, eggs or bones of subfossil or recent origin) or from apparently reliable paintings and/or travellers' accounts. Other forms, based either on less reliable evidence, or on specimens whose status and identity is more debatable, are not included. Dates of extinction (often approximate) are indicated against the extinct taxa. For completeness, a few taxa which became extinct before 1600 are considered (and listed at the end) if specimens exist in the collections examined.

This portion of the list is based on King (1978–1979), Orenstein (1985), Diamond (1987), Fuller (1987) and Stattersfield (1988) with modifications. We have included the names of a few species and subspecies whose taxonomic status is unclear but which may deserve further study. A number of these are based on single museum specimens.

ENDANGERED BIRDS

This portion of the list includes those taxa categorised as endangered by King (1978–1979) and Collar & Stuart (1985). A few African species listed by King have been omitted as they are now believed to be less threatened than was originally thought (Collar & Stuart 1985). One South American species (*Eulidia yarrellii*) was deleted from the list and one subspecies (*Geositta excelsior aricomae*) was added on the recommendation of Jon Fjeldså (pers. comm., 1987). Also included are some African species which may be extinct (Collar & Stuart 1985, p. 702). Nigel Collar kindly provided a manuscript list of additional species considered to be most threatened in the Americas (Collar in litt., November 1987). A few names have been removed from the published lists as they are now known to be synonyms, to represent hybrids, or otherwise to be invalid.

The important work by Collar & Andrew (1988a) includes many more species thought to be under threat than we have considered here, but their status was not categorised and too many forms were listed for cataloguing to have been practicable in the time available. Collar & Andrew (1988b) included categories but this only became available as the manuscript for the systematic list was nearing completion. The endangered taxa listed here and covered in the systematic accounts thus omit many forms now known to be severely endangered, and include a small number now thought to be less at risk.

Where every race of a polytypic species is endangered they are not listed separately.

SEQUENCE AND NOMENCLATURE

With only minor, unavoidable exceptions, we have followed the sequence and nomenclature of Morony *et al.* (1975) for recent species and Peters (1934–1987) for subspecies. These works are used for curation within The Natural History Museum, as they are in many large collections. Although scientific names may change from time to time they should easily be traced through these publications. Taxa not included in Morony *et al.* or Peters have been inserted in alphabetical order at the end of the relevant species (for unlisted subspecies), genus (for species) or family (for genera). Where possible, the names of sub-fossil forms follow Brodkorb (1963–1978).

Conventions and abbreviations used in the list:

*	taxa for which entries will be found in the catalogue
(1952)	approximate date of extinction
[1952]	bird perhaps still extant
1959–1982	bird became extinct sometime between given dates
1864,1970	last definite record, possible later record
>	became extinct sometime after the given date
<	became extinct sometime before the given date
?	date not known
AO	Atlantic Ocean
C	century (e, early; m, mid)
I(s)	Island(s)
IO	Indian Ocean
NZ	New Zealand
PO	Pacific Ocean

All taxa are known from recent specimens unless the evidence is otherwise indicated under **B/T**:

B	known from bones
T	known from travellers' accounts and/or paintings

LIST OF EXTINCT AND ENDANGERED BIRDS OF THE WORLD

Species/subspecies (* taxa with entries in this catalogue)	Last date if extinct	B/T	Range or former range
STRUTHIONIDAE			
**Struthio camelus syriacus* Rothschild	m20th C		Middle East
**Struthio indicus* Bidwell	?		India
**Psammornis rothschildi* Andrews	?		Algeria
RHEIDAE			
**Rhea nana* Lydekker	?		S. America
**Pterocnemia pennata tarapacensis* Chubb			S. America
DROMAIIDAE			
**Dromaius novaehollandiae diemenensis* Le Souef	e19th C		Tasmania
**Dromaius diemenianus* (Jennings)	>1802		Kangaroo I, Australia
Dromaius minor (Spencer)	>1802		King I, Australia
TINAMIDAE			
Tinamus solitarius pernambucensis Berla			Brazil
Crypturellus saltuarius Wetmore	>1943		Colombia

Additional Ratite species are given at the end of the list.

Species/subspecies	Last date if extinct	B/T	Range or former range
PODICIPEDIDAE			
**Tachybaptus rufolavatus* (Delacour)			Madagascar
**Podilymbus gigas* Griscom	1987–1991		L. Atitlan, Guatemala
Podiceps nigricollis andinus Meyer de Schauensee	1977		Colombia
**Podiceps taczanowskii* Berlepsch & Stolzmann			Peru
DIOMEDEIDAE			
**Diomedea albatrus* Pallas			Izu Is, n.w. PO
Diomedea amsterdamensis Roux *et al.*			Amsterdam I, IO
PROCELLARIIDAE			
Pterodroma aterrima (Bonaparte)			Mascarene Is
**Pterodroma hasitata caribbaea* Carte	(1880)		Jamaica, W. Indies
**Pterodroma cahow* (Nichols & Mowbray)			Bermuda, AO
**Pterodroma magentae* (Giglioli & Salvadori)			Chatham Is, New Zealand

Species/subspecies (* taxa with entries in this catalogue)	Last date if extinct	B/T	Range or former range
**Pterodroma mollis madeira* Mathews			Madeira, AO
**Pterodroma phaeopygia* (Salvin)			Galapagos & Hawaiian Is
**Pterodroma cookii* (G.R. Gray)			New Zealand
**Pterodroma axillaris* (Salvin)			Chatham Is, New Zealand
**Procellaria parkinsoni* G.R. Gray			New Zealand
HYDROBATIDAE			
**Oceanodroma macrodactyla* Bryant	(1912)		Guadalupe I, Mexico
SULIDAE			
**Sula abbotti* Ridgway			Christmas I, IO
PHALACROCORACIDAE			
**Phalacrocorax perspicillatus* Pallas	(1852)		Bering I, n. PO
ARDEIDAE			
**Nycticorax caledonicus crassirostris* Vigors	(1889)		Bonin Is, n.w. PO
Nycticorax mauritianus (Newton & Gadow)	17th C	B	Mauritius, Mascarene Is
Nycticorax megacephalus (Milne-Edwards)	<1761	B	Rodriguez, Mascarene Is
CICONIIDAE			
**Ciconia ciconia boyciana* Swinhoe			e. Asia
THRESKIORNITHIDAE			
**Geronticus eremita* (Linnaeus)			s.w. Palearctic
**Nipponia nippon* (Temminck)			e. Asia
Lampribis olivacea rothschildi Bannerman	1901		Principe I, e. AO
Lampribis olivacea bocagei Chapin			Sao Tome, e. AO
Borbonibis latipes Mourer-Chauvire & Moutou	?	B	Reunion, Mascarene Is
ANATIDAE			
Cygnus sumnerensis (Forbes)	1590-1690	B	NZ & Chatham I
Alopochen mauritianus (Newton & Gadow)	1693-1698	BT	Mauritius, Mascarene Is
**Tadorna cristata* (Kuroda)			e. Siberia, China
Anas strepera couesi (Streets)	(1876-1926)		Fanning Group, PO
Anas gibberifrons remissa Ripley	1959		Rennell I, Solomon Is
Anas georgica niceforoi Wetmore & Borrero	1952		Colombia
Anas theodori Newton & Gadow	1696	BT	Mauritius, Reunion, Mascarene Is
**Rhodonessa caryophyllacea* (Latham)	[1944]		India

Species/subspecies (* taxa with entries in this catalogue)	Last date if extinct	B/T	Range or former range
**Aythya innotata* (Salvadori)			Madagascar
**Camptorhynchus labradorius* (Gmelin)	1875		N. America
**Mergus australis* Hombron & Jacquinot	1905		Auckland Is, New Zealand
CATHARTIDAE			
Sarcoramphus sacer (Bartram)	>1741	T	Florida, ?S. Carolina, USA
**Gymnogyps californianus* (Shaw)			w. North America
ACCIPITRIDAE			
**Chondrohierax uncinatus mirus* Friedmann			Grenada, W. Indies
**Chondrohierax uncinatus wilsonii* (Cassin)			Cuba, W. Indies
**Haliaeetus vociferoides* Des Murs			Madagascar
**Haliaeetus leucocephalus leucocephalus* (Linnaeus)			N. America
**Haliaeetus pelagicus niger* Heude	[1887]		Korea
**Eutriorchis astur* Sharpe			Madagascar
Circus alphonsi (Newton & Gadow)	17th C	B	Mauritius, Mascarene Is
Accipiter princeps Mayr	(1933)		New Britain, Bismarck Archipelago
**Accipiter francesii pusillus* (Gurney)			Comoro Is, IO
Accipiter gundlachii Lawrence			Cuba, W. Indies
**Pithecophaga jefferyi* Ogilvie-Grant			Philippine Is
**Aquila heliaca adalberti* Brehm			s.w. Palearctic
FALCONIDAE			
**Polyborus lutosus* Ridgway	1900		Guadalupe I, Mexico
**Falco punctatus* Temminck			Mauritius, Mascarene Is
**Falco peregrinus tundrius* White			n. N. America
**Falco peregrinus anatum* Bonaparte			n. N. America
MEGAPODIIDAE			
**Megapodius andersoni* Gray	19th C		New Caledonia, PO
**Megapodius burnabyi* Gray	19th C		Hapace I, Tonga Is, PO
**Megapodius stairi* Gray	19th C		Samoa, PO
CRACIDAE			
**Ortalis erythroptera* Sclater & Salvin			Ecuador
**Penelope albipennis* Taczanowski			Peru
**Penelope perspicax* Bangs			Colombia
Aburria pipile pipile (Jacquin)			Trinidad, n. S. America
**Aburria jacutinga* (Spix)			Brazil

Species/subspecies (* taxa with entries in this catalogue)	Last date if extinct	B/T	Range or former range
**Oreophasis derbianus* G.R. Gray			Central America
**Crax mitu mitu* Linnaeus			Brazil
**Crax globulosa* Spix			S. America
**Crax blumenbachii* Spix			Brazil
PHASIANIDAE			
**Tympanuchus cupido cupido* (Linnaeus)	1932		N. America
**Dendrortyx barbatus* Gould			Mexico
Colinus virginianus insulans Howe	>1888		Florida, USA
**Colinus virginianus ridgwayi* Brewster			N. America
**Odontophorus strophium* (Gould)			Colombia
Francolinus ochropectus Dorst & Jouanin			Somalia
**Perdix perdix italica* Hartert			Italy
**Coturnix novaezelandiae* Quoy & Gaimard	1875		New Zealand
**Ophrysia superciliosa* (J.E. Gray)	[1868,1876]		n.w. Himalayas
**Tragopan melanocephalus* (J.E. Gray)			w. Himalayas
**Tragopan caboti* (Gould)			China
**Lophophorus lhuysii* Saint-Hilaire			China
**Crossoptilon mantchuricum* Swinhoe			China
**Catreus wallichii* (Hardwicke)			w. Himalayas
**Syrmaticus ellioti* (Swinhoe)			China
**Argusianus bipunctatus* (Wood)	19th C		?Tioman I, e. Malaya
**Agelastes meleagrides* Bonaparte			West Africa
GRUIDAE			
**Grus americana* (Linnaeus)			N. America
**Grus leucogeranus* Pallas			Siberia
RALLIDAE			
**Rallus elegans levipes* Bangs			N. America
**Rallus antarcticus* King			S. America
**Rallus philippensis macquariensis* Hutton	1880–1894		Macquarie I, New Zealand
**Rallus wakensis* (Rothschild)	1945		Wake Is, PO
**Rallus pacificus* Gmelin	18th C		Tahiti, PO
**Nesolimnas dieffenbachii* (G.R. Gray)	>1842		Chatham Is, New Zealand
**Cabalus modestus* (Hutton)	(1895–1900)		Chatham Is, New Zealand
Atlantisia elpenor Olson	>1656,19th C	B	Ascension I, AO
**Atlantisia podarces* (Wetmore)	?	B	St Helena, AO
**Tricholimnas lafresnayanus* (J. Verreaux & Des Murs)	[1904]		New Caledonia, PO
**Tricholimnas sylvestris* (Sclater)			Lord Howe I, Australia

Species/subspecies (* taxa with entries in this catalogue)	Last date if extinct	B/T	Range or former range
**Dryolimnas cuvieri abbotti* (Ridgway)	<1937		Assumption I, Seychelles, IO
**Amaurolimnas concolor concolor* (Gosse)	(1890)		Jamaica, W. Indies
**Aramides gutturalis* Sharpe	(>1943)		S. America
**Nesoclopeus poeciloptera* (Hartlaub)	(1890),1973		Fiji, PO
**Gymnocrex intactus* (Sclater)	1869		Solomon Is, PO
**Porzanula palmeri* Frohawk	1944		Laysan I, Hawaiian Is
**Pennula millsi* Dole	1884–1893		Hawaii, Hawaiian Is
**Pennula sandwichensis* (Gmelin)	1884–1893		Hawaii, Hawaiian Is
**Nesophylax niger* (Miller)	18th C	T	Tahiti, PO
Aphanolimnas monasa (Kittlitz)	(1828–1880)		Kusai I, Caroline Is, PO
**Sarothrura watersi* (Bartlett)			Madagascar
**Poliolimnas cinereus brevipes* (Ingram)	1911,1925		Iwo Jimo, Volcano Is, PO
**Gallinula chloropus sandvicensis* Streets			Hawaiian Is
**Porphyriornis nesiotis* (Sclater)	(1872)		Tristan da Cunha, AO
**Pareudiastes pacificus* Hartlaub & Finsch	1873		Samoan Is, PO
Porphyrio albus (White)	(1834)		Lord Howe I, Australia
**Notornis mantelli* Owen			New Zealand
Fulica newtoni Milne-Edwards	<1700	B	Mauritius, Reunion, Mascarene Is
Aphanapteryx bonasia (Selys-Longchamps)	(1693–1700)	T	Mauritius, Mascarene Is
Aphanapteryx leguati (Milne-Edwards)	(1730)	BT	Rodriguez, Mascarene Is
Nesotrochis debooyi Wetmore	?	B	W. Indies
Stictolimnas sharpii Büttikofer	18–19th C		unknown
RHYNOCHETIDAE			
**Rhynochetos jubatus* Verreaux & Des Murs			New Caledonia, PO
OTIDIDAE			
**Choriotis nigriceps* (Vigors)			India
HAEMATOPODIDAE			
**Haematopus ostralegus chathamensis* Hartert			Chatham Is, New Zealand
**Haematopus moquini meadewaldoi* Bannerman	[20th C]		Canary Is, AO
RECURVIROSTRIDAE			
**Himantopus novaezelandiae* Gould			New Zealand

Species/subspecies (* taxa with entries in this catalogue)	Last date if extinct	B/T	Range or former range
GLAREOLIDAE			
**Rhinoptilus bitorquatus* (Blyth)			India
CHARADRIIDAE			
**Vanellus macropterus* (Wagler)	[1940]		Indonesia
**Charadrius novaeseelandiae* Gmelin			New Zealand
SCOLOPACIDAE			
**Numenius borealis* (J.R. Forster)			N. America
**Prosobonia leucoptera* (Gmelin)	(1777)		Society Is, PO
**Coenocorypha aucklandica iredalei* Rothschild	1964		New Zealand
Coenocorypha aucklandica barrierensis Oliver	1870		New Zealand
Calidris cooperi (Baird)	1833		N. America
LARIDAE			
Sterna albifrons browni Mearns			North & Central America
ALCIDAE			
**Pinguinus impennis* (Linnaeus)	1844		n. AO
RAPHIDAE			
**Raphus cucullatus* (Linnaeus)	>1681		Mauritius, Mascarene Is
Raphus solitarius (Selys-Longchamps)	(1710)	T	Reunion, Mascarene Is
**Pezophaps solitaria* (Gmelin)	(1760)		Rodriguez, Mascarene Is
COLUMBIDAE			
Columba palumbus maderensis Tschusi	1904		Madeira, AO
**Columba vitiensis godmanae* (Mathews)	(1800)	T	Lord Howe I, Australia
**Columba versicolor* Kittlitz	1889		Bonin Is, n.w. PO
**Columba jouyi* (Stejneger)	[1936]		Ryukyu & Borodino Is, n.w. PO
Columba inornata wetmorei Peters			Puerto Rico, W. Indies
**Columba mayeri* Prevost			Mauritius, Mascarene Is
**Streptopelia picturata coppingeri* (Sharpe)	1884–1906		Glorioso I, IO
**Streptopelia picturata rostrata* (Bonaparte)	1975		Seychelles, IO
**Ectopistes migratorius* (Linnaeus)	1914		N. America
**Zenaida macroura graysoni* (Lawrence)			Socorro I, Mexico
**Leptotila wellsi* (Lawrence)			Grenada, W. Indies
**Leptotila ochraceiventris* Chapman			Ecuador & Peru

Species/subspecies (* taxa with entries in this catalogue)	Last date if extinct	B/T	Range or former range
**Caloenas nicobarica pelewensis* Finsch			Palau Is, Caroline Is, PO
**Gallicolumba ferruginea* (Wagler)	>1774	T	Tanna I, PO
Gallicolumba norfolciensis (Latham)	>1801	T	Norfolk I, Australia
**Microgoura meeki* Rothschild	[1904,1965]		Choiseul, Solomon Is, PO
Phapitreron amethystina frontalis (Bourns & Worcester)	1892		Cebu, Philippine Is
Ptilinopus mercierii mercierii (Des Murs & Prevost)	1849		Nukuhiva, Marquesas, PO
**Ptilinopus mercierii tristrami* (Salvadori)	1922		Hivaoa, Marquesas, PO
Alectroenas rodericana (Milne-Edwards)	(1730)	B	Rodriguez, Mascarene Is
Alectroenas nitidissima (Scopoli)	(1830)		Mauritius, Mascarene Is
Ducula oceanica teraokai (Momiyama)			Caroline Is, PO
**Ducula galeata* (Bonaparte)			Nukuhiva, Marquesas, PO
**Hemiphaga novaeseelandiae spadicea* (Latham)	(1833–1848)		Norfolk I, Australia
**Hemiphaga novaeseelandiae chathamensis* (Rothschild)			Chatham Is, New Zealand
LORIIDAE			
Chalcopsitta atra spectabilis van Oort	(1908)		New Guinea
Eos goodfellowi Ogilvie-Grant	(1907)		Obi I, Moluccas, E. Indies
Lorius amabilis Stresemann	>1931		New Britain, Bismarck Archipelago
**Lorius tibialis* Sclater	>1871		unknown (India)
Charmosyna toxopei (Siebers)	[1931]		Buru I, Moluccas, E. Indies
Charmosyna diadema (J. Verreaux & Des Murs)	1860,1970		New Caledonia, s.w. PO
PSITTACIDAE			
**Nestor productus* (Gould)	(1851)		Norfolk & Philip Is, Australia
**Psephotus pulcherrimus* (Gould)			Australia
**Cyanoramphus novaezelandiae subflavescens* Salvadori	1869		Lord Howe I, Australia
**Cyanoramphus novaezelandiae cookii* (G.R. Gray)			Norfolk I, Australia
**Cyanoramphus novaezelandiae erythrotis* (Wagler)	(1914)		Macquarie Is, New Zealand
**Cyanoramphus auriceps forbesi* Rothschild			Chatham Is, New Zealand
**Cyanoramphus malherbi* Souance			New Zealand

Species/subspecies (* taxa with entries in this catalogue)	Last date if extinct	B/T	Range or former range
**Cyanoramphus zealandicus* (Latham)	1844		Tahiti, PO
**Cyanoramphus ulietanus* (Gmelin)	>1773		Raiatea, Society Is, PO
**Eunymphicus cornutus uvaeensis* (Layard & Layard)			Uvea, Loyalty Is, PO
**Pezoporus wallicus flaviventris* North			Australia
**Lophopsittacus mauritianus* (Owen)	>1638	BT	Mauritius, Mascarene Is
Lophopsittacus bensoni Holyoak	(1764)	B	Mauritius, ?Reunion, Mascarene Is
Necropsittacus rodericanus (Milne-Edwards)	(1760)	BT	Rodriguez, Mascarene Is
Mascarinus mascarinus (Linnaeus)	(1834)		Reunion, Mascarene Is
**Coracopsis nigra barklyi* E. Newton			Seychelles, IO
**Loriculus philippensis chrysonotus* Sclater	1906		Cebu, Philippine Is
**Loriculus philippensis siquijorensis* Steere	1908		Siquijor, Philippine Is
**Loriculus flosculus* Wallace	>1862		Flores I, Indonesia
**Loriculus salvadorii* Hachisuka	1929		Mindanao, Philippine Is
**Psittacula wardi* (E. Newton)	1870–1906		Seychelles, IO
**Psittacula eques echo* (Newton & Newton)			Mauritius, Mascarene Is
Psittacula eques eques (Boddaert)	(<1830)		Reunion, Mascarene Is
Psittacula exsul (A. Newton)	1875		Rodriguez, Mascarene Is
**Psittacula intermedia* (Rothschild)			India
**Anodorhynchus glaucus* (Vieillot)	1955		Brazil & Uruguay
**Anodorhynchus leari* Bonaparte			Brazil
**Cyanopsitta spixii* (Wagler)			Brazil
**Ara caninde* (Wagler)			Bolivia
**Ara tricolor* Bechstein	1864,1885		Cuba & Isle of Pines, W. Indies
Aratinga chloroptera maugei (Souance)	>1892		Mona I & Puerto Rico, W. Indies
**Ognorhynchus icterotis* (Massena & Souance)			Colombia & Ecuador
Rhynchopsitta pachyrhyncha terrisi Moore			Mexico
**Conuropsis carolinensis carolinensis* (Linnaeus)	1914		N. America
**Conuropsis carolinensis ludovicianus* (Gmelin)	1912		N. America
Pyrrhura albipectus Chapman			Ecuador
Bolborhynchus ferrugineifrons (Lawrence)			Colombia
**Amazona vittata vittata* (Boddaert)			Puerto Rico, W. Indies
Amazona vittata gracilipes Ridgway	1899		Culebra I, W. Indies
**Amazona pretrei* (Temminck)			Brazil & Argentina

Species/subspecies (* taxa with entries in this catalogue)	Last date if extinct	B/T	Range or former range
**Amazona brasiliensis* (Linnaeus)			Brazil
**Amazona dufresniana rhodocorytha* (Salvadori)			Brazil
**Amazona vinacea* (Kuhl)			S. America
**Amazona versicolor* (P.L.S. Muller)			St Lucia, W. Indies
**Amazona arausiaca* (P.L.S. Muller)			Dominica, W. Indies
**Amazona guildingii* (Vigors)			St Vincent, W. Indies
**Amazona imperialis* Richmond			Dominica, W. Indies
**Strigops habroptilus* G.R. Gray			New Zealand
MUSOPHAGIDAE			
**Tauraco bannermani* (Bates)			Cameroon
CUCULIDAE			
**Neomorphus geoffroyi dulcis* Snethlage			Brazil
**Neomorphus radiolosus* Sclater & Salvin			Colombia & Ecuador
**Coua delalandei* (Temminck)	[1930]		Madagascar
TYTONIDAE			
**Tyto soumagnei* (Milne-Edwards)			Madagascar
STRIGIDAE			
**Otus ireneae* Ripley			Kenya
**Otus rutilus capnodes* (Gurney)			Anjouan, Comoro Is, IO
Otus commersoni (Oustalet)	1837	T	Mauritius, Mascarene Is
Otus scops botelensis Kuroda			Botel Tobago I, Philippine Is
**Ninox novaeseelandiae albaria* Ramsay	1940		Lord Howe I, Australia
'Ninox'/'Tyto' murivora (Milne-Edwards)	1726	BT	Rodriguez, Mascarene Is
Sceloglaux albifacies rufifacies Buller	1889		North I, New Zealand
**Sceloglaux albifacies albifacies* (G.R. Gray)	[1950]		South I, New Zealand
**Athene blewitti* (Hume)	[(1914,1968)]		central India
**Speotyto cunicularia amaura* Lawrence	19th C		W. Indies
Speotyto cunicularia guadeloupensis Ridgway	19th C		Marie Galante or Guadeloupe, W. Indies
AEGOTHELIDAE			
Aegotheles savesi Layard & Layard	1881		New Caledonia

Species/subspecies (* taxa with entries in this catalogue)	Last date if extinct	B/T	Range or former range
CAPRIMULGIDAE			
**Siphonorhis americanus* (Linnaeus)	1859		Jamaica, W. Indies
**Eleothreptus anomalus* (Gould)			S. America
TROCHILIDAE			
**Glaucis dohrnii* (Bourcier & Mulsant)			Brazil
Phaethornis margarettae Ruschi			n. S. America
Chlorostilbon ricordii bracei (Lawrence)	>1877		Bahama Is, W. Indies
Lepidopyga lilliae Stone			Colombia
**Eupherusa poliocerca* Elliot			Mexico
Eupherusa cyanophrys Rowley & Orr			Mexico
**Hylonympha macrocerca* Gould			Venezuela
**Eriocnemis nigrivestis* (Bourcier & Mulsant)			Ecuador
**Eriocnemis godini* (Bourcier)			Ecuador & Colombia
**Acestrura bombus* (Gould)			Ecuador & Peru
Acestrura berlepschi (Simon)			Ecuador
ALCEDINIDAE			
**Halcyon cinnamomina cinnamomina* Swainson			Guam, Mariana Is, PO
Halcyon gambieri gambieri Oustalet	1841–1922		Mangareva, Tuamotu Is, PO
Halcyon miyakoensis Kuroda	1887		Miyakoshima, Ryukyu Is, n.w. PO
MOMOTIDAE			
**Electron carinatum* (Du Bus)			Central America
GALBULIDAE			
**Jacamaralcyon tridactyla* (Vieillot)			Brazil
PICIDAE			
**Colaptes auratus rufipileus* Ridgway	1906		Guadalupe I, Mexico
**Dryocopus galeatus* (Temminck)			S. America
**Dryocopus javensis richardsi* Tristram			Korea
Campephilus principalis bairdii Cassin			Cuba, W. Indies
**Campephilus principalis principalis* (Linnaeus)			USA
**Campephilus imperialis* (Gould)	[1958]		Mexico
**Sapheopipo noguchii* (Seebohm)			Okinawa, Ryukyu Is, n.w. PO

Species/subspecies (* taxa with entries in this catalogue)	Last date if extinct	B/T	Range or former range
FURNARIIDAE			
**Geositta excelsior aricomae* (Carriker)			Peru & Bolivia
Leptasthenura pileata xenothorax Chapman			Peru
Synallaxis courseni Blake			Peru
Synallaxis cherriei Gyldenstolpe			e. S. America
Megaxenops parnaguae Reiser			Brazil
FORMICARIIDAE			
**Myrmotherula erythronotos* (Hartlaub)			Brazil
**Pyriglena atra* (Swainson)			Ecuador
**Grallaria gigantea* Lawrence			Ecuador & Colombia
Grallaria alleni Chapman			Colombia
Grallaria chthonia Wetmore & Phelps			Venezuela
**Grallaria rufocinerea* Sclater & Salvin			Colombia
**Grallaria milleri* Chapman	[1911]		Colombia
COTINGIDAE			
**Iodopleura pipra* (Lesson)			Brazil
TYRANNIDAE			
Empidonax euleri johnstonei Barbour	1910		Grenada, W. Indies
Ceratotriccus furcatus (Lafresnaye)			Brazil
Anairetes alpinus (Carriker)			w. S. America
XENICIDAE			
**Xenicus longipes stokesii* G.R. Gray	1955		New Zealand
**Xenicus longipes longipes* (Gmelin)	1972		New Zealand
**Xenicus longipes variabilis* Stead	[1967]		New Zealand
**Xenicus lyalli* (Rothschild)	1894		New Zealand
ATRICHORNITHIDAE			
**Atrichornis clamosus* (Gould)			Australia
ALAUDIDAE			
**Calandrella razae* (Alexander)			Razo, Cape Verde Is
MOTACILLIDAE			
Anthus chacoensis Zimmer			S. America
CAMPEPHAGIDAE			
**Coracina striata cebuensis* (Ogilvie-Grant)	1906		Cebu, Philippine Is

Species/subspecies (* taxa with entries in this catalogue)	Last date if extinct	B/T	Range or former range
**Coracina coerulescens altera* (Ramsay)	1906		Cebu, Philippine Is
**Lalage leucopyga leucopyga* (Gould)	1962		Norfolk I, Australia
PYCNONOTIDAE			
Hypsipetes siquijorensis monticola (Bourns & Worcester)	1906		Cebu, Philippine Is
LANIIDAE			
Lanius ludovicianus mearnsi Ridgway			N. America
**Lanius newtoni* Bocage			Sao Tome, e. AO
CINCLIDAE			
**Cinclus cinclus olympicus* Madarasz	1939		Cyprus
TROGLODYTIDAE			
Salpinctes obsoletus exsul Ridgway	1952		Revillagigedo Is, Mexico
**Hylorchilus sumichrasti* (Lawrence)			Mexico
**Thryomanes bewickii leucophrys* (Anthony)	1968		California, USA
**Thryomanes bewickii brevicauda* Ridgway	1897		Guadalupe I, Mexico
**Ferminia cerverai* Barbour			Cuba, W. Indies
Thryothorus nicefori Meyer de Schauensee			Colombia
Troglodytes troglodytes orii Yamashina	1938		Borodino Is, n.w. PO
**Troglodytes aedon guadeloupensis* (Cory)			Guadeloupe I, W. Indies
**Troglodytes aedon martinicensis* (Sclater)	1886		Martinique, W. Indies
**Troglodytes aedon mesoleucus* (Sclater)			St Lucia, W. Indies
MIMIDAE			
**Mimodes graysoni* (Lawrence)			Socorro I, Mexico
**Cinclocerthia ruficauda gutturalis* (Lafresnaye)			Martinique, W. Indies
**Ramphocinclus brachyurus* (Vieillot)			W. Indies
MUSCICAPIDAE			
Erithacus komadori subrufus (Kuroda)			Ryukyu Is, n.w. PO
**Alethe choloensis* Sclater			Malawi & Mozambique
**Copsychus sechellarum* Newton			Seychelles, IO
**Copsychus niger cebuensis* (Steere)			Cebu, Philippine Is
Myadestes elisabeth retrusus Bangs & Zappey	1934		Isle of Pines, Cuba, W. Indies
**Saxicola dacotiae murielae* Bannerman	1913		Canary Is, AO
Zoothera terrestris (Kittlitz)	>1830		Bonin Is, n.w. PO

Species/subspecies (* taxa with entries in this catalogue)	Last date if extinct	B/T	Range or former range
**Cichlherminia lherminieri sanctaeluciae* (Sclater)			St Lucia, W. Indies
**Phaeornis obscurus myadestinus* Stejneger			Kauai, Hawaiian Is
**Phaeornis obscurus oahensis* Wilson & Evans	(1825)		Oahu, Hawaiian Is
**Phaeornis obscurus rutha* Bryan			Molokai, Hawaiian Is
**Phaeornis obscurus lanaiensis* Wilson	1933		Lanai, Hawaiian Is
**Phaeornis palmeri* Rothschild			Kauai, Hawaiian Is
**Turdus helleri* (Mearns)			Kenya
**Turdus poliocephalus pritzbueri* Layard	[?]		Lifu, Loyalty Is, PO
**Turdus poliocephalus mareensis* Layard & Tristram	1939		Mare, Loyalty Is, PO
**Turdus poliocephalus poliocephalus* Latham			Norfolk I, Australia
**Turdus poliocephalus vinitinctus* (Gould)	1918		Lord Howe I, Australia
Turdus celaenops yakushimensis Ogawa	1904		Yakushima, Ryukyu Is, n.w. PO
**Turdus ravidus* (Cory)	1938		Grand Cayman I, W. Indies
*'*Turdus' ulietensis* Gmelin	18th C	T	Raiatea, Society Is, PO
**Moupinia altirostris altirostris* (Jerdon)	1914		Burma
Cettia diphone restricta (Kuroda)	1922		Borodino Is, n.w. PO
Acrocephalus luscinia astrolabii Holyoak & Thibault	?		Micronesia, ?Yap I, PO
**Acrocephalus familiaris* (Rothschild)	1904–1923		Laysan I, Hawaiian Is
**Acrocephalus caffer garretti* Holyoak & Thibault	19th C		Huahine, Society Is, PO
Acrocephalus caffer longirostris (Gmelin)			Moorea, Society Is, PO
Acrocephalus caffer aquilonis (Murphy & Mathews)			Eiao, Marquesas Is, PO
**Acrocephalus caffer musae* (J.R. Forster)	18–19th C		Raiatea, Society Is, PO
**Bebrornis rodericanus* (A. Newton)			Rodriguez, Mascarene Is
**Nesillas aldabranus* Benson & Penny	[1983]		Aldabra, IO
**Sylvia melanocephala norissae* Nicoll	>1939		Faiyum, Egypt
**Amaurocichla bocagii* Sharpe			Sao Tome, e. AO
Bowdleria punctata wilsoni Stead			New Zealand
**Bowdleria punctata rufescens* (Buller)	1900		Chatham Is, New Zealand
Trichocichla rufa Reichenow			Fiji Is, PO
**Dasyornis broadbenti litoralis* (Milligan)			Australia
**Gerygone insularis* Ramsay	>1918		Lord Howe I, Australia
Newtonia fanovanae Gyldenstolpe			Madagascar
**Petroica traversi* (Buller)			Chatham Is, New Zealand
**Platysteira laticincta* Bates			Cameroon

Species/subspecies (* taxa with entries in this catalogue)	Last date if extinct	B/T	Range or former range
Eutrichomyias rowleyi (Meyer)	[>1878,1978]		Great Sangi I, Indonesia
**Pomarea nigra nigra* (Sparrman)			Tahiti, PO
Pomarea nigra pomarea (Garnot)	1823		Maupiti, Society Is, PO
**Pomarea mendozae mendozae* (Hartlaub)			Marquesas Is, PO
**Pomarea mendozae nukuhivae* Murphy & Mathews			Marquesas Is, PO
**Rhipidura fuliginosa cervina* Ramsay	1924–1928		Lord Howe I, Australia
**Turnagra capensis tanagra* (Schlegel)	1955		New Zealand
**Turnagra capensis capensis* (Sparrman)	1963		New Zealand
PARIDAE			
Parus varius orii (Kuroda)	1938		Borodino Is, n.w. PO
DICAEIDAE			
Melanocharis arfakiana (Finsch)	1900		New Guinea
**Dicaeum quadricolor* (Tweeddale)	1906,1992		Cebu, Philippine Is
**Dicaeum trigonostigma pallidius* Bourns & Worcester	1910		Cebu, Philippine Is
NECTARINIIDAE			
Nectarinia afra prigoginei (Macdonald)			Zaire
ZOSTEROPIDAE			
**Zosterops everetti everetti* Tweeddale	1906		Cebu, Philippine Is
**Zosterops luteirostris luteirostris* Hartert			Gizo, Solomon Is, PO
**Zosterops strenua* Gould	>1918		Lord Howe I, Australia
**Zosterops albogularis* Gould			Norfolk I, Australia
**Zosterops mayottensis semiflava* E. Newton	1888		Seychelles, IO
**Zosterops modesta* E. Newton			Seychelles, IO
**Rukia ruki* (Hartert)			Truk, Caroline Is, PO
MELIPHAGIDAE			
**Meliphaga cassidix* (Gould)			Australia
**Apalopteron familiare familiare* (Kittlitz)	[(1980)]		Bonin Is, n.w. PO
**Moho braccatus* (Cassin)			Kauai, Hawaiian Is
**Moho bishopi* (Rothschild)			Molokai, Hawaiian Is
**Moho apicalis* Gould	1837		Oahu, Hawaiian Is
**Moho nobilis* (Merrem)	1934		Hawaii, Hawaiian Is
Chaetoptila angustipluma (Peale)	1859		Hawaii, Hawaiian Is
**Anthornis melanura melanocephala* G.R. Gray	1906		Chatham Is, New Zealand

Species/subspecies (* taxa with entries in this catalogue)	Last date if extinct	B/T	Range or former range
EMBERIZIDAE			
Zonotrichia melodia amaka (Gabrielson & Lincoln)			Alaska, USA
Zonotrichia melodia graminea (Townsend)	1967		California, USA
**Ammodramus maritimus nigrescens* Ridgway	1987		Florida, USA
**Ammodramus baileyi* (Bangs)			Mexico
Amphispiza belli clementeae Ridgway			California, USA
**Poospiza rubecula* Salvin			Peru
Loxigilla portoricensis grandis Lawrence	>1929		St Kitts, W. Indies
Pipilo erythrophthalmus consobrinus Ridgway	1897		Guadalupe I, Mexico
Atlapetes flaviceps Chapman			Colombia
Nemosia rourei Cabanis			Brazil
**Dacnis albiventris* (Sclater)			S. America
**Dacnis berlepschi* Hartert			Colombia & Ecuador
PARULIDAE			
**Vermivora bachmanii* (Audubon)	[1988]		s.e. USA
**Dendroica petechia petechia* (Linnaeus)			Barbados, W. Indies
**Dendroica kirtlandii* (Baird)			USA
**Catharopeza bishopi* (Lawrence)			St Vincent, W. Indies
**Leucopeza semperi* Sclater	[1972]		St Lucia, W. Indies
**Myioborus pariae* Phelps & Phelps Jr.			Venezuela
DREPANIDIDAE			
**Loxops sagittirostris* (Rothschild)	1900		Hawaii, Hawaiian Is
**Loxops maculata maculata* (Cabanis)			Oahu, Hawaiian Is
**Loxops maculata flammea* Wilson	[1963]		Molokai, Hawaiian Is
**Loxops maculata montana* (Wilson)	1937		Lanai, Hawaiian Is
**Loxops coccinea ochracea* Rothschild			Maui, Hawaiian Is
**Loxops coccinea rufa* (Bloxam)			Oahu, Hawaiian Is
Hemignathus obscurus ellisianus (G.R. Gray)	1837		Oahu, Hawaiian Is
**Hemignathus obscurus lanaiensis* Rothschild	1894		Lanai, Hawaiian Is
**Hemignathus obscurus obscurus* (Gmelin)	>1895, 1940		Hawaii, Hawaiian Is
**Hemignathus procerus* Cabanis	[1967]		Kauai, Hawaiian Is
**Hemignathus lucidus hanapepe* Wilson			Kaui, Hawaiian Is
Hemignathus lucidus lucidus Lichtenstein	1860–1893		Oahu, Hawaiian Is
**Hemignathus lucidus affinis* Rothschild			Maui, Hawaiian Is

Species/subspecies (* taxa with entries in this catalogue)	Last date if extinct	B/T	Range or former range
**Hemignathus wilsoni* (Rothschild)			Hawaii, Hawaiian Is
Dysmorodrepanis munroi Perkins	>1913		Lanai, Hawaiian Is
**Psittirostra psittacea* (Gmelin)			Hawaiian Is
**Psittirostra bailleui* Oustalet			Hawaii, Hawaiian Is
**Psittirostra palmeri* (Rothschild)	(1896,1937)		Hawaii, Hawaiian Is
**Psittirostra flaviceps* (Rothschild)	[1891]		Hawaii, Hawaiian Is
**Psittirostra kona* (Wilson)	1894		Hawaii, Hawaiian Is
**Himatione sanguinea freethii* Rothschild	1923		Laysan I, Hawaiian Is
**Ciridops anna* (Dole)	1892		Hawaii, Hawaiian Is
**Drepanis pacifica* (Gmelin)	1898		Hawaii, Hawaiian Is.
**Drepanis funerea* Newton	1907		Molokai, Hawaiian Is
ICTERIDAE			
**Icterus leucopteryx bairdi* Cory	1938		Grand Cayman, W. Indies
**Icterus bonana* (Linnaeus)			Martinique, W. Indies
**Quiscalus palustris* (Swainson)	1910		Rio Lerma, Mexico
FRINGILLIDAE			
Serinus flavigula Salvadori			Ethiopia
**Neospiza concolor* (Bocage)	[1888–1909]		Sao Tome, e. AO
**Carduelis cucullata* Swainson			n. S. America
**Carduelis siemiradzkii* (Berlepsch & Taczanowski)			Ecuador
**Carpodacus mexicanus mcgregori* Anthony	1938		Mexico
**Chaunoproctus ferreorostris* (Vigors)	1890		Bonin Is, n.w. PO
**Pyrrhula pyrrhula murina* Godman			Azores, AO
PLOCEIDAE			
**Ploceus golandi* (Clarke)			Kenya
Ploceus aureonucha Sassi			Zaire
**Malimbus ibadanensis* Elgood			Nigeria
**Foudia rubra* (Gmelin)			Mauritius, Mascarene Is
**Foudia flavicans* Newton			Rodriguez, Mascarene Is
STURNIDAE			
Aplonis pelzelni Finsch	(1930–35,1956)		Ponape, Caroline Is, PO
**Aplonis corvina* (Kittlitz)	1828–1880		Kusaie I, Caroline Is, PO
**Aplonis mavornata* Buller	e19th C		Mauke I, Cook Is, PO
**Aplonis fusca fusca* Gould	(1925)		Norfolk Is, Australia
**Aplonis fusca hulliana* Mathews	>1918		Lord Howe I, Australia
Necropsar leguati Forbes	18th C		?Rodriguez, Mascarene Is

Species/subspecies (* taxa with entries in this catalogue)	Last date if extinct	B/T	Range or former range
Necropsar rodericanus Slater	1726	B	Rodriguez, Mascarene Is
**Fregilupus varius* (Boddaert)	1850–1860		Reunion, Mascarene Is
**Leucopsar rothschildi* Stresemann			Bali, Indonesia
ORIOLIDAE			
**Oriolus xanthonotus assimilis* Tweeddale	1906		Cebu, Philippine Is
CALLAEIDAE			
**Callaeas cinerea cinerea* (Gmelin)			New Zealand
**Heterolocha acutirostris* (Gould)	1907		North I, New Zealand
CRACTICIDAE			
**Strepera graculina crissalis* Sharpe			Lord Howe I, Australia
CORVIDAE			
**Cyanolyca nana* Du Bus			Mexico
**Cyanolyca mirabilis* Nelson			Mexico
**Corvus kubaryi* Reichenow			Marianas Is, PO
**Corvus tropicus* Kerr			Hawaii, Hawaiian Is

Species presumed extinct prior to 1600, of which there are specimens at Tring

AEPYORNITHIDAE			
**Aepyornis maximus* Geoffroy	?		Madagascar
**Aepyornis hildebrandti* Burckh.	?		Madagascar
EMEIDAE			
**Megalapteryx didinus* (Owen)	?		New Zealand
**Emeus huttonii* (Owen)	?		New Zealand
**Eurapteryx gravis* (Owen)	?		New Zealand
DINORNITHIDAE			
**Dinornis maximus* Haast	?		New Zealand

CATALOGUE OF EXTINCT AND ENDANGERED BIRDS

CONVENTIONS AND ABBREVIATIONS USED IN THE CATALOGUE

The catalogue entries are prefaced by short introductory statements concerning the distribution and conservation status of the relevant species or subspecies, together with notes on nomenclature, if appropriate. Further information on most of these forms may be found in Collar & Andrew (1988a), Collar & Stuart (1985), Collar *et al.* (1992), Diamond (1987), Fuller (1988), Greenway (1967), King (1978–1979), Mountfort (1988) and Peters (1934–1987, for distribution).

Information on sequence and nomenclature is given on page 7. For consistency, the English names used here are largely based on those in Sibley & Monroe (1990), although many of the names would not be the first choice of either of us.

Because of constraints imposed by the computer database, particularly on the indexing of keywords, hyphens are rarely used, and diacritical marks have been omitted.

The copy from which most of the catalogue was printed was set up from the computer database. The catalogue entries are therefore presented in a format with fixed punctuation, for example (for a specimen of *Numenius borealis*):

Reg. no. 1891.10.20.1033. From Brownsville; Texas; United States. Date: 28/ 3/1889. Source: F.B. Armstrong; Salvin–Godman coll.. Material: adult F; whole/good skin.

A record with no data would take the form:
Reg. no. . From ; ; . Date: //. Source: Material: ; .

Explanation of data entries:

Reg. no.

Museum registration number.

Registration number prefixes:

- A spirit register
- E egg register
- G general vertebrate register
- N nest register
- S skeleton register
- VEL Vellum Catalogue

Numbers without prefixes are from the main series of bird registers.

From

Locality (three levels of information, separated by semicolons).

n., s., e., w., points of the compass

ft. feet
lr. lower
nr. near
( ) data as shown on the labels, in the register or in the *Catalogue of Birds*, if an alternative spelling or place–name is also given in this catalogue
[ ] inferred data (not necessarily correct)

Chinese localities for specimens are given in Giles–Wade spellings, but in the species headings are in Pinyin.

Date

Date of collection, or date of death of specimens from captivity. Presented in the format:

day/ month/ year.

Source

The name (if known) of the original collector, and the route (in chronological sequence) by which the specimen reached the Museum, different stages being separated by semicolons.

coll. private collection
[ ] inferred data

Material

The nature of the specimen:
For eggs, the number of eggs in the set (not necessarily a clutch unless so stated).
For other specimens, age and/or sex, if known:

chick
juv juvenile
imm immature
adult
M male
F female

Condition is given as whole/good unless damage is sufficiently serious as to impair the general usefulness of the specimen.

Nature of specimen:

alc specimens in spirit (alcohol)
egg(s) only a clutch if stated
painting or drawing
skeleton including subfossil bones
skin including loose feathers

Remarks

Where appropriate, these are given after **Material**. There are occasional references to the catalogue of type–specimens (Warren & Harrison, 1966–1973), the *History of the Collections* (*Hist. Colls.*; Sharpe, 1906), the *Catalogue of Birds* (*Cat. Bds.*; Sharpe & others, 1874–1898) and the *Catalogue of the Collection of Birds' Eggs in the British Museum (Natural History)* (*Cat. Eggs*; Oates 1901–1912). Reference to Seebohm's ms catalogue relates to a manuscript in the Museum at Tring, compiled in the late 1800s.

For each species or subspecies the specimens are presented in the sequence: skins, skeletons, paintings, nests, eggs and spirit specimens. Within each category they are arranged in the alphanumeric order of their registration numbers. This is not the same as being in registration order: see, e.g., the first two entries under *Struthio camelus syriacus*.

Information particularly relevant to the specimens listed in this catalogue is given in Knox & Walters (1992). The latter includes descriptions of the *Registers* and *Catalogues* of the Museum, accounts of the collections and explanations of many label annotations. It also describes some of the shortcomings of the data and indicates areas where caution should be exercised over its interpretation. For comment on the reliability of data associated with the Meinertzhagen collection, see Knox (1993).

Struthio camelus syriacus
Syrian Ostrich

Formerly occurred in the deserts from Syria through Arabia. Became extinct during the mid-20th century.

Reg. no. 1924.12.15.1. From ; Mesopotamia; Iraq. Date: //. Source: Mrs. Yetts. Material: ; part/good skin. Pieces of skin.

Reg. no. 1924.4.8.1. From Jauf, s. Syrian Desert; ; Saudi Arabia (Syria), Captivity. Date: / 4/1924. Source: Cox-Cheesman coll.. Material: M; whole/good skin. Died in captivity. Skeleton in collection; S1924.4.8.1.

Reg. no. 1939.12.9.1019. From Syrian Desert; ; [Saudi Arabia], Captivity. Date: //1918. Source: J. Aharoni; Rothschild coll.. Material: adult M; whole/good skin. Holotype. Chick received from the Syrian Desert. Reared in captivity near Tel Aviv (Jaffa), [Israel], by Aharoni. Killed 1918. Rothschild, *Bull. Brit. Orn. Club* 39 (1919): 81-3.

Reg. no. 1939.12.9.1020. From Syrian Desert; ; [Saudi Arabia] (Syria). Date: //. Source: Rothschild coll.. Material: ; whole/good skin.

Reg. no. 1939.12.9.1021. From Rhebot, nr. Tel Aviv (Jaffa); ; Israel (Syria). Date: //. Source: Rothschild coll.. Material: juv ; skin. Not found 1988.

Reg. no. 1939.12.9.1024. From Saltplains, s.e. Hauran; ; Syria. Date: / 1/1921. Source: Rothschild coll.. Material: ; skin. Not found 1988.

Reg. no. 1988.15.1. From Syrian Desert; ; [Saudi Arabia] (Syria). Date: //1931. Source: Aharoni coll.. Material: ; whole/good skin.

Reg. no. 1923.6.10.1. From Jauf; ; Saudi Arabia (Syria). Date: / 6/1923. Source: Cox-Cheesman coll.. Material: imm M; whole/good skeleton.

Reg. no. 1924.4.8.1. From Jauf; ; Saudi Arabia (Syria). Date: //. Source: Cox-Cheesman coll.. Material: adult M; part/good skeleton. Skull and wings missing, these form part of skin 1924.4.8.1.

Reg. no. 1912.11.13.1-2. From ; ; Syria. Date: //. Source: J. Aharoni; L.W. Rothschild. Material: 2 ; whole/good eggs.

Reg. no. E1922.9.5.1-2. From Wadi Abiadh, 300 miles s.w. of Baghdad; ; [Saudi Arabia]. Date: 31/ 1/1922. Source: R.E. Cheesman. Material: 2 ; whole/good eggs. Collected in the central Arabian Desert, halfway between Baghdad and Jerusalem. *Ibis* 1923: 208-211.

Reg. no. E1927.8.4.1-2. From Bir Naama, two days e. of Maan; ; [Jordan] (Arabia). Date: //. Source: Lt.Col. Bewsher. Material: 2 ; whole/good eggs.

Reg. no. E1951.6.1. From Empty Quarter; ; [Saudi Arabia]. Date: //. Source: R. Meinertzhagen. Material: 3 ; whole/poor eggs. Shell eroded.

Reg. no. . From Fao; [Mesopotamia]; [Iraq] (Persian Gulf). Date: //. Source: Possman; T. Wells; Jourdain coll.. Material: 1 ; egg. Given to Wells by Possman's widow, 27

August 1920. Not in the collection; probably not received with the Jourdain collection.

Reg. no. E1922.12.20.206. From Wadi Abiadh, 300 miles s.w. of Baghdad; ; [Saudi Arabia] (Arabia). Date: //. Source: R.E. Cheesman. Material: 1 ; whole/good egg. Collected between 31°–32°N. Locality data may not be correct.

Reg. no. E1922.12.20.207. From Jauf; ; Saudi Arabia. Date: //1922. Source: R.E. Cheesman. Material: 1 ; whole/good egg. Collected in spring 1922.

Reg. no. E1937.11.1.10. From Afalil; ; [Arabia]. Date: 30/ 9/1936. Source: H.St.J. Philby. Material: – ; part egg. Fragment found on 30 September 1936.

Reg. no. E1937.11.1.3. From Muassam; ; [Arabia]. Date: 15/ 1/1936. Source: H.St.J. Philby. Material: – ; part egg. Found, 15 January 1936, on a low sand ridge, at the edge of a salt mud–flat along the sea. Fragments of shell only.

Reg. no. E1937.11.1.4. From Shudhaif Valley; ; [Arabia]. Date: 31/ 7/1936. Source: H.St.J. Philby. Material: – ; part egg. Fragments found 31 July 1936.

Reg. no. E1937.11.1.5. From Akhbash; ; [Arabia]. Date: 27/ 7/1936. Source: H.St.J. Philby. Material: – ; part egg. Fragments of shell only.

Reg. no. E1937.11.1.6. From Abu Ka'b; ; [Arabia]. Date: 4/ 8/1936. Source: H.St.J. Philby. Material: – ; part egg. Shell fragments only. Found on 4 August 1936.

Reg. no. E1937.11.1.7. From Matau; ; [Southern Yemen]. Date: 3/ 8/1936. Source: H.St.J. Philby. Material: – ; part egg. Fragments found on 3 August 1936. Matau not traced; possibly in Southern Yemen, as Philby was active there at about this time.

Reg. no. E1937.11.1.8. From Ramlat Sab'atain; ; [Yemen] (Arabia). Date: 8/ 8/1936. Source: H.St.J. Philby. Material: – ; part egg. Fragments found on 8 August 1936, together with shells and flint flakes (tools?).

Reg. no. E1937.11.1.9. From Ramlat Al Fuwa, nr. Tukhman; ; [Southern Yemen]. Date: 20/ 9/1936. Source: H.St.J. Philby. Material: – ; part egg. Sub–fossil fragments of shell found on 20 September 1936. Locality not traced; possibly in Southern Yemen, as Philby was active there about this time.

Reg. no. E1938.12.1.10. From Shabwah (Shabwa); ; [Southern Yemen] (Arabia). Date: 22/ 9/1936. Source: H.St.J. Philby. Material: – ; part egg. Fragments only.

Reg. no. E1941.4.1.69. From Biseita; ; Saudi Arabia. Date: //1880. Source: C. Doughty; Lawrence; Meinertzhagen. Material: 1 ; whole/good egg. Inscribed: 'This egg was taken by Charles Doughty from a nest at Biseita, within 10 miles of Lat 30°N and Long 38°E, about 1880. Given by Doughty to Colonel Lawrence who gave it to me in 1922. R. Meinertzhagen.'

Reg. no. E1941.9.4.2638. From Syrian Desert; ; [Saudi Arabia]. Date: 8/ 4/1928. Source: J. Aharoni; Rothschild. Material: 1

; whole/good egg. Collected in the southernmost Syrian Desert. The locality is presumably the same as 1941.9.4.2639, which was collected by Aharoni on the same date.

Reg. no. E1941.9.4.2639. From Kaf, Syrian Desert; ; [Saudi Arabia]. Date: 8/ 4/1928. Source: J. Aharoni; Rothschild. Material: 1 ; whole/good egg.

Reg. no. E1941.9.4.2640. From Kasr Ezrak, s. of Djebel Druz and Djebel Kawen; ; [Saudi Arabia]. Date: 3/ 4/1928. Source: J. Aharoni; Rothschild. Material: 1 ; whole/good egg.

Reg. no. E1941.9.4.2641. From Syrian Desert; ; [Syria]. Date: //. Source: J. Aharoni; Rothschild. Material: 1 ; whole/good egg. Received by Rothschild 1912.

Reg. no. E1941.9.4.2642. From ; ; . Date: //. Source: J. Aharoni; Rothschild. Material: 1 ; whole/good egg.

Reg. no. E1941.9.4.2643. From Syrian Desert; ; [Syria]. Date: //1912. Source: J. Aharoni; Rothschild. Material: 1 ; whole/good egg.

Reg. no. E1951.6.2. From Wadi Sirham, Amman; ; Jordan (Trans Jordan). Date: 3/ 5/1920. Source: R. Meinertzhagen. Material: 1 ; whole/good egg. 'A fresh egg was brought to me at Amman in 1920 ...'. Meinertzhagen, 1954, *Birds of Arabia*, p. 574.

Reg. no. E1962.1.620. From Shuqalkat (Shuqayqat); ; Southern Yemen (Arabia). Date: 29/ 9/1936. Source: Material: – ; part egg. Fragment found at 15° 32'N 46° 48'E. Possibly collected by H.St.J. Philby, who was active in the area at this time.

Reg. no. E1974.31.1. From Rub Al Khali; Dhufar; Oman. Date: //1972. Source: P.A. Smith; Mrs F.E. Warr. Material: – ; part egg. Found at 18° 26'N 53° 08'E. Fragments (possibly sub-fossil) with slides of sections.

Struthio indicus

Indian Ostrich

Known only from eggshell fragments of uncertain age.

Reg. no. E1963.9.219. From Nullas, Ken River or Kain River, Banda; Uttar Pradesh or Madhya Pradesh; India. Date: //. Source: A. Carlyle; E. Bidwell; Rothschild coll.. Material: – ; part egg. Syntypes. Bidwell, *Ibis* 1910: 759–61. Two sub-fossil or fossil shell fragments, found sometime prior to 1905. *Bull. Brit. Orn. Club* 15 (1905): 72–3. Banda is near the border of Uttar Pradesh and Madhya Pradesh. There are three further syntypes in the Palaeontology Department.

Psammornis rothschildi

Rothschild's Ostrich

Originally described from eggshell fragments of uncertain age.

Reg. no. E1963.9.218. From between Touggourt and El Oued; ; Algeria. Date: / 4/1909. Source: W. Rothschild & E. Hartert. Material: – ; part egg. Syntypes. Andrews, 1911 (publ. 1912), *Proc. 5th*

Internat. Ornithol. Congr., Berlin, p. 169–74. Three fragments of shell collected in April 1909, about 20 miles east of Touggourt. The description was based on two fragments.

Rhea nana

Dwarf Rhea

A dubious species described and known only from a single egg.

Reg. no. 1894.11.28.1. From ; s. Patagonia; [Argentina]. Date: //. Source: R. Lydekker. Material: 1 ; whole/good egg. A wax cast of an egg in Museo de la Plata, which is the holotype of *R. nana*, believed to be a small species of rhea, but never subsequently rediscovered. See *Proc. zool. Soc. Lond.* 1894: 654.

Pterocnemia pennata tarapacensis

Puna Rhea

Previously found in the Puna zone of the eastern Andes from Peru south to Chile and Argentina. Probably no longer occurs over much of its former range, this subspecies is rare and declining in most places. (Includes *P.p. garleppi* of Peters 1931–1987).

Reg. no. 1892.2.10.1040. From Canchosa, Tarapaca; ; n. Chile. Date: 23/ 4/1890. Source: A.A. Lane; H.B. James. Material: imm M; whole/good skin. Holotype. Chubb, *Bull. Brit. Orn. Club* 33 (1913): 79. The original (worn) label bears the date listed here, but it was given as 2 February 1890, apparently by mistake, in the published description. See the note by Warren in the catalogue of type–specimens.

Dromaius novaehollandiae diemenensis

Tasmanian Emu

Formerly inhabited Tasmania, but became extinct in the 19th century.

Reg. no. 1838.1.15.203. From ; Tasmania; Australia. Date: //. Source: R. Gunn. Material: M; whole/good skin.

Reg. no. 1838.1.15.204. From ; Tasmania; Australia. Date: //. Source: R. Gunn. Material: F; whole/good skin.

Reg. no. 1894.8.21.1. From ; Tasmania; Australia. Date: //. Source: A.W. Franks. Material: ; part/good skin. No head or legs.

Reg. no. 1870.12.1.1. From ; ; Tasmania. Date: //. Source: R.J. Mercer. Material: 1 ; whole/good egg.

Reg. no. E1960.6.13. From ; ; Tasmania. Date: //. Source: Dr Milligen. Material: 1 ; whole/good egg.

Dromaius diemenianus/D. baudinianus

Kangaroo Island Emu

Formerly inhabited Kangaroo Island, but never seen after its discovery in 1802. The species should be called *D. baudinianus* (*Bull. Brit. Orn. Club* 104 (1984): 19–22).

Reg. no. E1938.6.1.1. From Nepean Bay, Kangaroo Island; South Australia; Australia. Date: //1830. Source: Dr J.D. Lang; G.D. Rowley; Mrs G.E. Mason. Material: 1 ; whole/good egg. See *Emu* 35 (1936): 331–2.

Additional Ratite species are given at the end of the list.

Tachybaptus rufolavatus

Alaotra Grebe

Known to breed mainly on Lake Alaotra, Madagascar. It was formerly common, but is now in the process of disappearing as a result of hybridization with the Little Grebe *T. ruficollis*, which became established in Madagascar in the present century.

Reg. no. 1931.8.18.1133. From Andreba; ; Madagascar. Date: 4/ 6/1929. Source: Archbold Vernay Expedition; American Museum of Natural History. Material: adult M; whole/good skin. Previously identified as *T. pelzelni*.

Reg. no. 1931.8.18.1134. From Andreba; ; Madagascar. Date: 30/ 5/1929. Source: Archbold Vernay Expedition; American Museum of Natural History. Material: adult M; whole/good skin. Previously identified as *T. pelzelni*.

Reg. no. 1931.8.18.1135. From Andreba; ; Madagascar. Date: 5/ 6/1929. Source: Archbold Vernay Expedition; American Museum of Natural History. Material: adult M; whole/good skin. Previously identified as *T. pelzelni*.

Reg. no. 1931.8.18.1136. From Andreba; ; Madagascar. Date: 30/ 5/1929. Source: Archbold Vernay Expedition; American Museum of Natural History. Material: adult F; whole/good skin. Laying. Previously identified as *T. pelzelni*.

Reg. no. 1931.8.18.1137. From Andreba; ; Madagascar. Date: 31/ 5/1929. Source: Archbold Vernay Expedition; American Museum of Natural History. Material: adult F; whole/good skin. Previously identified as *T. pelzelni*.

Reg. no. 1939.12.9.3389. From Lake Alaotra; ; Madagascar. Date: 13/ 8/1931. Source: Huschell Chauvin; Neumann; Rothschild coll.. Material: adult ; whole/good skin. Previously identified as *T. pelzelni*.

Reg. no. 1939.12.9.3390. From Lake Alaotra; ; Madagascar. Date: 24/ 8/1931. Source: Huschell Chauvin; Neumann; Rothschild coll.. Material: adult ; whole/good skin. Previously identified as *T. pelzelni*.

Reg. no. E1988.4.30. From ; ; Madagascar. Date: //. Source: Rothschild coll.. Material: 1 ; whole/good egg. No data or identification, other than locality; might be *T. rufolavatus*, *T. ruficollis* or *T. pelzelni*.

Podilymbus gigas

Atitlan Grebe

Known only from Lake Atitlan in Guatemala, where it is now extinct as a result of competition with and perhaps predation from introduced fish, changes in the lake levels following geological activity and competition or hybridization with the Pied-billed Grebe *P. podiceps*. Searches in April 1991 revealed no birds.

Reg. no. 1892.1.20.21. From Lake Atitlan; ; Guatemala. Date: 7/ 9/1862. Source: Salvin; Salvin-Godman coll.. Material: M; whole/good skin.

Reg. no. 1892.1.20.22. From Lake Atitlan; ; Guatemala. Date: / 2/1883. Source:

Salvin; Salvin–Godman coll.. Material: ; whole/good skin.

Reg. no. 1893.2.1.886. From Lake Atitlan; ; Guatemala. Date: /12/1890. Source: W.B. Richardson; Salvin–Godman coll.. Material: ; whole/good skin.

Reg. no. 1893.2.1.887. From Lake Atitlan; ; Guatemala. Date: /12/1890. Source: W.B. Richardson; Salvin–Godman coll.. Material: F; whole/good skin.

Podiceps taczanowskii

Puna Grebe

Known only from Lake Junin in Peru, where it was formerly common, but is rapidly declining, with 300 birds left in 1978 and only 100 in 1987. It is now almost extinct due to pollution and fluctuations in the level of the lake.

Reg. no. 1892.12.24.30. From Incapizea, Jezioro, Lake Junin; ; Peru. Date: 27/ 5/1890. Source: J. Kalinowski; Count Branicki; J. Stolzmann. Material: adult F; whole/good skin.

Reg. no. 1946.49.57. From Lake Junin; ; Peru. Date: 15/ 2/1938. Source: A. Morrison. Material: imm ; whole/good skin.

Reg. no. 1946.49.58. From Lake Junin; ; Peru. Date: 15/ 2/1938. Source: A. Morrison. Material: adult F; whole/good skin.

Reg. no. 1946.49.59. From Lake Junin; ; Peru. Date: 21/ 3/1938. Source: A. Morrison. Material: adult M; whole/good skin.

Reg. no. 1946.49.60. From Lake Junin; ; Peru. Date: 25/ 3/1938. Source: A. Morrison. Material: adult F; whole/good skin.

Reg. no. 1946.49.61. From Lake Junin; ; Peru. Date: 25/ 3/1938. Source: A. Morrison. Material: adult M; whole/good skin.

Reg. no. 1946.49.62. From Lake Junin; ; Peru. Date: 25/ 3/1938. Source: A. Morrison. Material: adult M; whole/good skin.

Reg. no. 1946.49.63. From Lake Junin; ; Peru. Date: 3/ 4/1938. Source: A. Morrison. Material: adult M; whole/good skin.

Reg. no. E1938.12.32.2–3. From Lake Junin; ; Peru. Date: 20/ 2/1938. Source: A.R.G. Morrison. Material: 2 ; whole/good eggs (clutch). Incubation, fresh. Collected as '*Colymbus chilensis* (?)' [= *Rollandia rolland*]; might be *R. rolland*, *P. taczanowskii* or *P. occipitalis*. See *Ibis* 1939: 645.

Reg. no. E1938.12.32.4. From Lake Junin; ; Peru. Date: 1/ 2/1938. Source: A.R.G. Morrison. Material: 1 ; whole/good egg. Collected as '*Colymbus chilensis* (?)' [= *Rollandia rolland*]; might be *R. rolland*, *P. taczanowskii* or *P. occipitalis*. See *Ibis* 1939: 645.

Reg. no. A1982.44.1. From Lake Junin, 14,000 ft.; ; Peru. Date: / 3/1938. Source: A.R.G. Morrison. Material: adult ; whole/good alc.

Diomedea albatrus

Short–tailed Albatross

Currently breeding only on Torishima in the Izu Islands. The species started to become rare about 100 years ago due to feather–hunters, and was considered extinct by 1949. It was rediscovered in 1951, when the population was less than 50 birds. By 1990/1991 there were about 500 birds, and the population was increasing by about 7% each year.

Reg. no. 1884.1.16.113. From Nagasaki; ; Japan. Date: 28/ 6/1881. Source: F. Ringer. Material: imm M; whole/good skin. Date believed to be correct, but label almost illegible.

Reg. no. 1887.10.2.10. From Kuril Islands; ; Russia. Date: 20/ 5/1883. Source: Janson. Material: imm ; whole/good skin. Either this bird or 1887.10.2.11 is from Iterup (Iteiruf) Island (*Cat. Bds.*).

Reg. no. 1887.10.2.11. From Kuril Islands; ; Russia. Date: //. Source: Janson. Material: imm ; whole/good skin. Either this bird or 1887.10.2.10 is from Iterup (Iteiruf) Island (*Cat. Bds.*).

Reg. no. 1887.10.2.8. From Kuril Islands; ; Russia. Date: //. Source: Janson. Material: adult ; whole/good skin.

Reg. no. 1887.10.2.9. From Kuril Islands; ; Russia. Date: //. Source: Janson. Material: adult ; whole/good skin.

Reg. no. 1888.5.18.160. From Chefoo; ; n. China. Date: 18/ 6/1873. Source: R. Swinhoe; Salvin–Godman coll.. Material: ; whole/good skin. Syntype of *D. derogata*. There is no date on the label or in the register, where this bird is claimed to be the type. This being so, the date and sex would have been the same as that for 1888.7.10.79. See *Proc. zool. Soc. Lond.* 1873: 784–6.

Reg. no. 1888.7.10.78. From Hakodate (Hakodadi); [Hokkaido]; Japan. Date: / 7/1873. Source: T. Blakiston; R. Swinhoe. Material: juv M; whole/good skin. *Ibis* 1874: 165.

Reg. no. 1888.7.10.79. From Chefoo; ; China. Date: 18/ 6/1873. Source: R. Swinhoe; Seebohm coll.. Material: M; whole/good skin. Syntype. Type of *D. derogata*. Swinhoe, *Proc. zool. Soc. Lond.* 1873: 784–6. There is another syntype in the collection.

Reg. no. 1888.7.10.80. From Hakodate; Hokkaido; Japan. Date: 7/ 7/1880. Source: T. Blakiston; Seebohm coll.. Material: juv M; whole/good skin.

Reg. no. 1888.7.10.83. From Hakodate; Hokkaido; Japan. Date: 4/ 4/1883. Source: T. Blakiston; Seebohm coll.. Material: M; whole/good skin.

Reg. no. 1888.7.10.84. From Amoy; ; China. Date: / 4/1866. Source: R. Swinhoe. Material: adult M; whole/good skin.

Reg. no. 1888.7.10.85. From Amoy; ; China. Date: / 4/1886. Source: R. Swinhoe. Material: adult F; whole/good skin.

Reg. no. 1891.5.3.15. From Hamilton Island; ; [Southern Korea]. Date: //. Source:

A.G. Wilday. Material: adult ; whole/good skin. Damaged: leg and wing loose.

Reg. no. 1898.7.4.154. From Hakodate, Yezo Island; [Hokkaido]; Japan. Date: 10/ 6/1897. Source: G.E.H. Barrett Hamilton. Material: juv ; whole/good skin.

Reg. no. 1898.7.4.155. From Bering Island, Bering Sea; ; United States. Date: //1896. Source: G.E.H. Barrett Hamilton. Material: imm F; whole/good skin.

Reg. no. 1902.8.5.439. From Foochow; ; s. China. Date: 23/ 5/1896. Source: C.B. Rickett. Material: juv ; whole/good skin.

Reg. no. 1902.8.5.440. From Foochow; ; s. China. Date: / 2/1891. Source: C.B. Rickett. Material: adult ; whole/good skin.

Reg. no. 1848.7.28.8. From Kamchatka; ; Russia. Date: //. Source: J.G.W. Brant. Material: ; whole/good skeleton. Previously registered as Skeleton Vellum Cat. 596a.

Reg. no. 1884.2.29.12. From ; ; [Pacific Ocean]. Date: //. Source: Challenger Expedition. Material: ; whole/good skeleton. Six specimens were collected on the expedition in June and July 1874, and in April 1875; this skeleton was presumably from one of them.

Reg. no. 1884.2.29.13. From ; ; [Pacific Ocean]. Date: //. Source: Challenger Expedition. Material: ; part/good skeleton. Skull only. Six specimens were collected on the expedition in June and July 1874, and in April 1875; this skull is presumably from one of them.

Reg. no. 1884.2.29.14. From ; ; [Pacific Ocean]. Date: //. Source: Challenger Expedition. Material: ; part/good skeleton. Skull only, without lower jaw. Six specimens were collected on the expedition in June and July 1874, and in April 1875; this skull is presumably from one of them.

Reg. no. E1901.1.1.1541–4. From Kater Island (Ejiga Sima); Bonin Islands; Japan. Date: /11/1890. Source: P.A. Holst; Seebohm coll.. Material: 4 ; whole/good eggs. Eggs collected in mid–November. One egg is missing and believed stolen during 1975–1979.

Reg. no. E1901.1.1.1545–8. From Kater Island (Ejiga Sima); Bonin Islands; Japan. Date: /11/1890. Source: P.A. Holst; Seebohm coll.. Material: 4 ; whole/good eggs. Eggs collected in mid–November. One egg is missing and believed stolen during 1975–1979.

Reg. no. E1901.1.1.1549–52. From Kater Island (Ejiga Sima); Bonin Islands; Japan. Date: /11/1890. Source: P.A. Holst; Seebohm coll.. Material: 4 ; whole/good eggs. Eggs collected in mid–November. Some of the eggs are not inscribed with data and may therefore be the H. Pryer coll. eggs refered to by Seebohm in *Ibis* 1890: 105. One egg is missing and believed stolen during 1975–1979.

Reg. no. E1914.10.1.3336. From ; Bonin Islands; Japan. Date: /11/1902. Source: W. Radcliffe–Saunders coll.. Material: 1 ; whole/good egg.

Reg. no. E1961.15.17. From ; Bonin Islands; Japan. Date: /11/1910. Source: A.

Owston; Freeth. Material: 1 ; whole/good egg (clutch).

Reg. no. E1973.1.770. From ; Bonin Islands; Japan. Date: /11/1912. Source: Kobayashi coll.; Jourdain coll.. Material: 1 ; whole/good egg.

Reg. no. E1973.1.771. From ; Bonin Islands; Japan. Date: 27/ 1/1910. Source: A. Owston; F.W. Proctor; F. Stoneham; Jourdain coll.. Material: 1 ; whole/good egg.

Reg. no. E1973.1.772. From ; Bonin Islands; Japan. Date: /10/1910. Source: A. Owston; Jourdain coll.. Material: 1 ; whole/poor egg. Egg badly cracked.

Reg. no. E1988.4.1. From ; ; . Date: //. Source: Material: 1 ; whole/good egg. One egg recovered from a series of thefts in 1975–1979, but it is not clear which of the stolen eggs it is.

Pterodroma hasitata caribbaea

Jamaican Petrel

Formerly bred in the Blue Mountains of Jamaica, and believed to have become extinct around 1880. Only about 7 specimens were ever collected.

Reg. no. 1868.7.31.1. From ; ; Jamaica. Date: //. Source: J. Taylor. Material: ; whole/good skin. Ex–mount.

Reg. no. 1888.5.18.108. From ; ; Jamaica. Date: / 7/1875. Source: J. Hill; J. Gould; Salvin–Godman coll.. Material: ; whole/poor skin. Collected prior to July 1875. Skin complete but extremely delicate.

Reg. no. 1888.5.18.109. From Cinchona plantations; ; Jamaica. Date: 17/11/1879. Source: E. Newton; Salvin–Godman coll.. Material: M; whole/good skin.

Pterodroma cahow

Bermuda Petrel

Having been considered extinct since the 17th century, it was rediscovered in 1951. The population has climbed from 18 to 35 pairs over the succeeding period, and its progress is intensively monitored. The species breeds only on five small islets in Castle Harbour, Bermuda.

Reg. no. S1959.14.3. From Cockroach Island, Pear Rocks, Castle Harbour; ; Bermuda. Date: //1958. Source: D.B. Wingate; W.R.P. Bourne. Material: ; part/good skeleton. Parts of incomplete skeletons.

Pterodroma magentae

Magenta Petrel

Breeds on the Chatham Islands in small numbers. Formerly known only from a single skin collected in 1867 on the voyage of the Magenta. In 1978 it was rediscovered and the population is currently thought to be 50–100 birds.

Reg. no. S1966.49.13–23. From ; ; Chatham Islands. Date: //1892. Source: H.O. Forbes coll.. Material: ; part/good skeleton. Sub–fossil: 2 incomplete skulls, 1 sternum, 2 coracoids, 6 humeri.

Reg. no. S1967.9.1. From ; ; . Date: //. Source: H.O. Forbes coll.. Material: ; part/good skeleton. One sub–fossil tibia.

Pterodroma mollis madeira

Madeira Petrel

Breeds only on the high mountain slopes of Madeira. The form was considered extinct from 1909 to 1940. In 1969 a very small breeding population (currently only about 20 pairs) was discovered, but the birds are heavily predated by rats. Now considered to be a distinct species, *P. madeira*.

Reg. no. 1906.7.26.1. From San Antonio; ; Madeira. Date: 13/ 6/1906. Source: Padre E. Schmitz. Material: adult M; whole/good skin. Holotype. Mathews, *Bull. Brit. Orn. Club* 54 (1934): 161. See also *Bol. Mus. Mun. Funchal* 38: 141–165.

Reg. no. 1906.7.26.2. From San Antonio; ; Madeira. Date: 12/ 6/1906. Source: Padre E. Schmitz. Material: adult ; whole/good skin. See also *Bol. Mus. Mun. Funchal* 38: 141–165.

Reg. no. . From Risco do Poio de Louro, Curral; ; Madeira. Date: 13/ 9/1905. Source: E. Schmitz; Jourdain coll.. Material: 1 ; egg. Not found <1987. *Ornithol. Jahrb.* 17 (1906): 25, 199–204. *Bull. Brit. Orn. Club* 19 (1907): 37–38. This egg was probably not received by the Museum.

Pterodroma phaeopygia phaeopygia

Galapagos Petrel

This race breeds in the Galapagos Islands, but has been greatly reduced by predation from introduced mammals.

Reg. no. 1850.1.31.20. From ; San Cristobal (Chatham Island); Galapagos Islands. Date: //. Source: Capt. Kellett & Lt. Wood, voyage of H.M.S. Herald. Material: adult ; whole/good skin. Syntype. Ex-mount. Salvin, *Trans. zool. Soc. Lond.* 9 (1876): 507–8; pl. 88. There is another syntype in the collection.

Reg. no. 1850.1.31.21. From ; San Cristobal (Chatham Island); Galapagos Islands. Date: //. Source: Capt. Kellett & Lt. Wood, voyage of H.M.S. Herald. Material: adult ; whole/good skin. Syntype.

Reg. no. 1899.9.1.539. From ; Wenman Island; Galapagos Islands. Date: 3/ 8/1897. Source: Webster Harris Expedition. Material: adult F; whole/good skin. Eyes brown, bill black, tarsi whitish, feet black and whitish.

Reg. no. 1899.9.1.540. From ; Albemarle Island; Galapagos Islands. Date: 22/11/1897. Source: Webster Harris Expedition. Material: adult ; whole/good skin. Eyes brown, bill black, tarsi bluish–flesh, feet flesh and black.

Reg. no. S1967.19.4. From ; Santa Cruz Island; Galapagos Islands. Date: / 5/1967. Source: Dr M.P. Harris. Material: adult ; part/good skeleton. Skull only.

Reg. no. S1973.1.7. From ; Santa Cruz Island; Galapagos Islands. Date: /10/1972. Source: Dr M.P. Harris. Material: adult ; whole/good skeleton.

Reg. no. S1973.6.1. From ; Santa Cruz Island; Galapagos Islands. Date: / 5/1966. Source: Dr M.P. Harris. Material: adult ; whole/good skeleton.

Pterodroma phaeopygia sandwichensis
Hawaiian Petrel

This race breeds in the Hawaiian Islands, where it too has been greatly reduced by predation from introduced mammals.

Reg. no. 1890.10.3.3. From ; Kaui; Hawaiian Islands. Date: //. Source: Knudsen; R. Collett. Material: adult ; whole/good skin.

Reg. no. 1915.12.19.1. From Olokui; Molokai; Hawaiian Islands. Date: 20/ 7/1914. Source: Bryan. Material: adult M; whole/good skin.

Reg. no. 1915.12.19.2. From Olokui; Molokai; Hawaiian Islands. Date: 20/ 7/1914. Source: Bryan. Material: adult F; whole/good skin.

Pterodroma cookii
Cook's Petrel

Breeds on Great and Little Barrier Islands off North Island, and on Codfish Island off Stewart Island, New Zealand. The species is now recovering after removal of predators from its nesting grounds.

Reg. no. 1842.9.29.14. From ; ; New Zealand. Date: //. Source: Dr E. Dieffenbach; New Zealand Company. Material: adult ; whole/good skin. Holotype. Ex-mount. G.R. Gray, in Dieffenbach, 1843, *Travels in New Zealand*, 2: 199–200. Iris golden-yellow.

Reg. no. 1888.5.18.123. From ; ; the seas off New Zealand. Date: //. Source: Gould; Salvin-Godman coll.. Material: adult ; whole/good skin.

Reg. no. 1892.10.31.20. From w. coast; ; New Zealand. Date: //. Source: H.O. Forbes. Material: adult ; whole/good skin. Identified as *P. axillaris* in the register, and said to be the 'type' of the species.

Reg. no. 1892.10.31.21. From w. coast; ; New Zealand. Date: //. Source: H.O. Forbes. Material: adult ; skin. Details taken from the register and *Cat. Bds.*. Not found 1988. Identified as *P. axillaris* in the register, and said to be the 'type' of the species.

Reg. no. 1959.33.11. From ; ; . Date: //. Source: Dominion Museum. Material: adult ; whole/good skin. 'Dr Falla says that large birds like this are typical of the southern colonies' (W.R.P. Bourne, on the label).

Reg. no. S1965.7.2. From Tauranga; ; New Zealand. Date: 7/ 2/1962. Source: P. Harper; Dominion Museum. Material: adult ; part/good skeleton. Skull only.

Pterodroma axillaris
Chatham Islands Petrel

Breeds on Rangatira Island, Chatham Islands, where it has apparently always been rare. The present population is estimated to be about 500 birds.

Reg. no. 1892.10.31.14. From ; S.E. Island; Chatham Islands. Date: 8/ 5/1892. Source: W. Hawkins; H.O. Forbes. Material: imm M; whole/good skin. Syntype. Salvin, *Ibis* 1893: 264. Sub-adult, though noted as an adult in *Cat. Bds.*. H.O. Forbes is erroneously listed as A.O. Forbes in the catalogue of type-specimens.

Reg. no. 1892.10.31.15. From ; ; Chatham Islands. Date: 8/ 5/1892. Source: H.O. Forbes. Material: adult F; whole/good skin. Syntype.

Reg. no. 1895.10.17.3. From ; ; Chatham Islands. Date: //1894. Source: W. Rothschild. Material: adult M; whole/good skin. Damaged: leg detached.

Reg. no. 1895.10.17.4. From ; ; Chatham Islands. Date: //1894. Source: W. Rothschild. Material: imm F; whole/good skin. Sub-adult.

Reg. no. 1945.54.1. From ; ; Chatham Islands. Date: //. Source: W.J.P. Wardlaw. Material: adult M; whole/good skin. Damaged: head detached. This specimen bears a collector's label identical in style to those on 1895.10.17.3 & 4.

Procellaria parkinsoni

Black Petrel

Known to breed on Great and Little Barrier Islands, New Zealand, and formerly on some high mountain ranges of North and South Islands. Introduced cats and other predators greatly reduced the numbers, but this is now being combated. The population is estimated to be 3,000–4,000 birds.

Reg. no. 1842.5.17.16. From ; ; New Zealand. Date: //. Source: Miss R. Stone, Excise Office, Cath. Dock. Material: adult ; whole/good skin. Holotype. G.R. Gray, *Ibis* 1862: 245–6. 'Taiko'; dark dun eye, black pupil.

Reg. no. 1902.2.10.52–5. From Cape Maria Van Dieman, Auckland; North Island; New Zealand. Date: 9/ 7/1887. Source: E.P. Seymour; Crowley. Material: 4 ; whole/good eggs.

Oceanodroma macrodactyla

Guadalupe Storm–Petrel

Bred on Guadalupe Island off Baja California. It was last seen in 1911, and is presumed extinct as a result of predation by feral cats.

Reg. no. 1898.7.12.232. From Guadalupe Island; ; Mexico. Date: 25/ 3/1897. Source: C.K. Worthen; Salvin–Godman coll.. Material: adult M; whole/good skin. See 1898.7.12.233, and the entry for the eggs collected by A.W. Anthony, 1897.11.10.48–9.

Reg. no. 1898.7.12.233. From Guadalupe Island; ; Mexico. Date: 25/ 3/1897. Source: C.K. Worthen; Salvin–Godman coll.. Material: F; whole/good skin. See 1898.7.12.232, and the entry for the eggs collected by A.W. Anthony, 1897.11.10.48–9.

Reg. no. 1897.11.10.48–9. From Guadalupe Island; ; Mexico. Date: 24/ 3/1897. Source: A.W. Anthony. Material: 2 ; whole/good eggs. Nest comprised a few oak leaves and pine needles at the end of a burrow about 3 feet long, 2,500 ft. above the sea. A note with the eggs refers to the adult birds being collected. The only skins in the collection are 1898.7.12.232–3, taken by C.K. Worthen on the following day.

Sula abbotti

Abbott's Booby

Now breeds only on Christmas Island (Indian Ocean) but subfossil remains from other islands suggest a distribution which was wider. The population is small and vulnerable: unlike most boobies, the species nests only on tall forest trees. Much of the forest on Christmas Island is threatened by phosphate mining.

Reg. no. 1898.9.16.49. From ; ; Christmas Island, Indian Ocean. Date: /10/1897. Source: C.W. Andrews; Sir J. Murray. Material: adult M; whole/good skin.

Reg. no. 1909.5.2.3. From ; ; Christmas Island, Indian Ocean. Date: 15/ 8/1906. Source: [C.W.] Andrews; Sir J. Murray. Material: adult M; whole/good skin.

Reg. no. 1909.5.2.4. From ; ; Christmas Island, Indian Ocean. Date: 14/ 8/1906. Source: C.W. Andrews; Sir J. Murray. Material: adult F; whole/good skin. Part of the skull is in the skeleton collection, S1974.2.1.

Reg. no. 1909.5.2.5. From ; ; Christmas Island, Indian Ocean. Date: 12/ 8/1906. Source: C.W. Andrews; Sir J. Murray. Material: adult F; whole/good skin.

Reg. no. 1909.5.2.6. From Murray Hill; ; Christmas Island, Indian Ocean. Date: 28/ 8/1908. Source: C.W. Andrews; Sir J. Murray. Material: juv F; whole/good skin.

Reg. no. S1974.2.1. From ; ; Christmas Island, Indian Ocean. Date: 14/ 8/1906. Source: C.W. Andrews; Sir J. Murray. Material: adult F; part/good skeleton. Part of a skull (cranium) from a skin in the collection, 1909.5.2.4.

Phalacrocorax perspicillatus

Pallas's Cormorant

Extinct. Known only from Bering Island in the Commander group, north Pacific, but may formerly have occurred more widely. It was common in 1741 and was probably exterminated for food by Aleut natives who settled on the island in 1826.

Reg. no. 1842.12.21.4. From Bering Island (Behring Island); ; North America. Date: //. Source: E. Belcher. Material: ; part/good skin. Tail missing.

Reg. no. 1858.2.3.1. From Bering Island (Behring Island); ; [North America]. Date: //. Source: J. Gould coll.. Material: ; whole/good skin.

Nycticorax caledonicus crassirostris

Bonin Night-Heron

Formerly occurred on Peel Island and Nakondo Shima in the Bonin Islands, where it nested in low trees near the coast. The last specimen was taken in 1889. The cause of extinction is unknown, but human interference is suspected.

Reg. no. 1857.10.16.66. From ; Bonin Islands; Japan. Date: //1827. Source: Collie, voyage of H.M.S. Blossom; J. Gould. Material: adult ; whole/good skin. Holotype. Vigors, in F.W. Beechey, 1839, *The Zoology of Captain Beechey's Voyage* ..., p. 27. Presumably collected in 1827.

Reg. no. 1897.10.30.32. From Nakondo Shima; Bonin Islands; Japan. Date: 15/ 6/1889. Source: P.A. Holst; Seebohm coll.. Material: adult F; whole/good skin. Date taken from the register.

Ciconia ciconia boyciana

Oriental Stork

Formerly bred widely over eastern Siberia, Korea and Japan, but now confined to the Amur basin and northeastern China, where the population is estimated to be between 400 and 600 birds. One of the main breeding areas is threatened with reclamation. Once abundant in Japan, a major decline occurred between 1868 and 1895, with a few surviving until 1969.

Reg. no. 1876.6.10.1. From ; ; Japan. Date: 4/ 4/1876. Source: R.H. Boyce; R. Swinhoe; Zoological Society. Material: adult M; whole/good skin. One of two birds received at London Zoo in December 1873; this one died 4 April 1876. See *Proc. zool. Soc. Lond.* 1874: 2, pl. 1. This is not one of the types, which died and the skins were not preserved, see *Proc. zool. Soc. Lond.* 1873: 512–14. See also 1892.4.2.385.

Reg. no. 1891.2.10.1. From ; ; Japan. Date: //. Source: H. Pryer; Seebohm coll.. Material: adult ; whole/good skin.

Reg. no. 1891.2.10.2. From ; ; Japan. Date: //. Source: H. Pryer; Seebohm coll.. Material: adult ; whole/good skin.

Reg. no. 1892.4.2.385. From ; ; Japan. Date: 15/ 1/. Source: [R.H. Boyce; R. Swinhoe; Zool. Soc.; Lord Walden;] Tweeddale coll.. Material: adult F; whole/good skin. Apparently, this is the second bird sent by Swinhoe to London Zoo in December 1873 (see 1876.6.10.1), this one dying in 1874, its skin being exhibited on 19 May that year, before passing to Lord Walden. See *Proc. zool. Soc. Lond.* 1874: 2, pl. 1, 306–7.

Reg. no. 1896.10.1.47. From Amur Bai, ie mouth of River Amur; Siberia; [Russia]. Date: 8/ 5/1894. Source: C. Dorries; Seebohm coll.. Material: adult F; whole/good skin.

Reg. no. 1897.10.30.137. From ; ; Korea. Date: //. Source: C.W. Campbell; Seebohm coll.. Material: adult F; whole/good skin.

Reg. no. 1908.1.8.443. From Shaweishan; Kiang Su Province; e. China. Date: 9/12/1901. Source: F.W. Styan. Material: adult F; whole/good skin.

Reg. no. E1972.13.85. From Tsapreayer; [e. Siberia]; [Russia]. Date: 4/ 6/1896. Source: O. Bamberg; Rothschild. Material: 1 ; whole/good egg. Location not traced.

Geronticus eremita

Waldrapp

Bred in Europe until the 1600s, in Syria until some time between 1910 and 1929 and, at one time, probably over much of North Africa and the Middle East. Now confined to a few scattered colonies in Morocco and perhaps the Middle East. Drainage and pesticides led to losses at the last known colony of the remnant eastern group in Turkey, where only three birds returned from migration in 1989 and two subsequently died in a storm. The eastern population was then thought to be extinct, but several birds were seen in 1990 and 1991 in Saudi Arabia. Following habitat degradation in Morocco, numbers had declined but there was a partial recovery to 430 birds by 1992.

Reg. no. 1860.8.24.3. From ; ; Algeria. Date: //. Source: Verreaux. Material: juv M; whole/good skin. This appears to be bird c. in *Cat. Bds.*, listed as 'Tunis', collected by L[ouis] Fraser. This has been altered in pencil to 'Algeria, purchased of Verreaux'. According to the register, this latter annotation is correct.

Reg. no. 1869.10.16.431. From 7,000 ft. at Antalo; Tigre; Ethopia (Abyssinia). Date: 11/ 5/1868. Source: W.T. Blanford. Material: adult F; whole/good skin. The reg. no. on the label (1869.10.17.431) is apparently erroneous, as 1869.10.17.– was never used. The correct reg. no. is given here.

Reg. no. 1879.4.19.4. From Mogador; ; Morocco. Date: //. Source: Howard Saunders coll.. Material: adult ; part/poor skin. Head only, tip of bill broken.

Reg. no. 1887.9.12.1. From Ain Tarsilt Pass; ; interior of Morocco. Date: //. Source: H.E. White. Material: adult F; whole/good skin. Nesting in a hole of a cliff, about 50 ft. high. Additional information taken from *Cat. Bds.*.

Reg. no. 1893.9.20.38. From nr. Suakin; ; Sudan (Red Sea). Date: //. Source: Capt. R.H. Penton. Material: adult ; whole/good skin.

Reg. no. 1904.4.12.54. From sea cliffs n. of Sallee, Atlas Mountains; ; Morocco. Date: 24/ 5/1901. Source: E.G.B. Meade Waldo. Material: juv ; whole/good skin.

Reg. no. 1905.10.11.72. From Karyatein, n. Syrian Desert; ; [? Syria]. Date: 18/ 5/1905. Source: D. Carruthers & J.H. Miller. Material: adult M; whole/good skin. Bill and feet dark red, iris orange, bare parts of head dark red, blue on top.

Reg. no. 1905.10.11.73. From Karyatein, n. Syrian Desert; ; [? Syria]. Date: 18/ 5/1905. Source: D. Carruthers & J.H. Miller. Material: adult F; whole/good skin. Bill and feet dark red, iris orange, bare parts of head dark red, blue on top.

Reg. no. 1905.12.22.386. From sea cliffs n. of Sallee; ; Morocco. Date: 24/ 5/1901. Source: E.G.B. Meade Waldo. Material: adult F; whole/good skin. Occiput orange, pink stripe down crown of head, lead coloured patch over eyes, neck and throat dark coral, legs coral, iris scarlet and gold.

Reg. no. 1919.12.11.56. From Azrou, Lesser Atlas; ; central Morocco. Date: 24/ 5/1919. Source: H. Lynes. Material: adult ; part/good skin. Head only. Found breeding in cliffs. See *Ibis* 1920: 276, 300.

Reg. no. 1927.1.31.1. From Wukr Ez Zbede, n. Syrian Desert; ; [? Syria]. Date: 25/ 5/1914. Source: Messrs Janson. Material: chick ; whole/good skin.

Reg. no. 1927.1.31.2. From Wukr Ez Zbede, n. Syrian Desert; ; [? Syria]. Date: 10/ 5/1914. Source: Messrs Janson. Material: chick ; whole/good skin.

Reg. no. 1953.45.1. From Chottel Chergui, in the desert s. of Oran; ; Algeria. Date: /11/1952. Source: Maj. W.H. Payn. Material: adult F; whole/good skin.

Reg. no. 1965.M.2454. From nr. Jidda; ; Saudi Arabia. Date: 3/ 3/1950. Source: Meinertzhagen coll.. Material: adult F; whole/good skin. Iris red, bill and feet dark red.

Reg. no. 1965.M.2455. From s. of Lawdar (Lodar), Aden; ; South Yemen (Arabia). Date: 13/12/1948. Source: R. Meinertzhagen. Material: adult M; whole/good skin.

Reg. no. 1988.20.1. From Mogador; ; Morocco. Date: //. Source: Howard Saunders coll.. Material: adult ; whole/good skin.

Reg. no. VEL.38.39a. From ; ; Ethopia (Abyssinia). Date: //. Source: Dr Ruppell. Material: adult ; whole/good skin.

Reg. no. E1941.1.4.73. From nr. Mogador; ; Morocco. Date: //1905. Source: F.W. Riggenbach coll.; Rothschild. Material: 3 ; whole/good eggs. Collected in spring 1905.

Reg. no. E1972.13.349. From River Neme; ; . Date: 20/ 5/. Source: J. Aharoni; Rothschild. Material: 2 ; whole/good eggs. Locality untraced (possibly in the Middle East; see 1973.1.427); spelling uncertain.

Reg. no. E1972.13.351. From El Hadjeb; ; Morocco. Date: 23/ 4/1928. Source: F.C.R. Jourdain; Rothschild. Material: 5 ; whole/good eggs. See 1973.1.428. *Ool. Rec.* 14 (1934): 2–5.

Reg. no. E1973.1.427. From El Hudriat, Syrian Desert; ; [Syria]. Date: 4/ 4/1931. Source: J. Aharoni; Jourdain coll.. Material: 4 ; whole/good eggs (clutch). From a nest on a high cliff.

Reg. no. E1973.1.428. From El Hadjeb, 30km from Mekness; ; Morocco. Date: 23/ 4/1928. Source: Jourdain coll.. Material: 3 ; whole/good eggs (clutch). Not very heavily incubated. From a colony of 30 pairs breeding in a deep recess of an overhung cliff. *Ool. Rec.* 14 (1934): 2–5. This clutch was missing and presumed stolen during 1975–1979. A clutch recovered subsequently was believed to be this one.

Reg. no. E1973.1.429. From El Hadjeb; ; Morocco. Date: 23/ 4/1928. Source: Jourdain coll.. Material: 3 ; whole/good eggs (clutch). Very hard set eggs, but with distinct differences in the stages of incubation. See 1973.1.428. *Ool. Rec.* 14 (1934): 2–5.

Reg. no. 1901.8.12.50. From Ain Oosera, 20 miles from Lagonat; ; Algeria. Date: / 6/1860. Source: Loche; Tristram; Crowley. Material: 1 ; whole/good egg. One of two eggs from Algeria given the same reg. no. in error.

Reg. no. 1901.8.12.50. From ; ; Algeria. Date: //. Source: Loche; Salvin–Godman coll.. Material: 1 ; whole/good egg. One of two eggs from Algeria given the same reg. no. in error.

Reg. no. E1972.13.350. From nr. El Hajeb; ; Morocco. Date: 24/ 5/1927. Source: ? Jourdain; Rothschild. Material: 1 ;

part/good egg. Egg broken and large end missing.

Reg. no. E1972.13.352. From ; ; Ethiopia. Date: //. Source: Rothschild. Material: 1 ; whole/good egg.

Reg. no. A1981.24.2. From ; ; Captivity. Date: 9/11/1980. Source: Jersey Wildlife Preservation Trust. Material: M; part/good alc. Bred in captivity. Eviscerated specimen.

Reg. no. A1981.24.3. From ; ; Captivity. Date: 10/ 3/1981. Source: Jersey Wildlife Preservation Trust. Material: F; part/good alc. Bred in captivity. Eviscerated specimen.

Reg. no. A1982.51.1. From ; ; Captivity. Date: 11/12/1981. Source: Jersey Wildlife Preservation Trust. Material: ; part/good alc. Bred in captivity. Eviscerated specimen.

Nipponia nippon

Crested Ibis

Formerly bred in Ussuriland, eastern Manchuria, China and Japan. Now breeds only in southern China (Shanxi province), where there were about 40 birds in 1992. The five remaining individuals from Sado Island, Japan, were taken into captivity in 1981 but failed to breed.

Reg. no. . From ; ; Japan. Date: //. Source: Material: adult ; skin. Mount or ex-mount. Specimen b. in *Cat. Bds.*. Not found 1988.

Reg. no. . From Shanghai; ; China. Date: //. Source: J. Webb. Material: adult ; skin. Mount. Details taken from *Cat. Bds.*. Not found 1988.

Reg. no. 1842.1.19.90. From ; ; Japan. Date: //. Source: Leyden Museum; Temminck; Leadbeater. Material: adult ; whole/good skin.

Reg. no. 1852.2.5.12. From ; ; . Date: //. Source: Material: ; whole/good skin. No data; reg. no. appears to be incorrect, as the register claims it was assigned to a different species.

Reg. no. 1891.10.19.20. From Soul; ; Korea. Date: //. Source: C.W. Campbell. Material: F; whole/good skin.

Reg. no. 1892.4.2.492. From Shanghai; ; China. Date: //. Source: Tweeddale coll.. Material: adult ; whole/good skin.

Reg. no. 1892.4.2.493. From Shapooshan dist.; Chekiang; China. Date: / 5/. Source: W. Pryer; Tweeddale coll.. Material: imm M; whole/good skin.

Reg. no. 1897.10.30.2. From ; ; Korea. Date: 1/ 2/1891. Source: C.W. Campbell. Material: adult F; skin. Information taken from a label, which may be from the specimen without data below (1988.17.1). Not found 1988.

Reg. no. 1897.10.30.3. From Hakodate (Hakodadi); [Hokkaido]; Japan. Date: / 4/. Source: Blakiston; Seebohm coll.. Material: F; whole/good skin. The reg. no. on the specimen is 1897.10.1.3, but the number does not exist in the registers. The correct number is believed to be 1897.10.30.3.

Reg. no. 1897.10.30.4. From Yokohama; ; Japan. Date: //. Source: H. Pryer. Material: adult ; whole/good skin.

Reg. no. 1897.10.30.5. From Ningpo; ; China. Date: 18/11/1872. Source: R. Swinhoe. Material: adult M; whole/good skin.

Reg. no. 1897.10.30.6. From Ningpo; ; China. Date: 18/11/1872. Source: Swinhoe coll.. Material: imm F; whole/good skin.

Reg. no. 1900.9.9.12. From ; Shenshi; n. China. Date: //. Source: Rev. Father Hugh. Material: ; whole/good skin.

Reg. no. 1900.9.9.13. From ; Shenshi; n. China. Date: //. Source: Rev. Father Hugh. Material: ; skin. Details taken from the register. Not found 1988.

Reg. no. 1908.1.5.20. From Anhuei; ; China. Date: 6/ 6/1901. Source: Styan coll.. Material: adult F; whole/good skin.

Reg. no. 1908.1.5.21. From Anhuei; ; China. Date: 2/ 6/1901. Source: Styan coll.. Material: adult M; whole/good skin.

Reg. no. 1908.1.5.22. From Anhuei; ; China. Date: 2/ 6/1901. Source: Styan coll.. Material: adult M; whole/good skin.

Reg. no. 1908.1.5.23. From Anhuei; ; China. Date: 4/ 6/1901. Source: Styan coll.. Material: M; whole/good skin.

Reg. no. 1908.1.5.24. From Anhuei; ; China. Date: 4/ 6/1901. Source: Styan coll.. Material: adult F; whole/good skin.

Reg. no. 1908.1.5.25. From Lu man Lsien; Shansi; China. Date: /11/1896. Source: Styan coll.. Material: adult M; whole/good skin.

Reg. no. 1908.1.5.26. From Lu man Lsien; Shansi; China. Date: /11/1896. Source: Styan coll.. Material: adult F; whole/good skin.

Reg. no. 1912.9.23.14. From Feng Hsiang; Shensi; China. Date: 1/ 1/1910. Source: Duke of Bedford coll.. Material: adult M; whole/good skin.

Reg. no. 1988.17.1. From ; ; [Asia]. Date: //. Source: Material: ; whole/good skin.

Reg. no. 1988.17.2. From ; ; Korea. Date: 15/ 1/1901. Source: Material: M; whole/good skin. Locality uncertain. Damaged: bill broken.

Tadorna cristata

Crested Shelduck

Status unknown. It is believed to have bred in eastern Siberia and China and to have migrated south to Korea and Japan. Although previously known only from 4 skins (not 3 as sometimes stated), and long presumed extinct, birds were seen alive in 1964, 1971 and probably 1990.

Reg. no. . From ; ; . Date: //. Source: Material: ; painting. No specimens, but there is a water-colour painting in the NHM library (Tring), of two birds, by Kobayashi.

Rhodonessa caryophyllacea

Pink-headed Duck

Formerly occurred in the Ganges and Brahmaputra basins from Bhutan to central Bihar, and along the eastern coast of India to Madras. Always rare and now probably extinct though it is possible that a few may still survive. For a number of years in the early 20th century birds were kept in captivity in Europe but attempts to breed the species were unsuccessful.

Reg. no. 1837.6.10.78. From ; ; n. India. Date: //. Source: Col. Cobbe. Material: adult ; whole/good skin. Ex-mount.

Reg. no. 1843.1.13.1274. From ; ; Nepal. Date: //. Source: B.H. Hodgson. Material: juv ; whole/good skin. This bird and 1843.1.13.1275 are identified in the register as *Branta*. The latter specimen was not found 1988, but may have been another of the same species.

Reg. no. 1880.1.1.1911. From ; ; Bhutan. Date: //. Source: Pemberton; India Museum. Material: adult ; whole/good skin. Ex-mount.

Reg. no. 1880.1.1.4783. From ; ; Bhutan. Date: //. Source: India Museum. Material: adult ; whole/good skin. Ex-mount.

Reg. no. 1880.1.1.4784. From ; ; Bhutan. Date: //. Source: Blyth. Material: juv ; whole/good skin.

Reg. no. 1889.4.3.11. From Calcutta; ; India. Date: / 1/1865. Source: H.J. Elwes. Material: adult F; whole/good skin.

Reg. no. 1889.4.3.12. From Purnea; ; India. Date: //1867. Source: H.J. Elwes. Material: juv ; whole/good skin. Date believed to be correct, but not clear on label.

Reg. no. 1894.6.1.129. From Calcutta; ; India. Date: 7/ 3/1879. Source: Hume coll.. Material: adult ; whole/good skin.

Reg. no. 1894.6.1.130. From Madras; ; India. Date: //1871. Source: Hume coll.. Material: imm F; whole/good skin.

Reg. no. 1896.4.2.1. From ; ; Captivity. Date: //. Source: Lord Lilford. Material: ; skin. Died in captivity. Details taken from the register. Not found <1986.

Reg. no. 1896.4.2.2. From ; ; Captivity. Date: 20/ 4/1896. Source: Lord Lilford. Material: F; whole/good skin. Died in captivity.

Reg. no. 1929.12.8.1. From ; ; Captivity. Date: 7/12/1929. Source: A. Ezra. Material: F; whole/good skin. Died in captivity. The skeleton of this bird is also in the collection, 1929.12.8.2.

Reg. no. 1933.4.2.1. From ; ; Captivity. Date: 4/ 4/1933. Source: A. Ezra. Material: F; whole/good skin. Died in captivity.

Reg. no. 1936.5.9.1. From ; ; Captivity. Date: 8/ 5/1936. Source: A. Ezra. Material: F; whole/good skin. Died in captivity.

Reg. no. 1949.WHI.1.6165. From ; ; n.e. India, Captivity. Date: 21/ 5/1931. Source: A. Ezra; H. Whistler. Material: M; whole/good skin. Died in captivity a few weeks after arriving from India.

Reg. no. 1949.WHI.1.6166. From ; ; Calcutta. Date: / 8/1894. Source: Material: F; skin. Not in the collection. Registered in error, property of the India Museum, Calcutta.

Reg. no. 1949.WHI.1.6167. From Tirhut, Baghownie; Bihar; India. Date: 23/ 8/1910. Source: C.M. Inglis; H. Whistler. Material: F; whole/good skin.

Reg. no. 1949.WHI.1.6168. From ; ; India. Date: 11/ 7/1905. Source: H. Whistler. Material: M; whole/good skin.

Reg. no. 1845.1.12.349. From ; ; Nepal. Date: //. Source: B.H. Hodgson. Material: ; part/good skeleton. Wings, legs and skull missing. Previously registered as Skeleton Vellum Cat. 457a.

Reg. no. 1929.12.8.2. From ; ; Captivity. Date: 7/12/1929. Source: A. Ezra. Material: adult F; whole/good skeleton. Died in captivity. The skin of this bird is also in the collection, 1929.12.8.1.

Reg. no. 1891.3.20.6019–22. From Dabeepoor Factory, Purnea (Purneah); ; India. Date: 3/ 7/1880. Source: T. Hill; Shillingford; Hume coll.. Material: 4 ; whole/good eggs (clutch). Four well incubated eggs from a clutch of nine. See Hume & Marshall, 1880, *The Game Birds of India, Burmah and Ceylon*, 3: 176–8. One egg has the date 3 July 1870 on it, presumably in error.

Reg. no. 1891.3.20.6018. From Zillah, Purnea (Purneah); ; India. Date: //. Source: F.A. Shillingford; Hume coll.. Material: 1 ; whole/good egg. See Hume & Marshall, 1880, *The Game Birds of India, Burmah and Ceylon*, 3: 176.

Reg. no. E1925.12.25.5391. From ; ; India. Date: //. Source: India Museum, Calcutta; Barnes coll.; Davidson. Material: 1 ; whole/good egg. Egg has 'Reilly' written on it in pencil.

Reg. no. E1925.12.25.5392. From Malda or Purneah; ; India. Date: //. Source: Hill; Irwin; Davidson. Material: 1 ; whole/good egg.

Reg. no. E1973.32.1. From Maldah; Bengal; India. Date: / 6/1898. Source: J. Shillingford; Stuart Baker coll.. Material: 1 ; whole/good egg. 1898 is the date on the egg, but Baker gives 1879 in his ms catalogue: '... and given to Irvine [?Irwin] from whom I purchased it with the rest of his colln.'.

Reg. no. 1932.10.8.1. From ; ; [India], Captivity. Date: 12/10/1932. Source: A. Ezra. Material: adult ; whole/good alc. Died in captivity.

Reg. no. 1936.1.16.3. From ; ; [India], Captivity. Date: 28/ 1/1936. Source: A. Ezra. Material: adult ; whole/good alc. Died in captivity.

Reg. no. 1936.2.17.1. From ; ; [India], Captivity. Date: 18/ 2/1934. Source: A. Ezra. Material: adult ; whole/good alc. Died in captivity. The register gives the date as 17 February 1936.

Aythya innotata

Madagascar Pochard

Confined to lakes and pools in the northern central plateau of Madagascar. Formerly common, especially on Lake Alaotra, since 1930 it has become increasingly rare, and may now be on the verge of extinction. It had not been seen for many years when a fisherman netted a male in August 1991. Shooting, and the introduction of exotic fish which upset the ecology have been blamed for its decline.

Reg. no. 1879.6.7.7. From Betsileo; ; Madagascar. Date: //. Source: Rev. W. Shaw. Material: adult M; whole/good skin. Holotype. Salvadori, *Bull. Brit. Orn. Club* 4 (1894): 2. There seems to be some confusion over the Rev. Shaw's initial: the label and the catalogue of type-specimens gives W., *Cat. Bds.* gives J. and the *History of the Collections* gives G.

Reg. no. 1931.8.18.1089. From Andreba; ; Madagascar. Date: 31/ 5/1929. Source: Archbold Vernay Expedition; American Museum of Natural History. Material: adult M; whole/good skin.

Reg. no. 1931.8.18.1090. From Andreba; ; Madagascar. Date: 5/ 6/1929. Source: Archbold Vernay Expedition; American Museum of Natural History. Material: adult M; whole/good skin.

Reg. no. 1931.8.18.1091. From Andreba; ; Madagascar. Date: 4/ 6/1929. Source: Archbold Vernay Expedition; American Museum of Natural History. Material: adult F; whole/good skin.

Reg. no. 1931.8.18.1092. From Andreba; ; Madagascar. Date: 2/ 6/1929. Source: Archbold Vernay Expedition; American Museum of Natural History. Material: adult M; whole/good skin.

Reg. no. 1931.8.18.1093. From Andreba; ; Madagascar. Date: 4/ 6/1929. Source: Archbold Vernay Expedition; American Museum of Natural History. Material: adult F; whole/good skin.

Reg. no. 1931.8.18.1094. From Andreba; ; Madagascar. Date: 31/ 5/1929. Source: Archbold Vernay Expedition; American Museum of Natural History. Material: adult M; whole/good skin.

Reg. no. 1931.8.18.1095. From Andreba; ; Madagascar. Date: 31/ 5/1929. Source: Archbold Vernay Expedition; American Museum of Natural History. Material: adult M; whole/good skin.

Reg. no. 1931.8.18.1096. From Andreba; ; Madagascar. Date: 31/ 5/1929. Source: Archbold Vernay Expedition; American Museum of Natural History. Material: adult M; whole/good skin.

Reg. no. 1931.8.18.1097. From Andreba; ; Madagascar. Date: 2/ 6/1929. Source: Archbold Vernay Expedition; American Museum of Natural History. Material: adult F; whole/good skin.

Reg. no. 1935.7.12.1. From ; ; Captivity. Date: 22/ 5/1935. Source: R. & N. Stevens. Material: chick ; whole/good skin.

Reg. no. 1931.10.5.9. From ; ; Captivity. Date: //. Source: J. Delacour. Material:

chick ; whole/good alc. Died in Cleres aviaries.

Reg. no. 1968.30.4. From Lac Alaotra; ; Madagascar. Date: //. Source: Rand Archbold Expedition. Material: ; whole/good alc.

Camptorhynchus labradorius

Labrador Duck

Formerly occurred on the eastern coast of North America in winter. Its breeding range was believed to be in Labrador or the islands to the north. The last known specimen was collected in 1875. The bird was apparently always rare, even in the 17th century.

Reg. no. 1863.12.15.27. From ; Labrador; Canada. Date: //. Source: Verreaux. Material: adult F; whole/good skin. Mount.

Reg. no. VEL.42.59a. From ; ; North America. Date: //. Source: Hudson's Bay Company. Material: adult M; whole/good skin. Mount.

Reg. no. . From ; ; [North America]. Date: //. Source: Material: 1 ; egg. Not in the collection. There is said to be an egg in a private collection in Scotland (correspondence in the Labrador Duck file in the NHM).

Reg. no. 1901.11.15.266. From Disco, Baffin Bay; ; Canada. Date: / 5/1855. Source: Material: – ; egg. Not in Oates *Cat. Eggs*, so presumably not found <1906.

Reg. no. E1962.1.559. From ; Labrador; Canada. Date: 8/ 6/. Source: Rodern coll.. Material: 1 ; whole/good egg. Inscribed 'Canard labrador' over a pencilled inscription 'Labrador 8 Juin'. See *Ibis* 1951: 305–6. This egg may not be correctly identified, but is probably either *C. labradorius* or *Anas rubripes*.

Reg. no. E1962.1.560. From n. of Baffin Bay; ; Canada. Date: //1855. Source: Piers [Tristram?]; Tristram coll.. Material: 1 ; whole/good egg. Collected by Piers (Tristram) of H.M.S. Investigator of the Franklin Search Expedition. In Tristram's cat. p. 668: 'brought from the north of Baffins bay'. In a letter it is recorded 'got as they were going out there lst spring'. Some doubt remains as to the correctness of the identification.

Mergus australis

Auckland Islands Merganser

Endemic to the Auckland Islands and now believed to be extinct, not having been seen since 1905. The species inhabited fresh and brackish waters, occasionally wandering to salt–water harbours. The extinction may have been due to the introduction of pigs, mice and rats to the islands in the 19th century.

Reg. no. 1875.11.6.14. From ; Auckland Islands; New Zealand. Date: /12/1874. Source: Baron von Hugel. Material: adult ; whole/good skin.

Reg. no. 1901.10.21.57. From McClelland's Inlet; Auckland Islands; New Zealand. Date: 4/ 1/1901. Source: Capt. Hutton; Earl of Ranfurly. Material: adult M; whole/good skin. Iris dark, culmen and top of lower mandible black, cutting edge of upper mandible and rest of lower mandible yellowish–orange, legs and feet orange, webs, joints and soles dusky.

Reg. no. 1901.10.21.58. From McClelland's Inlet; Auckland Islands; New Zealand. Date: 4/ 1/1901. Source: Capt. Hutton; Earl of Ranfurly. Material: adult F; whole/good skin. Iris dark, culmen and top of lower mandible black, cutting edge of upper mandible and rest of lower mandible yellowish–orange, legs and feet orange, webs, joints and soles dusky.

Reg. no. 1902.8.6.1. From ; Auckland Islands; New Zealand. Date: //. Source: K. Dixon. Material: adult M; whole/good skin. See the unregistered spirit specimen.

Reg. no. 1904.4.30.1. From ; Auckland Islands; New Zealand. Date: //. Source: A.J. Stewart. Material: ; whole/good skin.

Reg. no. 1904.8.4.1. From Darnley Harbour; Auckland Islands; New Zealand. Date: 9/ 7/1901. Source: J.P. Rolleston; Earl of Ranfurly. Material: adult F; whole/good skin. Iris brown, bill black on top, red underneath, legs and feet light red. The skeleton is also in the collection, 1904.8.4.2.

Reg. no. 1904.8.4.2. From Darnley Harbour; Auckland Islands; New Zealand. Date: 9/ 7/1901. Source: J.P. Rolleston; Earl of Ranfurly. Material: F; part/good skeleton. The skin is also in the collection, 1904.8.4.1.

Reg. no. 1904.8.4.3. From ; Auckland Islands; New Zealand. Date: 9/ 7/1901. Source: J.P. Rolleston; Earl of Ranfurly. Material: M; part/good skeleton. Day of collection not certain. From a mounted specimen. Parts of legs missing.

Reg. no. 1904.8.4.4. From ; Auckland Islands; New Zealand. Date: //. Source: J.P. Rolleston; Earl of Ranfurly. Material: F; whole/good skeleton. Skeleton from mounted specimen.

Reg. no. . From ; Auckland Islands; New Zealand. Date: //. Source: Material: adult M; part/good alc. Trunk only, possibly from skin 1902.8.6.1. Faded note in pencil in bottle reads 'Belong to the one standing on the upright. Tibia and [illegible] left in the stuffed specimen'.

Gymnogyps californianus

California Condor

Formerly occurred along the Pacific coast of North America from British Columbia to Baja California, and in prehistoric times across the southern USA to Florida. Widespread though not abundant during the 19th and early 20th centuries, the population gradually dwindled due to human interference and, latterly, poisoning from lead ingested with carrion. The last birds were taken into capivity in the late 1980s. In 1992, 50 birds were held in captivity and several pairs were breeding successfully. The first birds were reintroduced to the wild in January 1992.

Reg. no. 1888.10.10.156a. From Jolon; California; United States. Date: / 2/1885. Source: J. McCormack; Henshaw coll; Salvin–Godman coll.. Material: juv ; whole/good skin.

Reg. no. 1915.7.12.1. From San Luis Obispo Co.; California; United States. Date: 1/ 1/1892. Source: E.B. Towne jnr.; Rothschild. Material: M; whole/good skin.

Reg. no. 1939.12.9.3411. From San Luis Obispo Co.; California; United States. Date: 6/ 6/1897. Source: E.B. Towne jnr.; Rothschild coll.. Material: ; whole/good skin.

Reg. no. 1939.12.9.3414. From San Diego; California; United States. Date: / 7/1900. Source: E.B. Towne jnr.; Rothschild coll.. Material: adult M; whole/good skin.

Reg. no. 1955.6.N.20.134. From Monterey; California; United States. Date: //. Source: Taylor; J.H. Gurney. Material: adult M; part/good skin. Feet only; the rest of the specimen was destroyed in May 1919.

Reg. no. 1955.6.N.20.135. From Monterey; California; United States. Date: //. Source: Taylor; J.H. Gurney coll.. Material: adult M; whole/good skin.

Reg. no. 1955.6.N.20.136. From Monterey; California; United States. Date: //. Source: Taylor; J.H. Gurney coll.. Material: juv F; whole/good skin.

Reg. no. 1955.6.N.20.137. From Santa Lucia Mountains; California; United States. Date: //. Source: Taylor; J.H. Gurney coll.. Material: chick ; whole/good skin. From 5 to 7 days old; weighed 10 ounces. See *Ibis* 1859: 469–70; 1860: 278, pl. 9.

Reg. no. 1988.12.1. From Monterey; California; United States. Date: //. Source: Taylor; J.H. Gurney coll.. Material: adult ; whole/good skin. Ex-mount.

Reg. no. 1988.12.2. From Monterey; California; United States. Date: //. Source: Taylor; J.H. Gurney coll.. Material: juv ; whole/good skin.

Reg. no. 1988.21.32. From ; ; [United States]. Date: //. Source: Rothschild coll.. Material: adult ; whole/good skin. Mount in Tring gallery. Set up at 'The Jungle' (Rowland Ward).

Reg. no. VEL.5.2a. From ; California; United States. Date: //. Source: A. Menzies. Material: adult ; whole/good skin. Holotype. Shaw & Nodder, 1797, *Naturalist's Miscellany* 9: (1–5, unpaginated), pl. 301. This specimen is also listed in the Old Catalogue complied in the Museum by W.E. Leach in the early 1800s.

Reg. no. 1857.9.20.2. From ; California; United States. Date: //. Source: J.H. Gurney coll.. Material: ; whole/good skeleton. Previously registered as Skeleton Vellum Cat. 1381b. The register entries for this specimen and 1868.12.29.17 had been transposed.

Reg. no. 1868.12.29.17. From Monterey; California; United States. Date: //. Source: Zoological Society. Material: ; whole/good skeleton. Previously registered as Skeleton Vellum Cat. 1381a. The register entries for this specimen and 1857.9.20.2 had been transposed.

Reg. no. S1954.30.105. From ; [California]; [United States]. Date: //. Source: H.G.

Taylor. Material: ; part/good skeleton. Skull only.

Reg. no. E1941.1.6.760. From Santa Barbara Co.; California; United States. Date: 29/ 4/1897. Source: E.B. Towne; Rothschild coll. Material: 1 ; whole/good egg. The collection of what is believed to be this egg was described in *Museum* 4 (1898): 103, which also gives detailed notes of the nest site.

Reg. no. E1941.1.6.761. From Coast Range, Mountains of Ventura Co.; California; United States. Date: 2/ 4/1906. Source: Frank Arundel (Frank Fillmore); Rothschild coll.. Material: 1 ; whole/good egg. Incubation slightly advanced. The parent birds were previously observed at the nesting site and flushed from the nest at the time the egg was taken. Previously said to have been taken by Frank Fillmore, but this apparently refers to Frank Arundel of the town of Fillmore.

Reg. no. E1955.5.5. From San Rafel, Talarcitos Mountains, Carmel Valley,; California; United States. Date: 28/ 4/1859. Source: A.S. Taylor; J.H. Gurney coll.. Material: 1 ; whole/good egg. 'Weight 10.5 oz. avoir.. Colour, slightly tinged bluish.' *Ibis* 1859: 469–70; 1860: 278, pl. 8.

Chondrohierax uncinatus mirus
Grenada Hook-billed Kite

Found only on Grenada, Lesser Antilles, from where it was described in 1934, when it was already rare. Its decline is believed to have been due to hunting, a reduction in the giant snails on which it feeds, and destructive hurricanes. A survey in July 1987 estimated the population as 15–40 birds.

Reg. no. 1898.2.8.150. From ; ; Grenada. Date: 17/11/1891. Source: D.W Smith; Salvin–Godman coll.. Material: adult M; whole/good skin.

Chondrohierax uncinatus wilsoni
Cuban Hook-billed Kite

Formerly occurred over much of eastern and southern Cuba, but now restricted to a small area at the eastern end of the island. It occurs along rivers in forest and woodland where the tree–snails on which it feeds may be found. The snail shells are collected in large quantities as curios and this is considered the chief reason for the bird's decline. The present population is not known.

Reg. no. 1939.12.9.289. From La Patana, nr. Cape Marisi; ; e. Cuba. Date: / 4/1915. Source: Rothschild coll.. Material: adult F; whole/good skin.

Haliaeetus vociferoides
Madagascar Fish–Eagle

Now confined to rivers and coasts of the western side of central and northern Madagascar, but it may formerly have extended round more of the island. It was not rare in the 19th century and the reduction in numbers seems to have commenced in the 1930s. The reasons for its decline are not clear.

Reg. no. 1884.4.10.1. From ; ; Madagascar. Date: //. Source: Capt. Shelley. Material: ; whole/good skin.

Reg. no. 1894.6.16.332. From ; ; Madagascar. Date: //. Source: Bewsher; Shelley coll.. Material: ; whole/good skin.

Reg. no. 1931.8.18.15. From coast w. of Mt. D'Ambre; ; n. Madagascar. Date: 5/11/1930. Source: J. Delacour; Archbold Vernay Expedition; Rand & Archbold coll.. Material: M; whole/good skin.

Reg. no. 1931.8.18.16. From Anorontsangana; ; n. Madagascar. Date: 8/1/1931. Source: J. Delacour; Archbold Vernay Expedition; Rand & Archbold coll.. Material: F; whole/good skin.

Reg. no. 1931.8.18.17. From coast opp. Nossibe Island; ; n. Madagascar. Date: 9/12/1930. Source: J. Delacour; Archbold Vernay Expedition; Rand & Archbold coll.. Material: M; whole/good skin.

Reg. no. 1931.8.18.18. From Ampasimena; ; Madagascar. Date: 29/12/1930. Source: J. Delacour; Archbold Vernay Expedition; Rand & Archbold coll.. Material: M; whole/good skin.

Reg. no. 1931.8.18.19. From coast opp. Nossibe Island; ; n. Madagascar. Date: 7/12/1930. Source: J. Delacour; Archbold Vernay Expedition; Rand & Archbold coll.. Material: F; whole/good skin.

Reg. no. 1931.8.18.20. From Marotony, Prov. of Nossibe; ; n. Madagascar. Date: 7/1/1931. Source: J. Delacour; Archbold Vernay Expedition; Rand & Archbold coll.. Material: M; whole/good skin.

Reg. no. 1931.8.18.21. From Ampasimena; ; Madagascar. Date: 28/12/1930. Source: J. Delacour; Archbold Vernay Expedition; Rand & Archbold coll.. Material: F; whole/good skin.

Reg. no. 1931.8.18.22. From Lake Kinkony; ; n.w. Madagascar. Date: 21/3/1931. Source: J. Delacour; Archbold Vernay Expedition; Rand & Archbold coll.. Material: M; whole/good skin.

Reg. no. 1931.8.18.23. From coast opp. Nossibe Island; ; n. Madagascar. Date: 4/12/1930. Source: J. Delacour; Archbold Vernay Expedition; Rand & Archbold coll.. Material: M; whole/good skin.

Reg. no. 1955.6.N.20.671. From ; ; Mauritius. Date: //. Source: Sir E. Newton; Gurney coll.; Norwich Castle Museum. Material: imm ; part/good skin. Head and foot only.

Reg. no. 1955.6.N.20.672. From Majunga; ; Madagascar. Date: / 7/1893. Source: Boucard; Gurney coll.; Norwich Castle Museum. Material: imm ; whole/good skin.

Haliaeetus leucocephalus leucocephalus
Southern Bald Eagle

Formerly occurred widely through North America from latitude 40°N south to northern Mexico. Now breeds only in fragmented locations through its former range but, under protection, has made a remarkable recovery in numbers. The validity of the subspecies is questionable.

Reg. no. . From ; ; [North America]. Date: //. Source: Material: ; skeleton. There are four skeletons and two skulls in the collection, but they have not been identified to subspecies and bear no locality data.

Reg. no. E1955.5.65. From ; ; North America. Date: //. Source: J.H. Gurney coll.. Material: 2 ; whole/good eggs. Locality unknown; these eggs may not be from the endangered subspecies *leucocephalus*.

Reg. no. 1859.1.26.1. From Orcas Island; Washington; United States. Date: //. Source: Dr Lyall. Material: imm ; whole/good skin.

Reg. no. 1859.1.26.2. From Orcas Island; Washington; United States. Date: //. Source: Dr Lyall. Material: imm ; whole/good skin.

Reg. no. 1888.10.10.336. From Sing Sing; New York; United States. Date: 5/ 6/1881. Source: A.K. Fisher; Henshaw coll.; Salvin–Godman coll.. Material: adult F; whole/good skin. Year given as 1881 on the label, but as 1888 in the register.

Reg. no. 1888.10.10.337. From West Point; New York; United States. Date: 15/ 2/1878. Source: E.A. Mearns; Salvin–Godman coll.. Material: adult M; whole/good skin. Detailed bare parts descriptions available on the collector's label.

Reg. no. 1890.4.28.256. From Tarpon Springs; Florida; United States. Date: 30/10/1886. Source: W.E.D. Scott; Salvin–Godman coll.. Material: adult F; whole/good skin.

Reg. no. 1890.4.28.257. From Tarpon Springs; Florida; United States. Date: 11/12/1886. Source: Salvin–Godman coll.. Material: adult F; whole/good skin.

Reg. no. 1890.4.28.412. From Tarpon Springs; Florida; United States. Date: 30/11/1886. Source: W.E.D. Scott; Salvin–Godman coll.. Material: adult F; whole/good skin.

Reg. no. 1906.12.7.589. From San Rafael; California; United States. Date: 7/ 2/1886. Source: G.S. Miller. Material: imm ; whole/good skin.

Reg. no. 1906.12.7.590. From Oneida Lake; New York; United States. Date: 30/ 8/1886. Source: G.S. Miller. Material: imm F; whole/good skin.

Reg. no. 1906.12.7.591. From Geneva; New York; United States. Date: 3/ 2/1888. Source: G.S. Miller. Material: adult M; whole/good skin.

Reg. no. 1955.6.N.20.681. From ; Florida; United States. Date: //. Source: Smithsonian Institution; J.H. Gurney coll.. Material: chick ; whole/good skin.

Reg. no. 1955.6.N.20.682. From ; Florida; United States. Date: //. Source: Whitley; J.H. Gurney coll.. Material: chick ; whole/good skin.

Reg. no. 1955.6.N.20.683. From Atlantic Coast; ; United States. Date: //. Source: Gurney coll.. Material: imm M; whole/good skin.

Reg. no. 1955.6.N.20.686. From Monterey; California; United States. Date: //. Source:

A.S. Taylor; Gurney coll.. Material: imm ; whole/good skin.

Reg. no. 1955.6.N.20.687. From Scotch Key; Florida; United States. Date: / 1/1883. Source: Whitley; J.H. Gurney coll.. Material: imm F; whole/good skin.

Reg. no. 1955.6.N.20.688. From ; ; North America. Date: //. Source: J. Wolley; Gurney coll.. Material: ; skin. Not in the collection. Described as an old, dirty specimen and destroyed with the Trustees' approval in October 1959.

Reg. no. 1988.21.15. From ; ; North America. Date: //. Source: Earl of Derby. Material: adult M; whole/good skin. Ex-mount. Skin a. in the *Cat. Bds.*.

Reg. no. E1941.4.6.557. From ; Maryland; United States. Date: / 2/1846. Source: Dr Heermann; Buckley coll.. Material: 2 ; whole/good eggs. Two eggs from Maryland, collected separately, but registered together by mistake. The date refers to only one of the eggs; the other is undated. These eggs were stolen during 1975–1979, and subsequently recovered. Other eggs with faked registration numbers had been substituted in the meantime.

Reg. no. 1891.3.1.488. From Santa Cruz Island; California; United States. Date: //1875. Source: H.W. Henshaw coll.; Godman Salvin coll.. Material: 1 ; whole/good egg.

Reg. no. 1901.2.25.307. From Acron; Florida; United States. Date: //. Source: Leopold Field coll.; Crowley. Material: 1 ; whole/good egg.

Reg. no. 1902.10.25.166. From Chesterton; South Carolina; United States. Date: 8/ 1/1900. Source: Radcliffe–Saunders coll.. Material: 1 ; whole/good egg. Data taken from the register, no slip or label with the egg.

Reg. no. E1955.5.66. From ; e. Florida; United States. Date: 1/ 3/1876. Source: Leopold Field coll.; J.H. Gurney coll.. Material: 1 ; whole/good egg.

Reg. no. E1958.6.39. From Prince Island, Santa Barbara Co.; California; United States. Date: 8/ 3/1920. Source: S. Peyton; M.C. Badger; D. Moore Lindsay coll.. Material: 1 ; whole/good egg. Egg fresh. Nest of large sticks, lined with sea grass and sheep's wool, on the face of a cliff, 50 ft. down from the top.

Reg. no. E1988.4.2. From Chatham Bay; Florida; United States. Date: 19/ 1/1889. Source: Rothschild coll.. Material: 1 ; whole/good egg (clutch). One egg from a clutch of two.

Haliaeetus pelagicus niger

Heude's Sea–Eagle

Formerly bred in Korea, and perhaps also in Ussuriland and Dauria. It has not been seen since 1887 and is believed to be extinct.

Reg. no. 1939.12.9.3575. From ; ; Korea. Date: 7/12/1910. Source: T. Angelo; Rothschild coll.. Material: adult M; whole/good skin. Ex–mount.

Reg. no. 1955.6.N.20.660a. From ; ; Korea. Date: //. Source: Zoological Society;

Norwich Castle Museum. Material: adult M; whole/good skin. Presumably died in captivity.

Eutriorchis astur

Madagascar Serpent Eagle

Confined to the rainforests of eastern Madagascar, and formerly known from only 11 specimens collected more than 50 years ago. It had not definitely been recorded since 1930 until September 1988 when one was seen in the Marojejy Reserve. In February 1990 one was found dead in the Ambatovaky Special reserve.

Reg. no. 1875.2.1.19. From ; ; s. Madagascar. Date: //. Source: A. Crossley; W. Cutter. Material: adult ; whole/good skin. Holotype. Sharpe, *Proc. zool. Soc. Lond.* 1875: 73, pl. 13.

Reg. no. 1991.2.1–3. From Ambatovaky Special Reserve (49.08E 16.51S); Toamasina Province; Madagascar. Date: 23/2/1990. Source: Dr C.J. Raxworthy. Material: ; part/good skin. 7th, 8th & 9th primaries from the same individual, registered separately. The skull is also in the collection, S1991.11.1. See also *Bull. Brit. Orn. Club* 112: 108–111.

Reg. no. S1991.11.1. From Ambatovaky Special Reserve (49.08E 16.51S); Toamasina Province; Madagascar. Date: 23/2/1990. Source: Dr C.J. Raxworthy. Material: ; part/good skeleton. Skull only. There are also feathers from this bird in the collection, 1991.2.1–3. See also *Bull. Brit. Orn. Club* 112: 108–111.

Reg. no. 1968.30.25. From 40 km n.w. of Maroantsetra,; ; Madagascar. Date: 8/5/1930. Source: Rand Archbold Expedition. Material: adult F; part/good alc. Trunk only.

Accipiter francesii pusillus

Anjouan Goshawk

Known only from Anjouan Island in the Comoros, where it inhabits the evergreen forest. Once common, it has been seriously affected by deforestation, and is now confined to the two small areas of forest still remaining. The present population is unknown, but the bird is believed to be near extinction.

Reg. no. 1846.7.16.8. From ; [Anjouan Island]; [Comoro Islands] (Madagascar) (Java). Date: //. Source: Massena coll.; E. Wilson. Material: adult ; whole/good skin. Locality is given as Madagascar on the label; *Falco cuculoides*, Java, male, in the register.

Reg. no. 1846.7.16.9. From ; [Anjouan Island]; [Comoro Islands] (Madagascar) (Java). Date: //. Source: Massena coll.; E. Wilson. Material: adult F; skin. Listed as *Falco cuculoides*, Java, female, in the register. Not found 1988. See 1846.7.16.8.

Reg. no. 1873.10.7.1. From ; Anjouan Island (Joanna Island); Comoro Islands. Date: //. Source: Bartle Frere. Material: adult M; whole/good skin. The reg. no. was given as 1873.10.17.1 on the label in error.

Reg. no. 1888.5.4.35. From ; Anjouan Island; Comoro Islands. Date: //. Source:

G.A. Frank jnr.. Material: imm ; whole/good skin.

Reg. no. 1888.5.4.5. From ; Anjouan Island; Comoro Islands. Date: //. Source: G.A. Frank jnr.. Material: adult ; whole/good skin.

Reg. no. 1888.5.4.6. From ; Anjouan Island; Comoro Islands. Date: //. Source: G.A. Frank jnr.. Material: adult ; whole/good skin.

Reg. no. 1888.5.4.7. From ; Anjouan Island; Comoro Islands. Date: //. Source: G.A. Frank jnr.. Material: adult M; whole/good skin.

Reg. no. 1888.5.4.8. From ; Anjouan Island; Comoro Islands. Date: //. Source: G.A. Frank jnr.. Material: imm ; whole/good skin.

Reg. no. 1890.1.30.10. From ; Anjouan Island; Comoro Islands. Date: //. Source: A. Boucard. Material: imm F; whole/good skin.

Reg. no. 1890.1.30.5. From ; Anjouan Island; Comoro Islands. Date: //. Source: A. Boucard. Material: adult M; whole/good skin.

Reg. no. 1890.1.30.6. From ; Anjouan Island; Comoro Islands. Date: //. Source: A. Boucard. Material: adult M; whole/good skin.

Reg. no. 1890.1.30.7. From ; Anjouan Island; Comoro Islands. Date: //. Source: A. Boucard. Material: adult M; whole/good skin.

Reg. no. 1890.1.30.8. From ; Anjouan Island; Comoro Islands. Date: //. Source: A. Boucard. Material: adult M; whole/good skin.

Reg. no. 1890.1.30.9. From ; Anjouan Island; Comoro Islands. Date: //. Source: A. Boucard. Material: imm F; whole/good skin.

Reg. no. 1894.6.16.178. From ; Anjouan Island; Comoro Islands. Date: //. Source: J. Kirk; G.E. Shelley. Material: adult ; whole/good skin. Damaged: tail detached from body.

Reg. no. 1894.6.16.179. From ; Anjouan Island; Comoro Islands. Date: //. Source: J. Kirk; G.E. Shelley coll.. Material: adult ; whole/good skin.

Reg. no. 1955.6.N.20.3028. From ; Anjouan Island (Joanna Island); Comoro Islands. Date: //. Source: Dr Dickinson; J.H. Gurney coll.. Material: adult F; whole/good skin. Syntype. Presumed female.

Reg. no. 1955.6.N.20.3029. From ; Anjouan Island (Joanna Island); Comoro Islands. Date: //. Source: Dr Dickinson; J.H. Gurney coll.. Material: adult M; whole/good skin. Syntype. Gurney, *Ibis* 1875: 358–9. Presumed male. The other two syntypes are also in the collection.

Reg. no. 1955.6.N.20.3030. From ; ; Comoro Islands. Date: //. Source: C.E. Bewsher; J.H. Gurney coll.. Material: adult M; whole/good skin. Syntype.

Reg. no. 1955.6.N.20.3031. From ; Anjouan Island (Joanna Island); Comoro Islands. Date: //. Source: A. Boucard; J.H. Gurney coll.. Material: imm ; whole/good skin. Ex-mount.

Reg. no. 1955.6.N.20.3032. From ; Anjouan Island; Comoro Islands. Date: //. Source: G.A. Frank jnr.; J.H. Gurney coll. Material: imm ; whole/good skin. Ex-mount.

Pithecophaga jefferyi

Monkey-eating Eagle

Formerly widely distributed in primary forest in the Philippincs, but now greatly reduced in range and mainly confined to Leyte, Luzon, Samar and parts of Mindanao. The estimated total population is about 35 birds (of which 13 are in captivity) and the species is gravely threatened. The decrease has been very largely due to the clearing of forests. Recent attempts to protect the eagle's habitat have largely failed. The species continues to decline as a result of illegal persecution, in spite of being fully protected by law and having been nominated as the national bird of the Philippines.

Reg. no. . From ; ; . Date: //. Source: Material: ; whole/good skin. Mount in South Kensington gallery. Measurements suggest that this is a female. It had probably been in captivity for some time. This is presumably the missing specimen 1910.2.11.1 (*fide* C. Edelstam).

Reg. no. 1897.6.14.31. From Paranas; Samar; Philippines. Date: 15/ 6/1896. Source: J. Whitehead. Material: imm M; whole/good skin. Holotype. Ogilvie-Grant, *Bull. Brit. Orn. Club* 6 (1896): 17. In first moult. *Ibis* 1897: 214-20.

Reg. no. 1910.2.11.1. From ; Luzon; Philippines, [Captivity]. Date: 11/ 2/1910. Source: Zoological Society. Material: F; skin. Presumably died in captivity. Details taken from the register. Not found 1988. This was probably the bird from which the skeleton 1910.2.11.1a came.

Reg. no. 1953.48.1. From ; ; Philippines, Captivity. Date: 18/ 5/1953. Source: Zoological Society. Material: adult F; whole/good skin. Caught in the Philippines, died in captivity at London Zoo.

Reg. no. 1955.6.N.20.476. From ; ; Philippines. Date: //. Source: J.H. Gurney coll.. Material: imm ; whole/good skin. Ex-mount. In first moult.

Reg. no. 1910.2.11.1a. From ; Luzon; Philippines, [Captivity]. Date: 11/ 2/1910. Source: Zoological Society. Material: ; whole/good skeleton. Died in captivity. From a mounted specimen in the gallery.

Reg. no. S1955.22.5. From ; ; Philippines, Captivity. Date: 18/ 5/1953. Source: Zoological Society. Material: adult F; part/good skeleton. Caught in Philippines, died in captivity. The bird was received at the Zoo on 22 October 1939. Rough skeleton; trunk only, with flesh adhering. Previously preserved in formalin. The skin is also in the collection, 1953.48.1.

Reg. no. S1961.23.1. From ; ; Captivity. Date: /11/1961. Source: Zoological Society.

Material: ; whole/good skeleton. Died in captivity.

Reg. no. A1973.4.1. From ; ; Philippines, Captivity. Date: 26/10/1966. Source: Zoological Society. Material: ; part/good alc. Received by the Zoo 11 July 1963. Died in captivity. Eviscerated specimen.

Aquila heliaca adalberti

Spanish Imperial Eagle

Formerly widespread over the Iberian peninsula, northern Morocco and northeastern Algeria. Now restricted to southern Spain and Portugal where it is still threatened by poisoning, shooting and habitat destruction. About 100 pairs are thought to remain.

Reg. no. . From ; ; s. Europe. Date: //. Source: E. Wilson. Material: juv ; skin. Mount. Details taken from the *Cat. Bds.*. Not found 1988.

Reg. no. 1871.8.15.2. From Seville; ; Spain. Date: 24/ 9/1870. Source: Ruiz; E. Bartlett. Material: adult F; whole/good skin.

Reg. no. 1871.8.15.3. From Seville; ; Spain. Date: 15/10/1870. Source: Ruiz; E. Bartlett. Material: imm F; whole/good skin.

Reg. no. 1875.11.22.49. From Coto del Rey; ; Spain. Date: 12/ 4/1870. Source: H. Saunders. Material: adult F; skin. Details taken from the register. Not found 1988.

Reg. no. 1875.11.22.50. From Madre del Rocco; ; Spain. Date: 5/10/1869. Source: H. Saunders. Material: imm F; whole/good skin.

Reg. no. 1879.4.5.316. From Madrid; ; Spain. Date: //1868. Source: Salvin–Godman coll.. Material: imm ; whole/good skin.

Reg. no. 1887.11.1.287. From nr. Seville; ; Spain. Date: / 5/1876. Source: H. Saunders; Tweeddale coll.. Material: adult ; whole/good skin.

Reg. no. 1887.11.1.288. From nr. Seville; ; Spain. Date: / 6/1876. Source: H. Saunders; Tweeddale coll.. Material: imm F; whole/good skin.

Reg. no. 1895.5.1.5. From Coto Donana; ; Spain. Date: 4/ 2/1879. Source: Shelley coll.. Material: imm F; whole/good skin.

Reg. no. 1905.6.28.1274. From Coto del Rey, Seville; ; Spain. Date: 22/ 3/1872. Source: Lilford coll.; E.C. Cavendish Taylor. Material: adult F; whole/good skin.

Reg. no. 1934.1.1.1098. From Coto Donana; Andalucia; Spain. Date: 25/ 4/1898. Source: H.F. Witherby coll.. Material: adult F; whole/good skin.

Reg. no. 1934.1.1.1099. From Coto Donana; Andalucia; Spain. Date: 25/ 4/1898. Source: H.F. Witherby coll.. Material: chick ; whole/good skin.

Reg. no. 1934.1.1.1100. From Coto Donana; Andalucia; Spain. Date: 25/ 4/1898. Source: H.F. Witherby coll.. Material: chick ; whole/good skin.

Reg. no. 1955.6.N.20.250. From ; ; Spain. Date: //. Source: Dr R. Brehm; J.H. Gurney coll.. Material: imm M; whole/good skin.

Reg. no. 1955.6.N.20.251. From ; ; Spain. Date: //. Source: Dr R. Brehm; Gurney coll.. Material: imm F; whole/good skin.

Reg. no. 1955.6.N.20.252. From Seville; ; Spain. Date: 26/ 3/1874. Source: Lt. Col. L.H. Irby; Gurney coll.. Material: adult F; whole/good skin.

Reg. no. 1955.6.N.20.253. From Seville; ; Spain. Date: / 9/1869. Source: H. Saunders; W.M. Crowfoot; Gurney coll.. Material: imm F; whole/good skin. Collected in September–October.

Reg. no. 1955.6.N.20.254. From ; ; Spain. Date: //. Source: Lord Lilford; Anderson; W.M. Crowfoot; Gurney coll.. Material: adult F; whole/good skin.

Reg. no. 1901.2.25.457–8. From ; ; Spain. Date: 14/ 3/1888. Source: Leopold Field coll.; P. Crowley. Material: 2 ; whole/good eggs. Eggs missing, believed stolen during 1975–1979.

Reg. no. 1901.2.25.459–460. From ; ; Spain. Date: 10/ 4/1878. Source: Leopold Field coll.; P. Crowley. Material: 2 ; whole/good eggs. Eggs missing, believed stolen during 1975–1979. Missing by October 1977.

Reg. no. E1941.4.6.578. From Andalucia; ; Spain. Date: 18/ 3/1874. Source: Alberto Ruiz & Manuel Llano; M. Browne; Buckley coll.. Material: 2 ; whole/good eggs. One egg dated 18 March, and one the following day.

Reg. no. . From nr. El Palacio, Coto de Donana; Andalucia; Spain. Date: 12/ 4/1907. Source: Jourdain coll.. Material: 3 ; eggs (clutch). Incubation advanced. Bird flushed from the nest on climbing. A large nest of sticks, with flattened top, lined with wilted boughs of cork oak, among topmost branches of an enormous 'alcornoque'. Egg not found <1988; probably never received by the Museum.

Reg. no. . From Sierra de Las Cabras, nr. Jerez de La Frontera; Cadiz; Spain. Date: / 3/1906. Source: J.V. Benitez; Jourdain coll.. Material: 2 ; eggs (clutch). Eggs slightly incubated. Benitez took the eggs of this pair in 1905 and 1906. Not found <1988.

Reg. no. 1884.9.2.132,134. From Coto del Rey; ; Spain. Date: 4/ 4/1876. Source: Ruiz; Salvin–Godman coll.. Material: 2 ; whole/good eggs (clutch).

Reg. no. 1906.4.19.60–61. From Pinales de A; ; Spain. Date: 13/ 4/1870. Source: E.C. Cavendish Taylor. Material: 2 ; part/whole eggs (clutch). One egg is whole and in good condition, the other has a large portion of shell missing. The female was shot, but it is not in the collection.

Reg. no. E1901.1.1.7361–2. From Coto del Rey; ; Spain. Date: 12/ 4/1870. Source: H. Saunders; Seebohm coll.. Material: 2 ; eggs (clutch). Not found <1980.

Reg. no. E1901.1.1.7363–4. From Coto del Rey; ; Spain. Date: 18/ 4/1869. Source: H. Saunders; Seebohm coll.. Material: 2 ; whole/good eggs (clutch).

Reg. no. E1914.10.1.5288–9. From Mulahacen, Sierra Nevada; Granada; Spain.

Date: 29/ 3/1902. Source: O. Bamberg; Radcliffe–Saunders coll.. Material: 2 ; whole/good eggs (clutch).

Reg. no. E1919.12.1.5512–3. From Coto Donana; ; Spain. Date: / 4/1901. Source: Selous coll.. Material: 2 ; whole/good eggs (clutch).

Reg. no. E1919.12.1.5514–6. From Coto Donana; ; Spain. Date: 8/ 4/1900. Source: Selous coll.. Material: 3 ; whole/good eggs (clutch).

Reg. no. E1941.1.6.504. From Coto da Ana; ; Spain. Date: 6/ 4/1876. Source: C. Rothschild coll. Material: 2 ; whole/good eggs (clutch).

Reg. no. E1941.1.6.505. From Coto de Donana; ; Spain. Date: 27/ 4/1884. Source: ACS; C. Rothschild coll.. Material: 2 ; whole/good eggs (clutch).

Reg. no. E1941.1.6.508. From Coto del Rey, Seville; ; Spain. Date: 31/ 3/1908. Source: C. Rothschild coll.. Material: 3 ; whole/good eggs (clutch).

Reg. no. E1973.1.842. From nr. Choza de Vasquez, Coto de Donana; ; Spain. Date: 10/ 3/1903. Source: R. Vasquez; J.V. Benitez; Jourdain coll.. Material: 2 ; whole/good eggs (clutch). The eggs were nearly fresh. The nest was a huge structure in a pine tree among the sand dunes.

Reg. no. E1973.1.843. From Majada Real, Coto de Donana; Andalucia; Spain. Date: 21/ 2/1906. Source: R. Vasquez; R.H. Read; Jourdain coll.. Material: 3 ; whole/good eggs (clutch). Eggs nearly fresh, slight trace of incubation in one. The nest was right at the top of the tree, in a clump of three pines among the sandhills, about 200 yds from a 'corral'. Description of nest, including size, and a photograph of the tree available; also egg weights.

Reg. no. . From Forest of Zeid [Kef Oulsed Zeid]; ; Algeria. Date: 9/ 4/1857. Source: [La Tope; Tristram;] Salvin–Godman coll.. Material: 1 ; egg. Not in the collection since about 1906; apparently now in the US National Museum, Washington (*Bull. Brit. Orn. Club* 109 (1989): 86–93). [One of a clutch of two hard set eggs from a nest in a tree growing out of a rock.]

Reg. no. . From nr. El Palacio, Coto de Donana; ; Spain. Date: / 5/1906. Source: A. Ximenes; Jourdain coll.. Material: 1 ; egg. Egg fresh. Both birds were seen at the nest. Nest high up in a big cork oak. Found 3rd May, when it was lined with fresh green leaves and evidently just ready for eggs. A. Ximenes visited the place later on and took the egg, which was sent to Jourdain. Not found <1988; probably never received.

Reg. no. . From Coto de da Ana; ; Spain. Date: 5/ 3/1877. Source: Montague Browne; Buckley coll.. Material: 1 ; egg. Not found <1988.

Reg. no. 1884.9.2.133. From Coto del Rey; ; Spain. Date: 20/ 3/1875. Source: Ruiz; Salvin–Godman coll.. Material: 1 ; whole/good egg.

Reg. no. 1890.9.30.128. From Coto del Rey; ; Spain. Date: 28/ 3/1873. Source: Lord Lilford. Material: 1 ; whole/good egg.

Reg. no. 1901.2.25.462. From Madrid; ; Spain. Date: 14/ 4/1882. Source: Crowley. Material: 1 ; whole/good egg.

Reg. no. 1901.2.25.465. From Mount Edough, Bona; ; Algeria. Date: //1855. Source: Capt. Loche; Tristram coll; Crowley. Material: 1 ; whole/good egg. See *Bull. Brit. Orn. Club* 109 (1989): 86–93.

Reg. no. 1906.4.62.19. From Coto del Rey; ; Spain. Date: 28/ 3/1873. Source: E.C. Cavendish Taylor. Material: 1 ; whole/good egg.

Reg. no. E1901.1.1.7355. From Seville; ; Spain. Date: //. Source: Ruiz; Seebohm coll.. Material: 1 ; whole/good egg.

Reg. no. E1901.1.1.7357. From nr. Seville; ; Spain. Date: 2/ 4/1871. Source: Col. Irby; Seebohm coll.. Material: 1 ; whole/good egg.

Reg. no. E1901.1.1.7358. From Coto del Donana; ; Spain. Date: 8/ 4/1874. Source: Col. Irby; Seebohm coll.. Material: 1 ; whole/good egg.

Reg. no. E1901.1.1.7359. From Coto del Rey; ; Spain. Date: 12/ 4/1870. Source: H. Saunders; Seebohm coll.. Material: 1 ; whole/good egg. The female was shot, but it is not in the collection.

Reg. no. E1901.1.1.7360. From Coto del Rey; ; Spain. Date: 7/ 4/1869. Source: H. Saunders; Seebohm coll.. Material: 1 ; whole/good egg.

Reg. no. E1941.1.6.500. From Sierra Morena; ; Spain. Date: 14/ 4/1871. Source: C. Rothschild coll.. Material: 1 ; whole/good egg. This egg has '*Aq. Naevioides o rapax*' stroked out and *A. adalberti* substituted.

Reg. no. E1941.1.6.506. From Almonte; Huelva; Spain. Date: 28/ 6/1904. Source: C. Rothschild coll.. Material: 1 ; whole/good egg.

Reg. no. E1941.4.6.579. From ; ; . Date: / 2/1877. Source: M. Browne; Buckley coll.. Material: 1 ; whole/good egg.

Reg. no. E1955.5.55. From ; Granada; Spain. Date: / 3/1876. Source: Leopold Field coll.; J.H. Gurney coll.. Material: 1 ; whole/good egg.

Reg. no. E1962.1.1289. From Seville; ; Spain. Date: //1872. Source: Col. Irby coll.. Material: 1 ; whole/good egg. In Oates' *Cat. Eggs*, but not in Seebohm ms catalogue.

Reg. no. E1972.11.179. From ; ; . Date: 9/ 4/1875. Source: Rodern coll.. Material: 1 ; whole/good egg.

Reg. no. E1978.20.2. From ; ; Spain. Date: 17/ 3/1900. Source: L.S. Metcalfe. Material: 1 ; whole/good egg.

Reg. no. 1901.2.25.461. From Madrid; ; Spain. Date: 5/ 4/1883. Source: Crowley. Material: 1 ; whole/good egg (clutch). One egg from a clutch of three.

Reg. no. E1901.1.1.7356. From Cork Wood; ; Gibraltar. Date: 1/ 3/. Source: Col. Irby; Seebohm coll.. Material: 1 ; whole/good egg (clutch). Date given in

Oates' *Cat. Eggs* but not in Seebohm ms cat.. One egg from a probable clutch of three. Presumably collected about 1868–1874, when Irby was serving in Gibraltar.

Reg. no. E1941.1.6.503. From Coto de da Ana; ; Spain. Date: 5/ 3/1877. Source: C. Rothschild coll.. Material: 1 ; whole/good egg (clutch). One egg from a clutch of two.

Polyborus lutosus

Guadalupe Caracara

Formerly confined to Guadalupe Island, but extinct since 1900. It was deliberately exterminated by settlers who believed that it killed the kids of introduced goats. These goats have now despoiled the island and caused the extinction of several other endemic birds.

Reg. no. 1887.5.1.34. From Guadalupe Island; (California); Mexico. Date: //. Source: E. Palmer; Smithsonian Inst.; Salvin–Godman coll.. Material: adult ; whole/good skin.

Reg. no. 1887.5.1.35. From Guadalupe Island; (California); Mexico. Date: //. Source: E. Palmer; Smithsonian Inst.; Salvin–Godman coll.. Material: juv M; whole/good skin.

Reg. no. 1955.6.N.20.32. From Guadalupe Island; ; Mexico. Date: //. Source: E. Palmer; Smithsonian Inst.; Gurney coll.; Norwich Castle Museum. Material: adult ; whole/good skin.

Reg. no. 1955.6.N.20.33. From Guadalupe Island; (California); Mexico. Date: / 5/1875. Source: E. Palmer; Smithsonian Inst.; Gurney coll.; Norwich Castle Museum. Material: imm ; whole/good skin.

Falco punctatus

Mauritius Kestrel

Confined to Mauritius. Believed to have occurred formerly throughout the island, but now restricted to the remoter parts of the southwest. In the early 1970s the population was thought to be as low as 4 birds, but by the early 1980s this figure had climbed to 5 or 6 pairs with 10 additional birds in captivity. The decline was due largely to the reduction of the primary forest on which the species depends.

Reg. no. 1843.2.28.32. From ; ; Mauritius (Madagascar). Date: //. Source: Material: adult M; skin. Details taken from the register. Not found 1988.

Reg. no. 1843.2.28.33. From ; ; Mauritius. Date: //. Source: Material: adult F; whole/good skin. Ex–mount. Probably female.

Reg. no. 1874.5.13.1. From ; ; Mauritius. Date: 24/11/1866. Source: E. Newton. Material: ; whole/good skin. Damaged: head very loose.

Reg. no. 1894.6.16.407. From ; ; Mauritius. Date: //. Source: Shelley coll.. Material: ; whole/good skin.

Reg. no. 1955.6.N.20.1355. From ; ; Mauritius. Date: //. Source: Telfair; J.H. Gurney coll.; Norwich Castle Museum. Material: adult M; whole/good skin.

Reg. no. 1955.6.N.20.1356. From ; ; Mauritius. Date: //. Source: Cockburn; J.H. Gurney coll.. Material: M; whole/good skin.

Reg. no. 1955.6.N.20.1357. From ; ; Mauritius. Date: //. Source: F. Bedingfield; J.H. Gurney coll.. Material: M; whole/good skin.

Reg. no. 1955.6.N.20.1358. From ; ; Mauritius. Date: //1858. Source: F. Bedingfield; J.H. Gurney coll.. Material: F; part/good skin. Ex-mount. Tarsometatarsus removed in 1982, now in skeleton collection.

Reg. no. 1955.6.N.20.1359. From ; ; Mauritius. Date: //. Source: F. Bedingfield; J.H. Gurney coll.. Material: F; whole/good skin.

Reg. no. 1955.6.N.20.1360. From ; ; Mauritius. Date: //. Source: J.H. Gurney coll.. Material: adult M; whole/good skin.

Reg. no. 1955.6.N.20.1361. From ; ; Mauritius. Date: //. Source: J.H. Gurney coll.. Material: adult F; whole/good skin. Ex-mount.

Reg. no. 1955.6.N.20.1362. From ; ; Mauritius. Date: 1/ 1/1860. Source: E. Newton; A. Newton; J.H. Gurney coll., Norwich Castle Museum. Material: adult M; whole/good skin.

Reg. no. 1955.6.N.20.1363. From ; ; Mauritius. Date: //. Source: A. Newton; A. Newton; J.H. Gurney coll.; Norwich Castle Museum. Material: M; whole/good skin.

Reg. no. 1988.21.31. From ; ; [Mauritius] (Madagascar). Date: //. Source: Rothschild coll.. Material: adult F; whole/good skin. Mount in Tring gallery.

Reg. no. E1919.11.1.1. From ; ; Mauritius. Date: //. Source: Capt. C.K.G. Dick. Material: 1 ; whole/good egg.

Reg. no. A1993.9.1. From ; ; Mauritius, Captivity. Date: 18/ 9/1978. Source: Government of Mauritius *per* Prof. J.E. Cooper. Material: adult M; whole/good alc. Died in captivity in Mauritius.

Reg. no. A1993.9.2. From ; ; Mauritius, Captivity. Date: 4/ 1/1980. Source: Government of Mauritius *per* Prof. J.E. Cooper. Material: adult F; whole/good alc. Died in captivity in Mauritius.

Falco peregrinus anatum/tundrius
Peregrine Falcon

Formerly occurred throughout the boreal forests and tundras of North America, but now greatly reduced in range. Not as severely threatened as was once thought. The race *tundrius* was described only in 1968. We have not attempted to identify the specimens to subspecies.

Reg. no. . From ; ; Mexico. Date: //. Source: R.B. Sharpe. Material: adult M; skin. Details taken from *Cat. Bds.*. Not found 1988.

Reg. no. 1839.5.30.1. From ; ; [St.] Nevis. Date: //. Source: W. Cottle. Material: juv M; whole/good skin. The reg. no. was erroneously given on the label as 1839.6.6.1.

Reg. no. 1855.7.11.3. From Arctic coast; ; n.w. America. Date: //. Source: Capt. Collinson, Arctic Expedition of H.M.S. Enterprise. Material: M; whole/good skin.

Reg. no. 1855.7.11.4. From Arctic coast; ; n.w. America. Date: //. Source: Capt. Collinson, Arctic Expedition of H.M.S. Enterprise. Material: F; whole/good skin.

Reg. no. 1872.11.8.16. From ; ; Greenland. Date: //. Source: Gould coll.. Material: M; whole/good skin. The reg. no. was erroneously given on the label as 1872.11.8.6.

Reg. no. 1873.10.31.14. From ; Labrador; [Canada]. Date: //. Source: Gerrard. Material: juv M; whole/good skin. The reg. no. was erroneously given on the label as 1873.10.13.14.

Reg. no. 1873.5.28.13. From ; ; Mexico. Date: //. Source: E. Verreaux; R.B. Sharpe. Material: F; whole/good skin.

Reg. no. 1873.6.5.4. From Lichtenfels; ; Greenland. Date: 7/ 9/1871. Source: D.O. Frisch; R.B. Sharpe. Material: juv F; whole/good skin.

Reg. no. 1874.1.14.3. From Fiskenaesset; ; Greenland. Date: 12/ 7/1872. Source: Harrick; R.B. Sharpe. Material: adult F; whole/good skin.

Reg. no. 1874.1.14.4. From Fiskenaesset; ; Greenland. Date: 7/10/1872. Source: Harrick; R.B. Sharpe. Material: juv F; whole/poor skin. Damaged: skin rotten and in pieces.

Reg. no. 1876.4.15.16. From 49th parallel; ; North America. Date: //. Source: Mr Dawson, North American Boundary Commission; Foreign Office. Material: ; whole/good skin.

Reg. no. 1885.8.19.2239. From ; ; Greenland. Date: //. Source: Hume coll.. Material: imm M; whole/good skin.

Reg. no. 1887.5.1.798. From Duenas; ; Guatemala. Date: 21/ 2/1858. Source: O. Salvin; Salvin–Godman coll.. Material: juv F; whole/good skin.

Reg. no. 1887.5.1.799. From Veragua; ; . Date: //. Source: E. Arce; Salvin–Godman coll.. Material: M; whole/good skin.

Reg. no. 1887.5.1.800. From Mendoza; ; Argentina. Date: / 2/1871. Source: Weishaupt; Salvin–Godman coll.. Material: M; whole/good skin.

Reg. no. 1888.10.10.349. From Fort Reliance; Yukon; Canada. Date: 16/10/1878. Source: E.W. Nelson; Salvin–Godman coll.. Material: juv F; whole/good skin. The locality is given as Fort Reliance, Alaska, in the register, where it is also stated to be an adult.

Reg. no. 1888.10.10.350. From Chicago; Illinois; United States. Date: 9/10/1874. Source: E.W. Nelson; Salvin–Godman coll.. Material: imm F; whole/good skin.

Reg. no. 1888.10.10.351. From New Haven; Connecticut; United States. Date: 29/ 6/1871. Source: C. Hart Merriam. Material: adult F; whole/good skin.

Reg. no. 1888.2.25.1. From ; California; United States. Date: //. Source: W. Crewdson. Material: ; whole/good skin.

Reg. no. 1890.4.28.259. From Corpus Christi; Texas; United States. Date: 19/10/1889. Source: F.B. Armstrong; Salvin–Godman coll.. Material: M; whole/good skin.

Reg. no. 1890.4.28.261. From Corpus Christi; Texas; United States. Date: 6/10/1889. Source: F.B. Armstrong; Salvin–Godman coll.. Material: imm M; whole/good skin.

Reg. no. 1890.4.28.262. From Corpus Christi; Texas; United States. Date: 7/10/1889. Source: F.B. Armstrong; Salvin–Godman coll.. Material: imm M; whole/good skin.

Reg. no. 1890.4.28.398. From Brownsville; Texas; United States. Date: 18/10/1888. Source: F.B. Armstrong; Salvin–Godman coll.. Material: adult F; whole/good skin.

Reg. no. 1890.4.28.399. From Brownsville; Texas; United States. Date: 8/ 2/1889. Source: F.B. Armstrong; Salvin–Godman coll.. Material: imm F; whole/good skin.

Reg. no. 1892.6.29.9. From Buenos Aires; ; [Argentina]. Date: //1858. Source: D.F. Withington; Sclater coll.. Material: ; whole/good skin. Identified in the register as *Buteo albicaudatus*, female, collected on 28 March 1884.

Reg. no. 1896.1.11.15. From Lake Ontario; ; Canada. Date: //. Source: C.J. Phelps. Material: ; whole/good skin.

Reg. no. 1896.5.9.13. From Guadalajana, Santa Cruz del Valle; Jalisco; Mexico. Date: 3/ 9/1892. Source: A.C. Buller; Salvin–Godman coll.. Material: adult F; whole/good skin.

Reg. no. 1897.11.10.251. From Noursaak; ; Greenland. Date: / 6/1871. Source: Hargitt coll.. Material: juv ; whole/good skin.

Reg. no. 1897.11.14.168. From Espartillar; ; Argentina. Date: 17/ 2/1890. Source: A.H. Holland. Material: M; whole/good skin.

Reg. no. 1897.11.14.169. From Espartillar; ; Argentina. Date: 26/12/1890. Source: A.H. Holland. Material: juv M; whole/good skin.

Reg. no. 1897.3.16.1. From Buenos Aires; ; Argentina. Date: //. Source: H.M. Upcher. Material: ; whole/good skin.

Reg. no. 1898.2.8.146. From ; ; St. Vincent. Date: 1/ 2/1890. Source: D.W. Smith; Salvin–Godman coll.. Material: F; whole/good skin.

Reg. no. 1898.7.12.258. From San Benito Island; Baja California; [Mexico]. Date: 12/ 4/1897. Source: C.K. Worthen; Salvin–Godman coll.. Material: adult F; whole/good skin.

Reg. no. 1905.1.30.740. From Estancia Las Rosas; Prov. de Santa Fe; [Argentina]. Date: /12/1894. Source: Beales; Sclater coll.. Material: ; whole/good skin. Received by Sclater in 1895.

Reg. no. 1906.12.7.596. From Boston Market; Massachussetts; United States.

Date: 11/10/1884. Source: G.J. Goodale; G.S. Miller. Material: juv F; whole/good skin.

Reg. no. 1906.12.7.597. From North Truro; Massachussetts; United States. Date: 1/10/1889. Source: G.S. Miller. Material: juv F; whole/good skin.

Reg. no. 1913.11.10.71. From n. Chincha Island; ; w. Peru. Date: 23/12/1912. Source: H.O. Forbes. Material: M; whole/good skin.

Reg. no. 1913.11.10.72. From n. Chincha Island; ; w. Peru. Date: 16/ 2/1912. Source: H.O. Forbes. Material: adult F; whole/good skin.

Reg. no. 1913.5.15.9. From Grindrod; British Columbia; Canada. Date: 14/ 9/1911. Source: J. Wynne. Material: imm F; whole/good skin.

Reg. no. 1914.12.1.344. From Tampico; ; Mexico. Date: 6/ 4/1888. Source: P.R. Lowe. Material: F; whole/good skin. The year is almost illegible on the label; it is given as 1908 in the register.

Reg. no. 1920.5.20.18. From Ajo, Buenos Aires; ; Argentina. Date: 18/11/1917. Source: Mrs E. Gibson. Material: adult M; whole/good skin.

Reg. no. 1920.5.20.19. From Ajo, Buenos Aires; ; Argentina. Date: 5/12/1917. Source: Mrs E. Gibson. Material: juv F; whole/good skin.

Reg. no. 1920.9.30.82. From Los Yngleses, Ajo, Buenos Aires; ; Argentina. Date: 20/12/1909. Source: Mrs E. Gibson; C.H.B. Grant. Material: adult M; whole/good skin. Labelled as *cassini*.

Reg. no. 1934.11.1.1. From 69°N, Jakobshavn; ; w. Greenland. Date: 10/ 6/1934. Source: C.J. Dalgety. Material: adult F; whole/good skin. Eggs ready to lay. Bill yellowish–horn, black tip, yellow cere, orbital ring yellow, iris dark brown, feet yellow. See *Ibis* 1936: 580.

Reg. no. 1934.1.1.625. From Beaverhill Lake; Alberta; Canada. Date: 20/ 5/1924. Source: W. Rowan; Witherby coll.. Material: F; whole/good skin.

Reg. no. 1935.10.21.357. From Santiago; ; Chile. Date: 11/ 1/1933. Source: F.C. Platts. Material: imm F; whole/good skin. Eye black, legs yellow, cere greyish–yellow, bill slate, tip black.

Reg. no. 1937.6.4.1. From 20 miles out to sea; ; Jamaica. Date: //1937. Source: B.W. Williams. Material: adult F; whole/good skin.

Reg. no. 1938.3.19.1. From Humberstone, Lincolnshire; England; United Kingdom. Date: 28/ 9/1910. Source: J. Stubbs; G.H. Caton Haigh. Material: imm M; whole/good skin. Caught in a plover net by Stubbs. See *Brit. Birds* 5 (1912): 219–20.

Reg. no. 1938.9.10.1. From Great Whale River, Hudson's Bay; ; Canada. Date: 26/ 5/1915. Source: E. Renouf, British Museum. Material: M; whole/good skin.

Reg. no. 1946.36.28. From Restin Dist.; ; n. Peru. Date: 18/11/1940. Source: H.P. Manton. Material: ; whole/good skin. Bill pale blue to black tip, feet yellowish-grey, iris brown.

Reg. no. 1947.6.475. From White Island, Frozen Straits; Northwest Territories; Canada. Date: 30/ 7/1937. Source: T.H. Manning. Material: chick ; whole/good skin.

Reg. no. 1947.6.476. From White Island, Frozen Straits; Northwest Territories; Canada. Date: 30/ 7/1937. Source: T.H. Manning. Material: chick ; whole/good skin.

Reg. no. 1947.6.477. From mouth of Canyon River, Southampton Island; [Northwest Territories]; Canada. Date: 18/ 8/1937. Source: T.H. Manning. Material: juv ; whole/good skin.

Reg. no. 1947.6.478. From Seahorse Point, Southampton Island; [Northwest Territories]; Canada. Date: 16/ 6/1938. Source: T.H. Manning. Material: adult M; whole/good skin.

Reg. no. 1947.6.479. From Seahorse Point, Southampton Island; Northwest Territories; Canada. Date: 19/ 6/1938. Source: T.H. Manning. Material: adult M; whole/good skin.

Reg. no. 1947.6.480. From Seahorse Point, Southampton Island; [Northwest Territories]; Canada. Date: 10/ 6/1938. Source: T.H. Manning. Material: adult M; whole/good skin.

Reg. no. 1947.6.481. From Seahorse Point, Southampton Island; [Northwest Territories]; Canada. Date: 4/ 7/1938. Source: T.H. Manning. Material: adult M; whole/good skin.

Reg. no. 1947.6.482. From Seahorse Point, [Southampton Island]; [Northwest Territories]; [Canada]. Date: 4/ 7/1938. Source: T.H. Manning. Material: adult F; whole/good skin.

Reg. no. 1947.6.483. From mouth of Canyon River, Southampton Island; [Northwest Territories]; [Canada]. Date: 19/ 8/1937. Source: T.H. Manning. Material: F; whole/good skin.

Reg. no. 1947.6.484. From Koodlootook, Southampton Island; Northwest Territories; Canada. Date: 6/ 9/1934. Source: T.H. Manning. Material: juv F; whole/good skin.

Reg. no. 1947.6.485. From White Island, Frozen Straits; Northwest Territories; Canada. Date: 30/ 7/1937. Source: T.H. Manning. Material: chick ; whole/good skin.

Reg. no. 1953.68.34. From Mt. Cotopaxi; ; Ecuador. Date: 16/11/1941. Source: L. Gomez. Material: M; whole/good skin.

Reg. no. 1955.6.N.20.1855. From Hudson's Bay; ; [Canada]. Date: //. Source: J. Gould; Gurney coll.; Norwich Castle Museum. Material: imm ; whole/good skin.

Reg. no. 1955.6.N.20.1856. From Fort Churchill, Hudson's Bay; ; Canada. Date: //. Source: Capt. Herd; Leadbeater; Gurney coll.; Norwich Castle Museum. Material: F; whole/good skin.

Reg. no. 1955.6.N.20.1857. From Fort Churchill, Hudson's Bay; ; Canada. Date: //. Source: Capt. Herd; Leadbeater; Gurney coll.; Norwich Castle Museum. Material: imm F; whole/good skin.

Reg. no. 1955.6.N.20.1858. From Saskatchewan River; ; Canada (North America). Date: //. Source: Leadbeater; Gurney coll.; Norwich Castle Museum. Material: ; whole/good skin.

Reg. no. 1955.6.N.20.1859. From Port Kennedy, Billot Straits; Northwest Territories; Canada. Date: 29/ 8/1858. Source: Dr D. Walker, Fox Expedition; Gurney coll., Norwich Castle Museum. Material: ; whole/good skin. From 72°N 94°W.

Reg. no. 1955.6.N.20.1860. From ; ; Greenland. Date: //. Source: Holboll; Stevens; Gurney coll.; Norwich Castle Museum. Material: adult M; whole/good skin.

Reg. no. 1955.6.N.20.1861. From ; ; . Date: //. Source: Verreaux; Gurney coll.; Norwich Castle Museum. Material: imm M; whole/good skin.

Reg. no. 1955.6.N.20.1862. From ; ; Greenland. Date: //. Source: Holboll; Stevens; Gurney coll.; Norwich Castle Museum. Material: adult F; whole/good skin.

Reg. no. 1955.6.N.20.1863. From ; ; Canada. Date: //. Source: Stevens; Leadbeater; Gurney coll.; Norwich Castle Museum. Material: imm M; whole/good skin. Ex-mount.

Reg. no. 1955.6.N.20.1864. From ; ; Canada. Date: //. Source: Stevens; Gurney coll.; Norwich Castle Museum. Material: imm ; whole/good skin. Ex-mount.

Reg. no. 1955.6.N.20.1865. From ; ; Canada. Date: 27/ 9/1884. Source: Hadcraft; Gurney coll.; Norwich Castle Museum. Material: imm M; whole/good skin.

Reg. no. 1955.6.N.20.1866. From New York; ; United States. Date: //. Source: J.H. Backhouse; Leadbeater; Gurney coll.; Norwich Castle Museum. Material: F; whole/good skin.

Reg. no. 1955.6.N.20.1867. From Great Egg Harbour; New Jersey; United States. Date: //. Source: Cassin; Gurney coll.; Norwich Castle Museum. Material: imm F; whole/good skin.

Reg. no. 1955.6.N.20.1868. From ; California; [United States]. Date: //. Source: E. Verreaux; Gurney coll.; Norwich Castle Museum. Material: imm M; whole/good skin.

Reg. no. 1955.6.N.20.1869. From ; ; Mexico. Date: //. Source: Lefevre; Gurney coll.; Norwich Castle Museum. Material: imm F; whole/good skin.

Reg. no. 1955.6.N.20.1870. From coast of Yucatan; ; Mexico, Captivity. Date: 15/12/1881. Source: Zoological Society; Gurney coll.; Norwich Castle Museum. Material: adult F; whole/good skin. Received at the Zoo 31 October 1877, died 15 December 1881.

Reg. no. 1955.6.N.20.1871. From 124 miles from Colon; ; Isthmus of Panama. Date: //. Source: Rippon; Gurney coll.; Norwich Castle Museum. Material: imm ; whole/good skin. Collector erroneously given as Ripan on the Gurney label.

Reg. no. 1955.6.N.20.1874. From ; ; St. Croix. Date: 1/12/1862. Source: A. Newton; Gurney coll.; Norwich Castle Museum. Material: imm F; whole/good skin.

Reg. no. 1955.6.N.20.1875. From St. John's Estate; ; St. Croix. Date: 20/ 2/1861. Source: F.R.N. [?F.R. Nicholson]; A. Newton; Gurney; Norwich Castle Mus.. Material: imm F; whole/good skin. Ex–mount.

Reg. no. 1955.6.N.20.1884. From Okak; Labrador; [Canada]. Date: //. Source: Moravian Mission; Gurney coll.; Norwich Castle Museum. Material: adult M; whole/good skin.

Reg. no. 1955.6.N.20.1886. From ; ; Barbados. Date: 17/11/1888. Source: Col. Fielden; Gurney coll.; Norwich Castle Museum. Material: imm M; whole/good skin.

Reg. no. 1956.56.1. From 2 miles s. of Lompoc, Santa Barbara Co.; California; United States. Date: 30/ 6/1954. Source: T.J. Cade; University of California. Material: juv M; whole/good skin. One of four fledged young taken at a nesting cliff.

Reg. no. 1957.24.13. From Salinas; Guayas Prov.; Ecuador. Date: 3/11/1956. Source: S. Marchant. Material: F; whole/good skin. Bill blackish–grey, shading through grey to yellowish near cere, cere dirty yellow–white, iris brown, feet lemon–yellow.

Reg. no. 1965.M.1436. From Fishers' Island; New York; United States. Date: 9/10/1923. Source: L.C. Sanford; Meinertzhagen coll.. Material: M; whole/good skin.

Reg. no. 1965.M.1437. From 100 miles out to sea; ; [North Atlantic Ocean]. Date: /10/1920. Source: F. Guest; Meinertzhagen coll.. Material: M; whole/good skin. Carried by a ship from Newfoundland.

Reg. no. 1988.21.28. From ; ; North America. Date: //. Source: Hudson's Bay Company. Material: F; whole/good skin. Ex–mount.

Reg. no. 1988.21.29. From ; ; North America. Date: //. Source: Hudson's Bay Company. Material: M; whole/good skin.

Reg. no. 1988.21.30. From ; [Labrador]; North America. Date: //. Source: Sir E. Parry. Material: juv F; whole/good skin. The locality is given as North America on the label, and Labrador in *Cat. Bds.*.

Reg. no. 1860.7.22.12. From ; ; North America. Date: //. Source: Zoological Society coll.. Material: ; part/good skeleton. Skull missing. Previously registered as Skeleton Vellum Cat. 1274a. Subspecies not determined. This is the only North American skeleton of this species in the collection.

Reg. no. 1884.9.2.192–3. From Fort, Peel River; Yukon Territory; Canada. Date:

//1863. Source: J. Lockhart; Smithsonian; Salvin–Godman coll.. Material: 2 ; whole/good eggs. Two eggs from a clutch of three. One egg was damaged but has been repaired. The year 1863 is given on the data slip, but 1870 on the eggs. The latter is probably when Salvin received them.

Reg. no. 1902.10.25.688–9. From ; Utah; United States. Date: //1897. Source: Radcliffe–Saunders coll.. Material: 2 ; whole/good eggs.

Reg. no. E1901.1.1.7679–80. From Frederickshaab; ; Greenland. Date: //. Source: E. Muller; Hargitt coll.; Seebohm coll.. Material: 2 ; whole/good eggs.

Reg. no. E1948.1.1. From Seahorse Point, Southampton Island, Hudson Bay; Northwest Territory; Canada. Date: 5/ 7/1939. Source: T.H. Manning, British Canadian Arctic Expedition. Material: 4 ; whole/good eggs.

Reg. no. E1972.11.342. From ; Labrador; Canada. Date: //. Source: Rodern coll.. Material: 3 ; whole/good eggs.

Reg. no. E1952.10.1867. From Hawk's Nest, Delaware River, Orange Co.; New Jersey; United States. Date: 24/ 4/1921. Source: J.R. Gillin; Meiklejohn coll.. Material: 3 ; whole/good eggs (clutch). Incubation, 3/4; 2 eggs infertile. Nest easily approached from above, 30 ft. from the top of a cliff (250 ft.) facing the river, on a 4 ft. wide ledge, well protected by overhanging rocks and a good sized cedar. The female refused to fly out even after several revolver shots had been fired at the cliff.

Reg. no. E1952.10.1868. From Jordan River, Davis Co.; Utah; United States. Date: 29/ 4/1899. Source: N.H. Parker; T.H. Jackson; Meiklejohn coll.. Material: 4 ; whole/good eggs (clutch). Incubation, fresh. Eggs deposited on a shelving rock in face of a perpendicular cliff overhanging the river near its mouth. In slight depression in the sand and earth blown on the shelf, the only thing to keep the eggs from rolling off. (Sketch available.)

Reg. no. E1952.10.1869. From Monterey Co.; California; United States. Date: 19/ 4/1921. Source: P. Easton; Meiklejohn coll.. Material: 2 ; whole/good eggs (clutch).

Reg. no. E1958.6.49. From Peel River; ; n. Canada. Date: 30/ 5/1895. Source: Dr D. Moore Lindsay. Material: 3 ; whole/good eggs (clutch). Eggs fresh; nest looks like an old Raven's, in top of a balsam.

Reg. no. E1973.1.518. From Red Deer; Alberta; Canada. Date: 20/ 5/1928. Source: T.E. Randall; Jourdain coll.. Material: 4 ; whole/good eggs (clutch). Medium incubation, female on nest. Nest, on a ledge of a cliff overlooking river.

Reg. no. E1973.1.582. From Julianehaab; ; Greenland. Date: 13/ 6/1930. Source: L. Christinson; Pastor K. Barfod; Jourdain coll.. Material: 4 ; whole/good eggs (clutch). One egg damaged.

Reg. no. 1853.12.29.2. From ; ; North America. Date: //. Source: Stevens coll.. Material: 1 ; whole/good egg.

Reg. no. 1884.9.2.194. From Fort Yukon; Alaska; United States. Date: / 6/. Source: S. Jones; J. Lockhart; Smithsonian; Salvin-Godman coll.. Material: 1 ; whole/poor egg. One egg from a clutch of three. Damaged.

Reg. no. 1891.3.1.478. From Fort Yukon; Alaska; United States. Date: //. Source: J. Macdougall; Henshaw coll.; Salvin-Godman coll.. Material: 1 ; whole/good egg. One egg from a clutch of four.

Reg. no. 1901.2.25.345. From Fort Yukon; Alaska; United States. Date: //. Source: J. Sibbiston; Smithsonian; Crowley coll.. Material: 1 ; whole/good egg.

Reg. no. E1941.4.6.1512. From Mount Torn; Massachussetts; United States. Date: //. Source: C.W. Bennett; E.A. Samuels coll.; Buckley coll.. Material: 1 ; whole/good egg.

Reg. no. E1955.5.97. From ; ; Greenland. Date: //. Source: J.H. Gurney coll.. Material: 1 ; whole/good egg.

Reg. no. E1955.5.98. From ; Labrador; Canada. Date: //. Source: J.H. Gurney coll.. Material: 1 ; whole/good egg. The egg may or may not be the correct one: it bears the inscription 'Moravian Mission', but no reg. no.. It might have been moved from another part of the collection during a series of thefts in the 1970s.

Reg. no. E1988.4.29. From ; ; . Date: //1931. Source: Sawyer coll.. Material: 1 ; whole/good egg. One egg from a clutch of four.

***Megapodius* species**

Megapode species

The collections contain a number of megapode eggs said to be from places where megapodes no longer occur. It has been claimed that some of these are from distinct taxa that are now extinct, but this seems unlikely.

Reg. no. 1891.6.26.1,2,4,6. From ; ; Lord Howe Island. Date: //. Source: Dr Gunther. Material: 5 ; eggs. Not in the collection. The eggs were listed in Seebohm ms cat., but not in Oates' *Cat. Eggs*.

Reg. no. E1972.13.239. From Great Andaman; Andaman and Nicobar Islands; India. Date: //1878. Source: Groom Napier; Rothschild. Material: 2 ; whole/good eggs. Collected before 1878; purchased in that year. Megapodes are not supposed to have occured on the Andamans. This egg may be from an extinct population, or the data may be incorrect. [G. Young, 1981, *Slow Boats to China*, mentions a guest-house in Port Blair in the Andamans called the Megapode's Nest, commenting "the megapode is a big-footed bird about the size of a coot, found only in the Andamans".]

Megapodius andersoni

Reg. no. . From ; ; New Caledonia. Date: //. Source: Gould coll.. Material: 1 ; whole/good egg. Not in the collection. The egg was listed in Seebohm ms cat., but not in Oates' *Cat. Eggs*. See *Proc. zool. Soc. Lond.* 1861: 291.

Megapodius burnabyi

Reg. no. E1988.4.3. From Hapace Island; Tongatapu (Tongataboo); Tonga Islands. Date: //. Source: Lt. Burnaby. Material: 1 ; whole/good egg. Holotype. G.R. Gray, *Proc. zool. Soc. Lond.* 1861: 290; see also p. 296. The egg was received before 1861. Listed in Oates' *Cat. Eggs* under *M. pritchardi*. Type and only known specimen.

Megapodius stairi

Reg. no. 1847.10.11.5. From ; ; Samoa. Date: //1847. Source: Rev. J.B. Stair. Material: 1 ; whole/good egg. Holotype. G.R. Gray, *Proc. zool. Soc. Lond.* 1861: 290. The egg was obtained c. 1847. Listed in Oates' *Cat. Eggs* under *M. pritchardi*. Whitmee, *Ibis* 1875: 447, queried the location. Type and only known specimen.

Ortalis erythroptera

Rufous-headed Chachalaca

Confined to western Ecuador from the coast to the foothills of the Andes, and extreme northwestern Peru. Seriously at risk from habitat destruction.

Reg. no. 1889.6.1.352. From Balzar Mountains; ; Ecuador. Date: / 3/1880. Source: Illingworth; Salvin-Godman coll.. Material: adult ; whole/good skin. Figured in Gronvold, 1915, *Illustrations to the Birds of South America* by Brabourne & Chubb, pt. 1 (vol.2, pl. 9, no. 2).

Reg. no. 1889.6.1.353. From Babahoyo; ; Ecuador. Date: //1859. Source: L. Fraser; Salvin-Godman coll.. Material: adult ; whole/good skin. Syntype. P.L. Sclater & Salvin, *Proc. zool. Soc. Lond.* 1870: 540. There are other syntypes in Berlin and Paris.

Reg. no. 1946.36.12. From Lechugal Dist.; ; n. Peru. Date: 28/12/1944. Source: H.P. Manton. Material: adult F; whole/good skin. Bill and feet pale slate-grey, iris brownish.

Reg. no. 1953.68.45. From Jama, Manabi; ; Ecuador. Date: 20/ 4/1942. Source: L. Gomez. Material: adult M; whole/good skin.

Penelope albipennis

White-winged Guan

Restricted to the dry forest on the Pacific coast of extreme northwestern Peru. Previously known from 3 old specimens collected in 1876-1877, the species was not seen again for 100 years when about 30-40 birds were found. The present population is thought to number only a few hundred individuals.

Reg. no. 1940.3.3.2. From Hacienda de Pabur; ; n.w. Peru. Date: //. Source: Museo Xavier Prado; Lima Museum. Material: adult ; whole/good skin. *Ibis* 1948: 131.

Penelope perspicax

Cauca Guan

Formerly occurred throughout the middle of the Cauca river valley which has been almost totally deforested. The present range and population are not known.

Reg. no. 1889.6.1.208. From 'Bogota'; ; [Colombia]. Date: //. Source: Crowther; Salvin–Godman coll.. Material: ; whole/good skin. This specimen 'seems to approach *P. perspicax*, but it does not fully fit [the description] of that form; locality (if valid) is wrong for it, and I think it is just a rather reddish and/or foxed *P. purpurascens aequatorialis*' (note with the specimen by D. Goodwin).

Aburria jacutinga

Black-fronted Piping-Guan

Formerly occurred in southeastern Brazil from southern Bahia to Rio Grande do Sol, and adjacent areas of southeastern Paraguay and extreme northeastern Argentina. Deforestation has eliminated the species in much of the former range, and neither the present range nor the population is precisely known.

Reg. no. 1856.11.5.8. From ; ; Brazil. Date: //. Source: Verreaux. Material: adult ; whole/good skin.

Reg. no. 1856.11.5.9. From ; ; Brazil. Date: //. Source: Verreaux. Material: adult ; whole/good skin.

Reg. no. 1889.6.1.249. From ; ; Brazil, Captivity. Date: 16/ 8/1874. Source: H. Whitely; Salvin–Godman coll.. Material: adult F; whole/good skin. Date received in zoo: 5 September 1871; died at London Zoo: 16 August 1874.

Reg. no. 1889.6.1.250. From Sao Paulo; ; Brazil. Date: //. Source: J. Natterer; Salvin–Godman coll.. Material: adult ; whole/good skin.

Reg. no. 1889.6.1.251. From Rio Parana; [Goias]; Brazil. Date: //. Source: Rodgers (or Rogers); Salvin–Godman coll.. Material: adult ; whole/good skin.

Reg. no. 1895.4.1.1. From Rio de Janeiro; ; Brazil. Date: //. Source: A. Fry. Material: adult M; skin. Details taken from the register. Not found 1988.

Reg. no. 1895.4.1.2. From Rio de Janeiro; ; Brazil. Date: //. Source: A. Fry. Material: adult F; whole/good skin.

Reg. no. S1959.19.2. From ; ; Brazil. Date: 30/12/1959. Source: BBC; Zoological Society. Material: adult ; whole/good skeleton. Presented to the Zoo by the British Broadcasting Corporation. Received at the Zoo on 19 December 1958, died on 30 December 1959.

Oreophasis derbianus

Horned Guan

The Horned Guan, or Lord Derby's Mountain Pheasant, formerly occurred in moist mountain forest in southeastern Chiapas, and western Guatemala, and possibly also in Oaxaca. The species is now greatly reduced in range and numbers. It is good to eat and ridiculously tame and thus easily eliminated in any area which is opened up to hunters and settlers.

Reg. no. 1856.10.28.1. From ; ; Guatemala. Date: //. Source: J. Gould. Material: ; whole/good skin.

Reg. no. 1857.11.28.1. From ; ; Guatemala. Date: //. Source: J. Gould. Material: ;

whole/good skin. Ex-mount. Two birds bear this reg. no..

Reg. no. 1857.11.28.1. From ; ; Guatemala. Date: //. Source: J. Gould. Material: ; whole/good skin. Ex-mount. Two birds bear this reg. no..

Reg. no. 1889.6.1.354. From Volcan de Fuego; ; Guatemala. Date: //. Source: O. Salvin; Salvin-Godman coll.. Material: adult ; whole/good skin.

Reg. no. 1889.6.1.355. From Volcan de Fuego; ; Guatemala. Date: 9/ 1/1860. Source: O. Salvin; Salvin-Godman coll.. Material: adult M; whole/good skin.

Reg. no. 1889.6.1.356. From Volcan de Fuego; ; Guatemala. Date: /12/1861. Source: O. Salvin; Salvin-Godman coll.. Material: M; whole/good skin.

Reg. no. 1889.6.1.357. From Volcan de Fuego; ; Guatemala. Date: 10/ 1/1860. Source: O. Salvin; Salvin-Godman coll.. Material: adult F; whole/good skin.

Reg. no. 1891.10.21.290. From Volcan de Fuego; ; Guatemala. Date: / 8/1873. Source: O. Salvin; Salvin-Godman coll.. Material: adult F; whole/good skin.

Reg. no. 1891.10.21.291. From 8,000-9,000 ft. on Volcan de Fuego; ; Guatemala. Date: / 8/1873. Source: O. Salvin; Salvin-Godman coll.. Material: adult M; whole/good skin.

Reg. no. 1891.10.21.292. From Volcan de Fuego; ; Guatemala. Date: //. Source: O. Salvin; Salvin-Godman coll.. Material: ; whole/good skin.

Reg. no. 1891.10.21.293. From Volcan de Fuego; ; Guatemala. Date: //. Source: O. Salvin; Salvin-Godman coll.. Material: adult ; whole/good skin.

Reg. no. 1891.10.21.294. From Volcan de Fuego; ; Guatemala. Date: //. Source: Salvin-Godman coll.. Material: adult ; whole/good skin.

Reg. no. . From ; ; . Date: //. Source: Material: ; part skeleton. Not found 1988 and 1993. Two complete and two incomplete skeletons are listed in *Bull. Br. Mus. nat. Hist. (Zool.)* 34 (1979): 136. One of the incomplete skeletons could not be located.

Reg. no. 1860.6.18.23. From ; ; Guatemala. Date: //. Source: Salle; Cuming. Material: ; whole/good skeleton. Previously registered as Skeleton Vellum Cat. 1207a.

Reg. no. 1891.7.20.86. From Volcan de Fuego; ; Guatemala. Date: / 8/1873. Source: Salvin-Godman coll.. Material: M; whole/good skeleton.

Reg. no. S1952.3.68. From ; ; [South America]. Date: //. Source: Material: ; part/good skeleton. Trunk only, removed from a stuffed specimen. Previously registered as Skeleton Vellum Cat. 1207b.

Crax mitu

Razor-billed Curassow

The taxonomic status of the threatened population *C.m. mitu* is unsettled. It has been treated as specifically or subspecifically distinct from *C.m. tuberosa*, or not distinguishable at all. The nominate form is said to occur in a very restricted area of the Alagoas districts of Brazil, but may now only survive in a single private breeder's aviaries.

Reg. no. . From ; ; Brazil. Date: //. Source: Material: ; whole/good skin. There are several specimens of *C. mitu* in the collection listed as being from Brazil, usually without further qualification. The subspecies have not been determined. They may not belong to the endangered subspecies *mitu*.

Reg. no. . From ; ; [Brazil]. Date: //. Source: Material: – ; eggs. There are several eggs of this species in the collection. Most were laid by birds in captivity, but one came from Brazil. None had been identified to subspecies.

Crax globulosa

Wattled Curassow

Found on riverine islands in Colombia, Ecuador, Peru, Brazil and Bolivia. It is threatened by hunting and habitat loss over much of its range and possibly survives only in a few localities.

Reg. no. 1868.2.15.7. From ; ; South America. Date: //. Source: Zoological Society. Material: adult M; whole/good skin.

Reg. no. 1889.6.1.296. From ; ; s.e. Brazil. Date: 15/ 1/1878. Source: Zoological Society; Salvin–Godman coll.. Material: adult M; whole/good skin. Received at London Zoo on 2 January 1875, died 15 January 1878.

Reg. no. 1889.6.1.297. From ; ; South America (Tropical America). Date: 26/ 6/1875. Source: Zoological Society; Salvin–Godman coll.. Material: adult F; whole/good skin. Received at London Zoo on 25 May 1870, died 26 June 1875. Part of the skeleton is attached to the specimen.

Reg. no. 1889.6.1.365. From Samiria; ; [Peru]. Date: 21/11/1881. Source: J. Hauxwell; H. Whitely; Salvin–Godman coll.. Material: adult F; whole/good skin. Iris brown. The locality is given as Samaria in the register.

Reg. no. 1988.21.27. From ; ; South America. Date: //. Source: Material: adult M; whole/good skin.

Reg. no. 1849.3.26.62. From ; ; Guiana. Date: //. Source: Warwick. Material: ; whole/good skeleton.

Reg. no. 1867.7.12.1. From ; ; . Date: //. Source: Zoological Society. Material: ; whole/good skeleton. Previously registered as Skeleton Vellum Cat. 1362a.

Reg. no. G1854.3.29.2. From ; ; South America. Date: //. Source: Warwick. Material: ; whole/good skeleton. Previously registered as Skeleton Vellum Cat. 1092a.

Reg. no. . From ; ; . Date: //. Source: Material: ; alc. Two incomplete spirit specimens. Details taken from *Bull. Br. Mus. nat. Hist. (Zool.)* 34 (1979): 136. Not found 1988.

Crax blumenbachii

Red-billed Curassow

Formerly occurred in east Brazil in southeastern Bahia, eastern Minas Gerais, Espirito Santo and Rio de Janeiro. Now almost extinct and surviving only in a few reserves. The decline was due to a combination of hunting and forest destruction.

Reg. no. 1843.11.27.32. From ; Bahia; Brazil. Date: //. Source: Sudbury Museum. Material: adult M; whole/good skin.

Reg. no. 1889.6.1.293. From ; ; Brazil, Captivity. Date: 13/ 1/1879. Source: J. Natterer. Material: adult F; whole/good skin. Died in London Zoo. Iris red.

Tympanuchus cupido cupido

Heath Hen or Greater Prairie-chicken

Formerly occurred over the eastern coast of North America from Massachusetts to Washington DC, possibly north to Maine and south to the Carolinas. In its last years confined to the island of Martha's Vineyard where the last bird died in 1932.

Reg. no. 1893.3.1.1. From Martha's Vineyard; Massachussetts; United States. Date: 27/ 1/1893. Source: Southwick & Critchley. Material: adult M; whole/good skin.

Reg. no. 1893.3.1.2. From Martha's Vineyard; Massachussetts; United States. Date: 27/ 1/1893. Source: Southwick & Critchley. Material: adult F; whole/good skin.

Reg. no. 1939.12.9.551. From ; Illinois; United States. Date: //. Source: Rothschild coll.. Material: adult M; skin. Mount. Details taken from the register. Not found 1988.

Reg. no. 1939.12.9.552. From ; ; [United States]. Date: //. Source: Rothschild coll.. Material: adult M; skin. Mount. Details taken from the register. Not found 1988.

Dendrortyx barbatus

Bearded Wood-Partridge

Found in the mountain forests of eastern Mexico where it is threatened by hunting and deforestation.

Reg. no. . From Jalapa; ; Mexico. Date: //. Source: J. Verreaux. Material: ; skin. Mount. Details taken from *Cat. Bds.*. Not found 1988.

Reg. no. 1857.6.10.9. From Jalapa; ; Mexico. Date: //. Source: J. Verreaux. Material: imm ; whole/good skin.

Reg. no. 1881.5.1.5495. From Jalapa; ; Mexico. Date: //1872. Source: de Oca; Gould coll.. Material: adult ; whole/good skin.

Reg. no. 1889.6.1.147. From Jalapa; ; Mexico. Date: //1881. Source: de Oca; T.C. Eyton; Salvin-Godman coll.. Material: adult ; whole/good skin.

Reg. no. 1889.6.1.148. From Jalapa; ; Mexico. Date: //. Source: de Oca; Salvin-Godman coll.. Material: adult ; whole/good skin.

Reg. no. 1889.6.1.149. From Jalapa; ; Mexico. Date: //1872. Source: de Oca; Salvin–Godman coll.. Material: adult ; whole/good skin.

Colinus virginianus ridgwayi

Masked Bobwhite

Formerly occurred from southern Arizona and northern and central Sonora, Mexico. As a result of overgrazing by cattle it is now confined to a tiny area in the Benjamin Hill–Carbo area, north of Hermosillo.

Reg. no. 1887.1.27.82. From 18 miles s.w. of Sasabe; Sonora; Mexico. Date: 11/8/1884. Source: F. Stephens; G.F. Morcom. Material: adult M; whole/good skin. Holotype. Brewster, *Auk* 2 (1885): 196–200.

Reg. no. 1888.10.10.2795. From Barboquivari Mountains; Arizona; United States. Date: //. Source: Brown; Henshaw coll.; Salvin–Godman coll.. Material: adult M; whole/good skin.

Reg. no. 1888.10.10.2796. From Barboquivari Mountains; Arizona; United States. Date: //. Source: Brown; Henshaw coll.; Salvin–Godman coll.. Material: adult ; whole/good skin. Damaged: head loose.

Reg. no. 1913.5.16.1. From Las Arenas; Sonora; Mexico. Date: 5/ 1/1906. Source: W.W. Brown; J.E. Thayer Exp.; Mus. Comp. Zool., Cambr., Mass.. Material: adult M; whole/good skin. Damaged: tail detached.

Reg. no. 1913.5.16.2. From Las Arenas; Sonora; Mexico. Date: 9/12/1905. Source: W.W. Brown; J.E. Thayer Exp.; Mus. Comp. Zool., Cambr., Mass.. Material: adult M; whole/good skin. Damaged: head loose.

Reg. no. 1913.5.16.3. From Las Arenas; Sonora; Mexico. Date: 9/12/1905. Source: W.W. Brown; J.E. Thayer Exp.; Mus. Comp. Zool., Cambr., Mass.. Material: adult F; whole/good skin.

Reg. no. 1913.5.16.4. From Las Arenas; Sonora; Mexico. Date: 18/12/1905. Source: W.W. Brown; J.E. Thayer Exp.; Mus. Comp. Zool., Cambr., Mass.. Material: adult F; whole/good skin.

Reg. no. 1939.12.9.290. From Sonora la Bonancita; ; Mexico. Date: 15/12/1905. Source: W.W. Brown; Thayer Exp.; Bangs; Mus. Comp. Zool.; Rothschild coll.. Material: M; whole/good skin.

Odontophorus strophium

Gorgeted Wood–Quail

Formerly known from 8 specimens from the eastern Andes of Colombia. Survives only in Santander, 200 km north east of Bogota, where 9 groups were found in 1988 with the help of a local farmer.

Reg. no. 1844.12.12.11. From ; ; [Colombia] (Mexico). Date: //. Source: Gould coll.. Material: adult ; whole/good skin. Holotype. Ex–mount. Gould, *Proc. zool. Soc. Lond.* 1843 (publ. 1844): 134. From the southern countries of Mexico.

Reg. no. 1846.7.16.52. From Santa Fe de Bogota; ; Colombia. Date: //. Source: E. Wilson. Material: adult ; whole/good skin.

Reg. no. 1988.13.1. From Santa Fe de Bogota; ; Colombia. Date: //. Source: Gould coll.. Material: imm ; whole/good skin.

Reg. no. 1988.7.1. From Virolin; Santander; Colombia. Date: 6/ 3/1988. Source: M. de L. Brooke. Material: adult F; part/good skin. Wing only.

Reg. no. E1975.3.3. From Carare Region; ; Colombia. Date: //. Source: C. Allen. Material: 2 ; whole/good eggs. Identity uncertain: possibly *O. strophium*, *O. hyperythrus* or *O. atrifrons variegatus*.

Perdix perdix italica

Italian Grey Partridge

Formerly widely distributed through the Italian peninsula. Now confined to a few areas in the centre of the country, mainly as a result of hunting, changes in agricultural practice and genetic swamping by introduced birds of other races. Violani *et al.* (*Bull. Brit. Orn. Club* 108 (1988): 22–27) consider this race to be indistinguishable from *P.p. perdix*.

Reg. no. 1907.12.20.5. From Naples; ; Italy. Date: //. Source: H. Saunders. Material: adult M; whole/good skin.

Reg. no. 1965.M.2147. From Florence; ; Italy. Date: 7/12/1944. Source: R. Meinertzhagen coll.. Material: adult M; whole/good skin.

Reg. no. 1965.M.2148. From Florence; ; Italy. Date: 7/12/1944. Source: R. Meinertzhagen coll.. Material: adult F; whole/good skin.

Coturnix novaezelandiae

New Zealand Quail

Formerly common throughout New Zealand but declined rapidly during the mid–19th century and last seen in 1875. The cause of its sudden extinction is not known, but is believed to be due to diseases brought in by introduced gamebirds.

Reg. no. 1845.1.13.18. From ; ; New Zealand. Date: //. Source: P. Earl. Material: M; whole/good skin. Near adult.

Reg. no. 1847.1.8.32. From ; ; New Zealand. Date: //. Source: G. Grey. Material: M; whole/good skin. Label bears reg. no. 1847.1.8.38 in error.

Reg. no. 1847.1.8.33. From ; ; New Zealand. Date: //. Source: G. Grey. Material: adult M; whole/good skin.

Reg. no. 1849.12.12.19. From Port Cooper; ; New Zealand. Date: //. Source: F. Strange; Capt. Stokes. Material: imm M; whole/good skin.

Reg. no. 1849.12.12.20. From Port Cooper; ; New Zealand. Date: //. Source: F. Strange; Capt. Stokes. Material: adult F; whole/good skin.

Reg. no. 1855.12.19.111. From ; ; New Zealand. Date: //. Source: A. Dundas; Zoological Society. Material: adult F; whole/good skin.

Reg. no. 1855.12.19.112. From ; ; New Zealand. Date: //. Source: A. Dundas; Zoological Society. Material: imm M; whole/good skin.

Reg. no. 1881.5.1.5515. From Port Cooper; ; New Zealand. Date: //. Source: F. Strange; J. Gould coll.. Material: adult F; whole/good skin.

Reg. no. 1849.12.12.46–7. From Plains of Port Cooper; ; New Zealand. Date: 24/4/1849. Source: F. Strange; Capt. Stokes. Material: 2 ; whole/good eggs.

Reg. no. 1901.11.1.276–9. From ; ; New Zealand. Date: //1870. Source: T.H. Potts; Leopold Field coll.. Material: 4 ; whole/good eggs. One egg marked 10.3.70 (10 March 1870) and one egg marked 28.1.70 (28 January 1870).

Reg. no. E1960.6.54. From ; ; New Zealand. Date: //. Source: J. Gould. Material: 2 ; whole/good eggs.

Reg. no. E1960.6.55. From ; ; New Zealand. Date: //. Source: Forbes coll.. Material: 2 ; whole/good eggs.

Ophrysia superciliosa

Himalayan Quail

Known only from the western Himalayas, where it was last seen in 1868 or 1876. Some authorities consider that it may still exist.

Reg. no. 1868.4.27.2. From Budraj, Mussooree; Uttar Pradesh; India. Date: /11/1865. Source: K. Mackinnon; L'Estrange. Material: adult M; whole/good skin.

Reg. no. 1889.5.10.814. From Jerepani, Mussooree; Uttar Pradesh; India. Date: 28/11/1869. Source: T. Hutton; Hume coll.. Material: adult M; whole/good skin.

Reg. no. 1889.5.10.815. From Jerepani, Mussooree; Uttar Pradesh; India. Date: //. Source: T. Hutton; Hume coll.. Material: juv M; skin. Not in the collection; exchanged with the American Museum of Natural History.

Reg. no. 1889.5.10.816. From Jerepani, Mussooree; Uttar Pradesh; India. Date: /6/1870. Source: T. Hutton; Hume coll.. Material: imm M; whole/good skin. In moult.

Reg. no. 1889.5.10.817. From Jerepani, Mussooree; Uttar Pradesh; India. Date: 28/11/1869. Source: T. Hutton; Hume coll.. Material: adult F; whole/good skin.

Reg. no. 1889.5.10.818. From Sherkadanda, Naini Tal; ; India. Date: /12/1876. Source: Carwithen; Hume coll.. Material: adult F; whole/good skin.

Tragopan melanocephalus

Western Tragopan

Formerly occurred widely in the western Himalayas, but now greatly reduced, and occurring only in a few localities. The present population is estimated at 1600–4800 birds, the majority in small isolated populations, few of which are thought to be viable, all of which are threatened and none of which is adequately protected.

Reg. no. . From Kotegurh; ; India. Date: //. Source: Hume coll.. Material: adult F; skin. Not in the collection; exchanged with the Zoological Society.

Reg. no. 1838.7.9.1. From Himalayas; ; India. Date: //. Source: R. Ferguson. Material: M; skin. Details taken from the register. Not found 1987.

Reg. no. 1838.7.9.2. From Himalayas; ; India. Date: //. Source: R. Ferguson. Material: adult M; whole/good skin.

Reg. no. 1838.7.9.3. From Himalayas; ; India. Date: //. Source: R. Ferguson. Material: adult F; skin. Details taken from the register. Also registered as Vel.34.23d. Not found 1987.

Reg. no. 1845.1.9.132. From ; Punjab; India. Date: //. Source: B.H. Hodgson. Material: adult M; whole/good skin.

Reg. no. 1880.1.1.2941. From ; ; [India]. Date: //. Source: India Museum. Material: adult F; whole/good skin.

Reg. no. 1880.1.1.2946. From ; Punjab; India. Date: //. Source: India Museum. Material: adult M; whole/good skin.

Reg. no. 1881.7.30.49. From Simla; ; India. Date: //. Source: Capt. S. Pinwill. Material: imm M; whole/good skin.

Reg. no. 1881.7.30.50. From Simla; ; India. Date: //. Source: Capt. S. Pinwill. Material: adult M; whole/good skin. The label carried the reg. no. 1881.7.3.50 in error.

Reg. no. 1889.5.10.1863. From Kulu (Kooloo); Punjab; India. Date: //. Source: Hume coll.. Material: adult F; whole/good skin.

Reg. no. 1889.5.10.1864. From Kulu (Kooloo); Punjab; India. Date: / 2/1870. Source: Hume coll.. Material: adult M; whole/good skin.

Reg. no. 1889.5.10.1865. From Kulu (Kooloo); Punjab; India. Date: / 2/1870. Source: Hume coll.. Material: adult F; whole/good skin.

Reg. no. 1889.5.10.1866. From Taralla, Kulu (Kooloo); Punjab; India. Date: / 2/1869. Source: Hume coll.. Material: imm M; whole/good skin.

Reg. no. 1889.5.10.1867. From Kulu (Kooloo); Punjab; India. Date: / 4/1868. Source: Hume coll.. Material: adult M; whole/good skin.

Reg. no. 1889.5.10.1868. From Kulu (Kooloo); Punjab; India. Date: / 4/1868. Source: Hume coll.. Material: imm M; whole/good skin.

Reg. no. 1889.5.10.1869. From Kulu (Kooloo); Punjab; India. Date: / 4/1868. Source: Hume coll.. Material: adult F; whole/good skin. Label bears reg. no 1889.5.10.1860 in error.

Reg. no. 1889.5.10.1870. From Kotegurh, nr. Simla; ; India. Date: //1868. Source: Hume coll.. Material: adult M; whole/good skin.

Reg. no. 1889.5.10.1871. From Kotegurh; ; India. Date: //. Source: Hume coll.. Material: adult F; skin. Details taken from the register. Not found 1987.

Reg. no. 1889.5.10.1872. From Kotegurh; ; India. Date: //. Source: Hume coll.. Material: adult F; skin. Not in the collection; given to Mr Flemming.

Reg. no. 1889.5.10.1873. From Kotegurh; ; India. Date: / 3/1870. Source: Hume coll.. Material: adult M; whole/good skin.

Reg. no. 1889.5.10.1874. From Kotegurh; ; India. Date: / 3/1870. Source: Hume coll.. Material: juv M; whole/good skin.

Reg. no. 1889.5.10.1875. From Kotegurh; ; India. Date: / 3/1870. Source: Hume coll.. Material: adult F; whole/good skin.

Reg. no. 1889.5.10.1876. From Kotegurh; ; India. Date: / 3/1870. Source: Maj. Jones; Hume coll.. Material: adult F; whole/good skin.

Reg. no. 1889.5.10.1877. From Kotegurh; ; India. Date: 12/10/1870. Source: Hume coll.. Material: juv M; whole/good skin.

Reg. no. 1889.5.10.1878. From Kotegurh; ; India. Date: 12/10/1870. Source: Hume coll.. Material: adult F; whole/good skin.

Reg. no. 1889.5.10.1879. From Kotegurh; ; India. Date: / 4/1869. Source: Hume coll.. Material: adult M; whole/good skin.

Reg. no. 1889.5.10.1880. From Kotegurh; ; India. Date: //. Source: Hume coll.. Material: adult M; skin. Not in the collection; exchanged with the Zoological Society.

Reg. no. 1889.5.10.1881. From Kotegurh; ; India. Date: //. Source: Hume coll.. Material: adult M; skin. Details taken from the register. Not found 1987.

Reg. no. 1889.5.10.1882. From Kotegurh; ; India. Date: //. Source: Hume coll.. Material: adult M; whole/good skin.

Reg. no. 1889.5.10.1883. From Kotegurh; ; India. Date: //1871. Source: Hume coll.. Material: adult M; whole/good skin.

Reg. no. 1889.5.10.1884. From Kotegurh; ; India. Date: //. Source: Hume coll.. Material: adult M; skin. Details taken from the register. Not found 1987.

Reg. no. 1889.5.10.1885. From Kotegurh; ; India. Date: //1872. Source: Hume coll.. Material: adult M; whole/good skin.

Reg. no. 1889.5.10.1886. From Kotegurh; ; India. Date: //. Source: Hume coll.. Material: adult M; skin. Details taken from the register. Not found 1987.

Reg. no. 1889.5.10.1887. From Kotegurh; ; India. Date: //. Source: Hume coll.. Material: juv M; skin. Details taken from the register. Not found 1987.

Reg. no. 1889.5.10.1888. From Kotegurh; ; India. Date: //1871. Source: Hume coll.. Material: juv M; whole/good skin.

Reg. no. 1889.5.10.1889. From Kotegurh; ; India. Date: //. Source: Hume coll..

Material: adult F; skin. Details taken from the register. Not found 1988.

Reg. no. 1889.5.10.1890. From Simla; ; India. Date: //. Source: Hume coll.. Material: adult F; whole/good skin.

Reg. no. 1889.5.10.1891. From Snow Range, Himalayas; ; India. Date: 31/11/1868. Source: Hume coll.. Material: adult F; whole/good skin. Collector's label gives female, but Museum label says male. Specimen appears to be a female.

Reg. no. 1889.5.10.1892. From Snow Range, Himalayas; ; India. Date: 30/11/1868. Source: Hume coll.. Material: adult M; whole/good skin.

Reg. no. 1889.5.13.459. From the interior of the Simla Hills; ; India. Date: / 7/1866. Source: R.C. Beavan; Tweeddale coll.. Material: adult M; whole/good skin.

Reg. no. 1889.5.13.460. From n.w. of Simla; ; India. Date: //. Source: Capt. S. Pinwill; Tweeddale coll.. Material: adult F; whole/good skin.

Reg. no. 1889.5.13.461. From ; Kashmir; India. Date: //. Source: J. Biddulph; Tweeddale coll.. Material: imm M; whole/good skin. Label formerly had reg. no. 1889.5.13.462 in error.

Reg. no. 1949.WHI.1.6378. From the hills behind Dharmsala; Punjab; India. Date: 18/ 2/1923. Source: Whistler coll.. Material: adult M; whole/good skin. Irides dark brown, bill black, horns and throat skin blue, legs whitish–flesh, mottled with pink.

Reg. no. 1949.WHI.1.6379. From Kulu; Punjab; India. Date: 10/ 2/1923. Source: Whistler coll.. Material: adult M; whole/good skin.

Reg. no. 1949.WHI.1.6380. From Kangra; Punjab; India. Date: 26/12/1922. Source: Whistler coll.. Material: adult M; whole/good skin.

Reg. no. 1949.WHI.1.6381. From Baghnala Palumpur, Kangra; Punjab; India. Date: 27/12/1922. Source: Whistler coll.. Material: F; whole/good skin.

Reg. no. 1949.WHI.1.6382. From Kangra; Punjab; India. Date: 27/12/1922. Source: Whistler coll.. Material: adult F; whole/good skin.

Reg. no. 1949.WHI.1.6383. From Kulu; Punjab; India. Date: 5/ 2/1923. Source: Whistler coll.. Material: imm M; whole/good skin. Bill black, irides brick–red, legs fleshy mottled with dark brown.

Reg. no. 1949.WHI.1.6384. From nr. Kotegar, Simla Hills; ; India. Date: / 1/1918. Source: Whistler coll.. Material: adult F; whole/good skin.

Reg. no. 1949.WHI.1.6385. From the outer slopes of Duala Dar Range, nr. Dharmsala; Punjab; India. Date: 1/ 3/1923. Source: Whistler coll.. Material: F; whole/good skin. Irides dark brown, bill black, legs fleshy brown, claws dark brown.

Reg. no. 1965.M.2247. From 8,400 ft. on the hills w. of Wular; Kashmir; India. Date: 19/ 3/1925. Source: R. Meinertzhagen coll.. Material: M; whole/good skin.

Reg. no. 1965.M.2248. From 8,400 ft. on the hills w. of Wular; Kashmir; India. Date: 19/ 3/1925. Source: R. Meinertzhagen coll.. Material: imm M; whole/good skin.

Reg. no. 1988.14.1. From ; ; [India]. Date: //. Source: Anderson & Crowfoot coll.. Material: adult M; whole/good skin.

Reg. no. 1988.14.2. From ; ; [India]. Date: //. Source: Anderson & Crowfoot coll.. Material: adult M; whole/good skin.

Reg. no. VEL.34.23a. From Almorah, Himalayas; ; [India]. Date: //. Source: Maj. Gen. Hardwicke. Material: adult M; whole/good skin. Syntype. J.E. Gray, in Cuvier & Griffith, 1829, *The Animal Kingdom* ..., 8: 29. Listed as type in *Cat. Bds.* but, according to Warren in the catalogue of type-specimens, the bird does not exactly fit the original description.

Reg. no. VEL.34.23b. From Himalayas; ; [India]. Date: //. Source: Material: ; skin. Details taken from the register. Not found 1988.

Reg. no. VEL.34.23c. From ; ; [India]. Date: //. Source: Material: ; skin. Details taken from the register. Not found 1988.

Reg. no. 1891.3.20.2323–6. From Himalayas; Hazara; Pakistan. Date: 25/ 3/1869. Source: Capt. Lautour; Hume coll.. Material: 4 ; whole/good eggs. Four eggs from a clutch of six. See Baker, 1935, *The Nidification of Birds of the Indian Empire*, vol. 4, p. 222–4.

Reg. no. E1925.12.25.3377–9. From ; Chamba [Chambra]; India. Date: //. Source: H.N. Coltart; Davidson coll... Material: 3 ; whole/good eggs (clutch).

Reg. no. E1988.4.4. From about 7,000 ft., Pu Panjal; Kashmir; India. Date: / 6/1901. Source: D. Donald; R.H. Rattray; Stuart Baker coll.. Material: 2 ; whole/good eggs (clutch).

Reg. no. 1901.11.1.321. From ; ; Captivity. Date: //. Source: Dr Parker; Crowley. Material: 1 ; whole/good egg. Laid in capitivity.

Reg. no. 1902.10.25.60. From Himalayas; ; . Date: //1884. Source: Radcliffe–Saunders coll.. Material: 1 ; whole/good egg.

Reg. no. E1925.12.25.3375. From ; ; . Date: //1897. Source: Schluter; Davidson. Material: 1 ; whole/good egg.

Reg. no. E1925.12.25.3376. From ; Kashmir; India. Date: 24/ 5/1899. Source: Schluter; Davidson. Material: 1 ; whole/good egg.

Reg. no. E1941.1.3.153. From ; ; Captivity. Date: //1882. Source: Bartlett coll.; Munt coll.. Material: 1 ; whole/good egg. Laid in captivity.

Reg. no. E1988.4.5. From ; ; Captivity. Date: //1907. Source: Zoological Gardens, Berlin; M. Kuschel; Stuart Baker coll.. Material: 1 ; whole/good egg.

Reg. no. E1988.4.6. From Himalayas; ; . Date: //. Source: [Wilson]; Stoate coll.; Stuart Baker coll.. Material: 1 ; whole/good egg. Baker believed this egg to have been taken by Wilson.

Reg. no. E1988.4.7. From ; Hazara; Pakistan. Date: 25/ 5/1869. Source: Capt. Lautour; Hume; W.E. Brooks; Stoate coll.;Stuart Baker coll.. Material: 1 ; whole/good egg. Said to be from the same clutch as the Hume eggs, 1891.3.20.2323-6.

Tragopan caboti

Cabot's Tragopan

Confined to parts of southeastern China, mainly in evergreen mountain forest. The population has been drastically reduced by deforestation and hunting, so that it is now confined to a few areas. No birds were found during a recent search of an area formerly believed to have been a stronghold.

Reg. no. 1866.12.19.1. From ; ; n. China. Date: //. Source: L. Fraser. Material: adult M; whole/good skin.

Reg. no. 1898.1.28.87. From Kuatun; Fujian (Fukien); China. Date: //1896. Source: C.B. Rickett. Material: F; whole/good skin.

Reg. no. 1899.4.1.3. From Kuatun; Fujian (Fukien); China. Date: 27/10/1898. Source: C.B. Rickett. Material: adult F; whole/good skin.

Reg. no. 1899.4.1.4. From Kuatun; Fujian (Fukien); China. Date: 27/10/1898. Source: C.B. Rickett. Material: M; whole/good skin.

Reg. no. 1899.4.1.5. From Kuatun; Fohkien; China. Date: 7/11/1898. Source: C.B. Rickett. Material: adult F; whole/good skin.

Reg. no. 1900.6.23.48. From Kuatun; Fujian (Fukien); China. Date: 27/10/1898. Source: C.B. Rickett. Material: F; whole/good skin.

Reg. no. 1900.6.23.49. From Kuatun; Fujian (Fukien); China. Date: /11/1896. Source: C.B. Rickett. Material: F; whole/good skin.

Reg. no. 1900.6.23.50. From Kuatun; Fujian (Fukien); China. Date: /11/1896. Source: C.B. Rickett. Material: F; whole/good skin.

Reg. no. 1900.6.23.51. From Kuatun; Fujian (Fukien); China. Date: /11/1896. Source: C.B. Rickett. Material: F; whole/good skin.

Reg. no. 1900.6.23.52. From Kuatun; Fujian (Fukien); China. Date: //. Source: C.B. Rickett. Material: M; skin. Not in the collection. Exchanged with the Zoological Society.

Reg. no. 1901.2.18.1. From ; ; China. Date: 20/ 2/1901. Source: W.H.E. St.Quintin. Material: imm M; whole/good skin.

Reg. no. 1902.8.5.152. From Kuatun; Fujian (Fukien); China. Date: 4/12/1896. Source: C.B. Rickett. Material: adult M; whole/good skin.

Reg. no. 1902.8.5.153. From Kuatun; Fohkien; China. Date: 14/11/1896. Source: C.B. Rickett. Material: M; whole/good skin.

Reg. no. 1902.8.5.154. From Kuatun; Fujian (Fukien); China. Date: 28/10/1896. Source: C.B. Rickett. Material: imm F; whole/good skin.

Reg. no. 1902.8.5.155. From Kuatun; Fujian (Fukien); China. Date: 29/10/1898. Source: C.B. Rickett. Material: F; whole/good skin.

Reg. no. 1902.8.5.156. From Kuatun; Fujian (Fukien); China. Date: 27/10/1898. Source: C.B. Rickett. Material: F; whole/good skin.

Reg. no. 1903.10.23.1. From ; ; [Captivity]. Date: //. Source: W.H. St.Quintin. Material: juv ; whole/good skin.

Reg. no. 1905.12.24.1002. From Kuatun; Fujian (Fukien); China. Date: 23/10/1898. Source: C.B. Rickett. Material: adult M; whole/good skin.

Reg. no. 1905.12.24.1003. From Kuatun; Fujian (Fukien); China. Date: 23/10/1898. Source: C.B. Rickett. Material: F; whole/good skin.

Reg. no. 1905.12.24.1004. From Kuatun; Fujian (Fukien); China. Date: 5/11/1898. Source: C.B. Rickett. Material: imm M; whole/good skin. A later label bears the date 15 November 1898 in error.

Reg. no. 1905.12.24.1005. From Kuatun; Fujian (Fukien); China. Date: //1896. Source: C.B. Rickett. Material: F; whole/good skin. Collected during winter.

Reg. no. 1907.12.31.127. From Kuatun; n.w. Fokien; China. Date: //. Source: Styan coll.. Material: adult F; skin. Not in the collection. Exchanged with the Zoological Society.

Reg. no. 1907.12.31.128. From Kuatun; Fujian (Fukien); China. Date: / 3/1898. Source: Styan coll.. Material: adult M; whole/good skin.

Reg. no. 1907.12.31.129. From Kuatun; Fujian (Fukien); China. Date: 29/ 4/1899. Source: Styan coll.. Material: adult M; whole/good skin.

Reg. no. 1907.12.31.130. From Kuatun; Fujian (Fukien); China. Date: 7/11/1901. Source: Styan coll.. Material: adult F; whole/good skin.

Reg. no. 1907.12.31.131. From Kuatun; n.w. Fokien; China. Date: 28/10/1898. Source: Styan coll.. Material: adult F; whole/good skin.

Reg. no. 1907.12.31.132. From Kuatun; n.w. Fokien; China. Date: /12/1896. Source: Styan coll.. Material: adult F; whole/good skin.

Reg. no. 1907.12.31.133. From ; ; China. Date: / 4/. Source: Styan coll.. Material: adult M; whole/good skin.

Reg. no. 1907.12.31.134. From Kuatun; Fujian (Fukien); China. Date: 11/11/1901. Source: Styan coll.. Material: imm M; whole/good skin.

Reg. no. 1918.6.25.250. From ; Fujian (Fukien); China. Date: /11/. Source: W.R. Ogilvie Grant. Material: adult M; whole/good skin.

Reg. no. 1918.6.25.251. From ; Fujian (Fukien); China. Date: /12/. Source: W.R. Ogilvie Grant. Material: adult M; whole/good skin.

Reg. no. 1918.6.25.252. From ; Fujian (Fukien); China. Date: /12/. Source: W.R. Ogilvie Grant. Material: adult M; whole/good skin.

Reg. no. 1918.6.25.253. From ; Fujian (Fukien); China. Date: /12/. Source: W.R. Ogilvie Grant. Material: adult M; whole/good skin.

Reg. no. 1918.6.25.254. From ; Fujian (Fukien); China. Date: / 1/. Source: W.R. Ogilvie Grant. Material: adult M; whole/good skin.

Reg. no. 1918.6.25.255. From ; Fujian (Fukien); China. Date: /11/. Source: W.R. Ogilvie Grant. Material: adult F; whole/good skin.

Reg. no. 1918.6.25.256. From ; Fujian (Fukien); China. Date: /12/. Source: W.R. Ogilvie Grant. Material: adult F; whole/good skin.

Reg. no. 1918.6.25.257. From ; Fujian (Fukien); China. Date: / 8/. Source: W.R. Ogilvie Grant. Material: imm M; whole/good skin.

Reg. no. 1934.9.21.67. From ; ; [Captivity]. Date: //. Source: R. Haig Thomas. Material: M; whole/good skin.

Reg. no. 1938.6.3.3. From ; ; Captivity. Date: 4/ 6/1938. Source: J. Spedan Lewis. Material: chick ; whole/good skin. Bred in captivity.

Reg. no. 1939.5.7.2. From ; ; Captivity. Date: 18/ 5/1939. Source: J. Spedan Lewis. Material: chick ; whole/good skin. Bred in captivity. Ten days old.

Reg. no. 1941.5.30.9466. From ; ; [Captivity]. Date: / 5/1923. Source: W. Shore Baily (Shore Bailey); Ticehurst coll.. Material: chick ; whole/good skin. One day old.

Reg. no. 1901.6.10.1–2. From ; ; Captivity. Date: / 5/1901. Source: W. St.Quintin. Material: 2 ; whole/good eggs. Laid in captivity at Scampston, Yorkshire.

Reg. no. 1903.10.23.2–3. From ; ; Captivity. Date: //. Source: W. St.Quintin. Material: 2 ; whole/good eggs. Laid in captivity.

Reg. no. 1907.10.26.278–9. From Kuatun; ; China. Date: 17/ 5/1898. Source: J.D. La Touche. Material: 2 ; whole/good eggs. Three eggs with the same data were registered as 1907.10.26.278–9 and 1901.5.15.1.

Reg. no. E1988.4.10. From ; ; Captivity. Date: 4/ 6/1924. Source: W. Shore Baily coll.; Stuart Baker coll.. Material: 4 ; whole/good eggs (clutch). Laid in captivity at Westbury, Wiltshire. The nest was in thick grass.

Reg. no. E1988.4.11. From ; ; Captivity. Date: 4/ 5/1925. Source: W. Shore Baily coll.; Stuart Baker coll.. Material: 4 ; whole/good eggs (clutch). Laid in captivity at Westbury avairies, Wiltshire. The nest was a mass of sticks and weeds on top of a bush, 5 ft. from the ground.

Reg. no. E1988.4.9. From ; ; Captivity. Date: 30/ 4/1928. Source: W. Shore Baily coll.; Stuart Baker coll.. Material: 4 ; whole/good eggs (clutch). Laid in captivity at Westbury, Wiltshire. The nest was in a bed of nettles.

Reg. no. 1901.11.1.327. From ; ; Captivity. Date: //. Source: Dr Kutter Carsel; Crowley. Material: 1 ; whole/good egg.

Reg. no. 1901.5.15.1. From Kuatun; ; China. Date: 17/ 5/1898. Source: J.D. La Touche. Material: 1 ; whole/good egg. Three eggs with the same data were registered as 1907.10.26.278–9 and 1901.5.15.1.

Reg. no. E1941.3.1.151. From ; ; Captivity. Date: 25/ 5/1901. Source: Munt coll.. Material: 1 ; whole/good egg. Laid in captivity.

Reg. no. E1941.3.1.152. From ; ; Captivity. Date: 26/ 4/1902. Source: Munt coll.. Material: 1 ; whole/good egg. Laid in captivity.

Reg. no. E1988.4.8. From ; ; . Date: //. Source: Rothschild. Material: 1 ; whole/poor egg. Egg damaged.

Reg. no. A1989.15.1. From ; ; . Date: //. Source: Material: adult ; whole/good alc.

Lophophorus lhuysii

Chinese Monal

Confined to the mountains of central China in scrub and meadows above tree level. Now greatly reduced due to hunting, but precise numbers unknown.

Reg. no. 1867.5.20.1. From Moupin Mountains; w. Szechuen; China. Date: //. Source: M. Dabry; J. Verreaux. Material: adult M; whole/good skin. Syntype. St.–Hilaire, *Bull. Soc. imp. zool. Acclim.*, series 2, 3 (1866): 221–3; Verreaux, *Bull. Soc. imp. zool. Acclim.*, series 2, 4 (1867): 705–11.

Reg. no. 1867.5.20.2. From Moupin Mountains; Szechuen; China. Date: //. Source: M. Dabry; J. Verreaux. Material: imm F; whole/good skin. Syntype.

Reg. no. 1896.1.1.646. From Wantung, nr. Taschien; ; China. Date: / 6/1890. Source: Pratt; H. Seebohm. Material: adult M; whole/good skin.

Reg. no. 1896.1.1.647. From ; Szechuen; China. Date: //1890. Source: Pratt; H. Seebohm. Material: adult M; part/good skin. Head missing.

Reg. no. 1939.12.9.3549. From Janchienlu; Szechuen; China. Date: //. Source: Pratt; Rothschild coll.. Material: adult M; whole/good skin. Died in captivity.

Reg. no. 1939.12.9.3550. From Jacheinlu; Szechuen; China. Date: //. Source: Pratt;

Rothschild coll.. Material: adult F; whole/good skin.

Reg. no. 1954.59.5. From Chin Liang Shan, Ts'ao Po area, n.w. of Kwanhsien; w. Szechuen; China. Date: / 8/1938. Source: F. Tangier Smith. Material: ; whole/good skin. From 31° 18'N 103° 15'E.

Reg. no. 1954.59.6. From Chin Liang Shan, Ts'ao Po area, n.w. of Kwanhsien; w. Szechuen; China. Date: / 8/1938. Source: F. Tangier Smith. Material: M; whole/good skin. From 31° 18'N 103° 15'E.

Reg. no. 1954.59.7. From [n.w. of Kwanhsien]; w. Szechuen; China. Date: //. Source: F. Tangier Smith. Material: ; whole/good skin. Almost certainly from n.w. of Kwanhsien.

Reg. no. 1965.M.2269. From the hills n.w. of Kuanhsien, upper Minho; ; China. Date: /12/1914. Source: Dr. Weigold; R. Meinertzhagen coll.. Material: M; whole/good skin.

Crossoptilon mantchuricum

Brown Eared–Pheasant

Formerly inhabited the mountains of northern China, but declined in numbers with deforestation, and is now found only in three reserves.

Reg. no. 1864.6.9.1. From the mountains n.w. of Peking; ; China. Date: //. Source: F.W.A. Bruce. Material: M; whole/good skin. Ex–mount.

Reg. no. 1864.6.9.2. From the mountains n.w. of Peking; ; China. Date: //. Source: F.W.A. Bruce. Material: F; whole/good skin. Two birds bear this reg. no..

Reg. no. 1864.6.9.2. From Peking; ; China. Date: //. Source: F.W.A. Bruce. Material: adult F; whole/good skin. Two birds bear this reg. no..

Reg. no. 1867.7.22.5. From ; ; Captivity. Date: //. Source: Zoological Society. Material: chick ; whole/good skin. Ex–mount. Hatched 3 May 1867.

Reg. no. 1867.7.22.6. From ; ; Captivity. Date: //. Source: Zoological Society. Material: chick ; whole/good skin. Hatched 3 May 1867.

Reg. no. 1868.2.24.3. From ; ; Captivity. Date: //. Source: Zoological Society. Material: chick ; whole/good skin. Ex–mount.

Reg. no. 1881.5.1.6282. From ; ; China. Date: //. Source: J. Gould. Material: adult M; whole/good skin. Probably from n.w. China.

Reg. no. 1896.1.1.614. From Peking; ; China. Date: //. Source: A. David; R. Swinhoe; Seebohm coll.. Material: M; whole/good skin.

Reg. no. 1898.4.14.1. From ; Manchuria; China. Date: //. Source: H. Fothergill. Material: M; skin. Mount, from which the trunk was taken. Not found 1988. The trunk is in the skeleton collection with the no. 1898.4.14.2.

Reg. no. 1907.12.31.108. From Peking; ; China. Date: /12/1886. Source: F.W. Styan. Material: adult M; whole/good skin.

Reg. no. 1907.12.31.109. From Peking; ; China. Date: /12/1886. Source: F.W. Styan. Material: adult F; whole/good skin.

Reg. no. 1930.6.12.1. From ; ; Captivity. Date: 16/ 6/1930. Source: J. Spedan Lewis. Material: chick ; whole/good skin. Died in captivity.

Reg. no. 1939.6.1.1. From ; ; Captivity. Date: 2/ 6/1939. Source: J.D. Scale. Material: chick ; whole/good skin.

Reg. no. 1947.57.1. From ; ; Captivity. Date: 3/ 7/1947. Source: Prof. A. Ghigi. Material: adult M; whole/good skin. Hybrid: *mantchurium* x *Lophura (nycthemera beli x leucomelana lineata)*. The latter was previously noted as *Gennaeus (beli x lineatus)*.

Reg. no. 1947.57.2. From ; ; Captivity. Date: 3/ 7/1947. Source: Prof. A. Ghigi. Material: adult F; whole/good skin. Hybrid: *mantchurium* x *Lophura (nycthemera beli x leucomelana lineata)*. The latter was previously noted as *Gennaeus (beli x lineatus)*.

Reg. no. 1952.56.67. From ; ; Captivity. Date: //. Source: Zoological Society. Material: adult M; whole/good skin. Hybrid: male *mantchurium* x female *Lophura leucomelana hamiltoni (Gennaeus hamiltoni)*. Hatched at Whipsnade, 21 August 1938; date of death not noted.

Reg. no. 1952.56.68. From ; ; Captivity. Date: //. Source: Zoological Society. Material: adult M; whole/good skin. Hybrid: male *mantchurium* x female *Lophura leucomelana hamiltoni (Gennaeus hamiltoni)*. Hatched at Whipsnade, 21 August 1938; date of death not noted.

Reg. no. 1952.56.69. From ; ; Captivity. Date: //. Source: Zoological Society. Material: adult F; whole/good skin. Hybrid: male *mantchurium* x female *Lophura leucomelana hamiltoni (Gennaeus hamiltoni)*. Hatched at Whipsnade, 21 August 1938; date of death not noted.

Reg. no. 1955.3.127. From ; ; Captivity. Date: 7/ 8/1922. Source: W.G. Maginess; G.E. Lodge coll.. Material: M; whole/good skin. From captivity in Lisburn, N. Ireland.

Reg. no. 1897.3.13.1. From ; Manchuria; China. Date: //. Source: H. Fothergill. Material: F; whole/good skeleton. Reg. no. untraced. The no. appears in the register of skeletons which was compiled from the General Registers, but cannot now be found in the latter, where this number was allocated to a mammal. The label bears the inscription 'received 18.3.98', a year later than the specimen was registered.

Reg. no. 1898.4.14.2. From ; Manchuria; China. Date: //. Source: H. Fothergill. Material: M; part/good skeleton. Trunk only. See 1898.4.14.1 (same species).

Reg. no. S1972.1.61. From ; ; Captivity. Date: //. Source: Rothschild coll.. Material: ; whole/good skeleton.

Reg. no. 1891.6.22.32–4. From ; ; Captivity. Date: //. Source: Zoological Society. Material: 3 ; whole/good eggs. Laid in captivity at London Zoo.

Reg. no. E1941.3.1.81. From ; ; Captivity. Date: 16/ 5/1903. Source: Munt coll.. Material: 2 ; whole/good eggs. Laid in captivity.

Reg. no. E1972.13.425. From ; ; Captivity. Date: / 4/1916. Source: Rothschild. Material: 5 ; whole/good eggs. Laid in captivity at Tring, Hertfordshire.

Reg. no. E1972.13.427. From ; ; Captivity. Date: / 5/1918. Source: Rothschild. Material: 5 ; whole/good eggs. Laid in captivity at Tring, Hertfordshire.

Reg. no. E1988.4.12. From ; ; Captivity. Date: //. Source: Bartlett coll.; Stuart Baker coll.. Material: 2 ; whole/good eggs.

Reg. no. E1988.4.13. From ; ; Captivity. Date: 29/ 5/1904. Source: Berlin Zoo; Hon. M. Kuschel; H.N. Coltart; Stuart Baker coll.. Material: 3 ; whole/good eggs (clutch). Labelled '*tibetanum*' = *C. crossoptilon*. Identity uncertain.

Reg. no. E1988.4.14. From ; ; Captivity. Date: 1/ 6/1917. Source: Leach; Stuart Baker coll.. Material: 7 ; whole/good eggs (clutch). Laid in captivity at Epsom, Surrey, between 1 – 10 June 1917. Another hen in the same aviary is said to have laid an egg each day until she had 32 eggs.

Reg. no. E1941.3.1.81a. From ; ; Captivity. Date: //. Source: Munt coll.. Material: 1 ; whole/good egg. Laid in captivity.

Reg. no. E1963.9.189. From ; ; Captivity. Date: //1915. Source: W. Rothschild. Material: 17 ; whole/good egg. Laid in captivity between 1915 and 1918, at Tring Park, Hertfordshire.

Reg. no. E1972.13.423. From ; ; [Captivity]. Date: / 4/1867. Source: Rothschild. Material: 1 ; whole/good egg. No history, but presumed laid in captivity. No data except on egg. Identity not certain.

Reg. no. E1972.13.424. From ; ; Captivity. Date: / 5/1915. Source: Rothschild. Material: 1 ; whole/good egg. Laid in captivity at Tring, Hertfordshire.

Reg. no. E1972.13.426. From ; ; Captivity. Date: / 5/1917. Source: Rothschild. Material: 1 ; whole/good egg. Laid in captivity at Tring, Hertfordshire.

Reg. no. E1972.13.428. From ; ; Captivity. Date: / 4/1916. Source: Rothschild. Material: 1 ; whole/good egg. Laid during April or May 1916, at Tring, Hertfordshire.

Reg. no. 1929.3.11.5. From ; ; Captivity. Date: //. Source: J. Delacour. Material: chick ; whole/good alc.

Reg. no. A1969.38.1. From ; ; Captivity. Date: 6/12/1969. Source: P. Wayre. Material: adult ; whole/good alc.

Catreus wallichii

Cheer Pheasant

Occurred through much of the western Himalayas but greatly reduced by hunting and habitat destruction, and now survives only in scattered areas. A vigorous reintroduction programme is under way.

Reg. no. . From ; ; Nepal. Date: //. Source: B.H. Hodgson. Material: imm M; skin. Details taken from *Cat. Bds.*. Not found 1988.

Reg. no. . From ; ; Nepal. Date: //. Source: B.H. Hodgson. Material: imm M; skin. Details taken from *Cat. Bds.*. Not found 1988.

Reg. no. 1843.1.13.1086. From ; ; Nepal. Date: //. Source: B.H. Hodgson. Material: imm M; whole/good skin.

Reg. no. 1845.1.9.140. From ; ; Nepal. Date: //. Source: B.H. Hodgson. Material: F; whole/good skin.

Reg. no. 1867.3.16.100. From ; ; Captivity. Date: //. Source: J. Gould. Material: adult M; whole/good skin. Hybrid: *C. wallichii* x *Phasianus colchicus*, from England.

Reg. no. 1873.9.10.12. From ; ; Captivity. Date: //. Source: J.J. Stone. Material: chick ; whole/good skin. Hybrid: *Tragopan temminckii* x *C. wallichii*.

Reg. no. 1880.1.1.1909. From Himalayas; ; India. Date: //. Source: Bax; India Museum. Material: imm M; whole/good skin. Ex-mount.

Reg. no. 1880.1.1.2475. From Himalayas; ; Nepal. Date: 18/11/1833. Source: B.H. Hodgson; India Museum. Material: M; whole/good skin.

Reg. no. 1881.7.30.142. From n.w. Himalayas; ; . Date: //. Source: Capt. S. Pinwill. Material: F; whole/good skin.

Reg. no. 1881.7.30.51. From n.w. Himalayas; ; . Date: //. Source: Capt. S. Pinwill. Material: ; skin. Details taken from the register. Not found 1988.

Reg. no. 1881.7.30.52. From n.w. Himalayas; ; . Date: //. Source: Capt. S. Pinwill. Material: F; whole/good skin.

Reg. no. 1881.7.30.53. From n.w. Himalayas; ; . Date: //. Source: Capt. S. Pinwill. Material: ; skin. Details taken from the register. Not in the collection; to Nottingham Museum in 1895.

Reg. no. 1881.7.30.54. From n.w. Himalayas; ; . Date: //. Source: Capt. S. Pinwill. Material: ; skin. Details taken from the register. Not in the collection; to Dundee Museum in 1892.

Reg. no. 1889.5.10.1588. From Kumaon; Punjab; [India]. Date: //1869. Source: Hume coll.. Material: adult M; whole/good skin.

Reg. no. 1889.5.10.1589. From Kooloo; Punjab; [India]. Date: 29/ 6/1869. Source: Hume coll.. Material: adult M; whole/good skin.

Reg. no. 1889.5.10.1590. From Koteghur, nr. Simla; ; [India]. Date: /10/1868. Source: Hume coll.. Material: adult M; whole/good skin. Damaged: head detached and bill broken.

Reg. no. 1889.5.10.1591. From Koteghur; ; [India]. Date: //. Source: Hume coll.. Material: adult M; whole/good skin.

Reg. no. 1889.5.10.1592. From Koteghur; ; [India]. Date: //. Source: Hume coll.. Material: ; skin. Details taken from the register. Not in the collection; to Christiana Museum (via Seebohm) in 1892.

Reg. no. 1889.5.10.1593. From Koteghur; ; [India]. Date: / 4/1871. Source: Hume coll.. Material: adult M; whole/good skin.

Reg. no. 1889.5.10.1594. From Koteghur; ; [India]. Date: //. Source: Hume coll.. Material: adult M; skin. Details taken from the register. Not found 1988.

Reg. no. 1889.5.10.1595. From Koteghur; ; [India]. Date: /12/1870. Source: Hume coll.. Material: adult M; whole/good skin.

Reg. no. 1889.5.10.1596. From Koteghur; ; [India]. Date: //1872. Source: Hume coll.. Material: adult M; whole/good skin.

Reg. no. 1889.5.10.1597. From Koteghur; ; [India]. Date: / 4/1871. Source: Hume coll.. Material: adult M; whole/good skin.

Reg. no. 1889.5.10.1598. From Koteghur; ; [India]. Date: //. Source: Hume coll.. Material: adult M; whole/good skin.

Reg. no. 1889.5.10.1599. From Koteghur; ; [India]. Date: //1871. Source: Hume coll.. Material: adult M; whole/good skin.

Reg. no. 1889.5.10.1600. From Koteghur, nr. Simla; ; [India]. Date: /10/1868. Source: Hume coll.. Material: adult M; whole/good skin.

Reg. no. 1889.5.10.1601. From Koteghur; ; [India]. Date: //1871. Source: Hume coll.. Material: adult F; whole/good skin.

Reg. no. 1889.5.10.1602. From Koteghur; ; [India]. Date: //. Source: Hume coll.. Material: adult F; skin. Details taken from the register. Not found 1988.

Reg. no. 1889.5.10.1603. From Koteghur; ; [India]. Date: / 4/1871. Source: Hume coll.. Material: adult F; whole/good skin.

Reg. no. 1889.5.10.1604. From Koteghur; ; [India]. Date: / 4/1871. Source: Hume coll.. Material: adult F; whole/good skin.

Reg. no. 1889.5.10.1605. From Koteghur; ; [India]. Date: //. Source: Hume coll.. Material: adult F; skin. Details taken from the register. Not found 1988.

Reg. no. 1889.5.10.1606. From Koteghur; ; [India]. Date: /12/1870. Source: Hume coll.. Material: adult F; whole/good skin.

Reg. no. 1889.5.10.1607. From Bhagirati Valley; ; India. Date: //. Source: Hume coll.. Material: adult M; whole/good skin.

Reg. no. 1889.5.10.1608. From Bhagirati Valley; ; India. Date: //. Source: Hume coll.. Material: adult M; whole/good skin.

Reg. no. 1889.5.10.1609. From Bhagirati Valley; ; India. Date: //. Source: Hume coll.. Material: adult M; whole/good skin.

Reg. no. 1889.5.10.1610. From Bhagirati Valley; ; India. Date: //. Source: Hume coll.. Material: adult M; whole/good skin.

Reg. no. 1889.5.10.1611. From Bhagirati Valley; ; India. Date: //. Source: Hume coll.. Material: adult M; whole/good skin.

Reg. no. 1889.5.10.1612. From Bhagirati Valley; ; India. Date: //. Source: Hume coll.. Material: adult M; skin. Details taken from the register. Not found 1988.

Reg. no. 1889.5.10.1613. From Bhagirati Valley; ; India. Date: //. Source: Hume coll.. Material: adult M; whole/good skin.

Reg. no. 1889.5.10.1614. From Bhagirati Valley; ; India. Date: //. Source: Hume coll.. Material: adult F; whole/good skin.

Reg. no. 1889.5.10.1615. From Bhagirati Valley; ; India. Date: //. Source: Hume coll.. Material: adult F; whole/good skin.

Reg. no. 1889.5.10.1616. From Bhagirati Valley; ; India. Date: //. Source: Hume coll.. Material: adult F; whole/good skin.

Reg. no. 1889.5.10.1617. From Bhagirati Valley; ; India. Date: //. Source: Hume coll.. Material: adult F; whole/good skin.

Reg. no. 1889.5.10.1618. From Bhagirati Valley; ; India. Date: //. Source: Hume coll.. Material: adult F; skin. Details taken from the register. Not found 1988.

Reg. no. 1889.5.10.1619. From Bhagirati Valley; ; India. Date: //. Source: Hume coll.. Material: adult F; whole/good skin.

Reg. no. 1949.WHI.1.6353. From Chakrata, Dehra Dun; [Uttar Pradesh]; India. Date: //1893. Source: H. Whistler coll.. Material: adult F; whole/good skin. Date probably 1893.

Reg. no. 1949.WHI.1.6354. From Kangra; Punjab; India. Date: 30/ 1/1921. Source: H. Whistler coll.. Material: M; whole/good skin.

Reg. no. 1949.WHI.1.6355. From Kangra; Punjab; India. Date: 30/ 1/1921. Source: H. Whistler coll.. Material: F; whole/good skin.

Reg. no. 1949.WHI.1.6356. From Kangra; Punjab; India. Date: 26/12/1926. Source: H. Whistler coll.. Material: ; whole/good skin. Weight, 3.25 lbs.

Reg. no. 1965.M.2289. From Bodiya, Chakrata, Himalayas; ; . Date: 16/ 9/1912. Source: Meinertzhagen coll.. Material: F; whole/good skin.

Reg. no. 1969.41.1114. From ; ; India. Date: //. Source: Hewitt coll.. Material: M; whole/good skin.

Reg. no. 1988.21.21. From Himalayas; ; . Date: //. Source: Material: ; whole/good skin.

Reg. no. 1988.21.22. From ; ; . Date: //. Source: Material: ; whole/good skin. No history.

Reg. no. 1988.21.23. From ; ; . Date: //. Source: Material: ; whole/good skin. No history.

Reg. no. VEL.34.13a. From [Almorah Hills], Himalayas; ; [India]. Date: //. Source: Gen. T. Hardwicke. Material: M; whole/good skin. Syntype.

Reg. no. VEL.34.13b. From Almorah Hills, Himalayas; ; n.e. India. Date: //. Source: Gen. T. Hardwicke. Material: adult M; whole/good skin. Syntype. Ex-mount. Hardwicke, *Trans. Linn. Soc. Lond.* 15 (1826): 166–7. There are another two syntypes in the collection.

Reg. no. VEL.34.13c. From [Almorah Hills], Himalayas; ; [India]. Date: //. Source: Gen. T. Hardwicke. Material: ; whole/good skin. Syntype.

Reg. no. VEL.34.13d. From Himalayas; ; . Date: //. Source: R. Ferguson. Material: ; whole/good skin.

Reg. no. 1845.1.12.293. From ; ; Nepal. Date: //1859. Source: B.H. Hodgson. Material: ; part/good skeleton. Trunk, rough skeleton only. Previously registered as Skeleton Vellum Cat. 334a.

Reg. no. 1859.2.27.2. From ; ; Nepal. Date: //1859. Source: Zoological Society. Material: ; whole/good skeleton. Taken from the gallery at South Kensington.

Reg. no. 1861.6.3.2. From ; ; . Date: //. Source: Zoological Society. Material: ; whole/good skeleton. Previously registered as Skeleton Vellum Cat. 334b.

Reg. no. S1954.20.1. From ; ; . Date: 2/ 8/1954. Source: T. Sowerbutts. Material: chick ; whole/good skeleton. One day old chick. The label also bears the inscription 'Golden Pheasant chick'.

Reg. no. . From ; ; . Date: //. Source: Major Cathwler (sic) ; R.H. Rattray coll.; Jourdain coll.. Material: 2 ; eggs. Bought by Jourdain in 1914. Not received with Jourdain collection.

Reg. no. 1859.6.5.10. From ; ; India. Date: //. Source: Gould coll.. Material: 2 ; whole/good eggs.

Reg. no. 1891.3.20.3479–82. From [Kotegurh, Himalayas]; ; [India]. Date: //. Source: Hume coll.. Material: 4 ; whole/good eggs. Locality given as 'Himalayas' in Oates' *Cat. Eggs*, and as 'Kotegurh' in Seebohm ms cat..

Reg. no. 1901.11.1.431–3. From ; ; [Captivity]. Date: 4/ 5/1858. Source: London Zoo; Tristram coll.; Crowley. Material: 3 ; whole/good eggs.

Reg. no. E1925.12.25.3129–30. From ; ; Nepal. Date: //1887. Source: Schluter; Davidson. Material: 2 ; whole/good eggs.

Reg. no. E1925.12.25.3132–3. From Mussoorie Hills, Himalayas; ; Nepal. Date: 6/ 5/1899. Source: Davidson coll.. Material: 2 ; whole/good eggs.

Reg. no. E1962.1.738. From ; ; Captivity. Date: //. Source: H. Munt. Material: 4 ; whole/good eggs. Laid in captivity.

Reg. no. E1988.4.19. From ; ; Captivity. Date: 19/ 6/1900. Source: Hermanne Rolle; Stuart Baker coll.. Material: 2 ; whole/good eggs. Doubtfully a clutch. Laid in captivity in Berlin. Rolle informed Baker that the birds 'probably came from Kashmir, certainly not (underlined) China', as is given on the data slips pasted into the Baker catalogue.

Reg. no. E1988.4.20. From ; ; [Captivity]. Date: //. Source: London Zoo; Godman Salvin coll.. Material: 2 ; whole/good eggs.

Reg. no. E1941.1.7.1288. From Dharamsala; Punjab; India. Date: //1891. Source: Butler coll.. Material: 2 ; whole/good eggs (clutch).

Reg. no. E1988.4.15. From 8,000 ft., Bodyar, Chakrata; Uttar Pradesh; India. Date: 3/ 6/1918. Source: B.B. Osmaston; Stuart Baker coll.. Material: 8 ; whole/good eggs (clutch). 'Nest in long dry grass in open glade in Deodar forest – fairly steep slope. Found by my spaniel bitch "Rami" who put bird off nest' (note by Osmaston).

Reg. no. E1988.4.16. From Garhwal, (Garwhal); Uttar Pradesh; India. Date: 15/ 5/1906. Source: S.L. Whymper; Stuart Baker coll.. Material: 8 ; whole/good eggs (clutch). Six eggs stolen during 1975–1979, subsequently recovered. The nest was a few leaves, partly sheltered by a ledge of rock, at about 7,000 ft..

Reg. no. E1988.4.17. From ; ; [Captivity]. Date: 1/ 9/. Source: London Zoo; Bartlett coll.; H.N. Coltart; Stuart Baker coll.. Material: 3 ; whole/good eggs (clutch).

Reg. no. 1891.3.20.3478. From [Kotegurh, Himalayas]; ; [India]. Date: //. Source: Hume coll.. Material: 1 ; egg. Locality given as 'Himalayas' in Oates' *Cat. Eggs*, and as 'Kotegurh' in Seebohm ms cat.. Not in the collection.

Reg. no. E1925.12.25.3131. From Himalayas; ; . Date: //. Source: Davidson coll.. Material: 1 ; whole/good egg.

Reg. no. E1988.4.18. From 6,000 ft. in the Valley of Nepal; ; India. Date: 27/ 5/1894. Source: Stuart Baker coll.. Material: 1 ; whole/good egg. Some doubt about date and locality. Baker's catalogue contains the information 'Brought down to Darjeeling with the birds for sale. The Nepalese said the hen had laid the egg in the basket'.

Reg. no. E1988.4.21. From ; ; [Captivity]. Date: 18/ 4/1858. Source: London Zoo; Godman Salvin coll.. Material: 1 ; whole/good egg.

Reg. no. E1988.4.22. From ; ; Captivity. Date: //1861. Source: Rothschild. Material: 1 ; whole/good egg. Probably laid in captivity in France in 1861 or 1881. 'Faison I. Wallich'.

Reg. no. E1988.4.23. From Himalayas; ; . Date: 3/ 6/1900. Source: Rothschild. Material: 1 ; whole/good egg.

Syrmaticus ellioti

Elliot's Pheasant

Found in forested ravines in the hills of eastern China, where it is severely threatened by deforestation and has now become exceedingly scarce.

Reg. no. 1896.1.1.743. From Ningpo; Che Kiang; s.e. China. Date: 23/11/1871. Source: W.G.H. Cable; Swinhoe coll.;

Seebohm coll.. Material: adult M; whole/good skin. Syntype. Ex-mount. Swinhoe, *Proc. zool. Soc. Lond.* 1872: 550.

Reg. no. 1896.1.1.744. From Ningpo; Che Kiang; China. Date: 18/12/1873. Source: R. Swinhoe; Seebohm coll.. Material: adult F; whole/good skin.

Reg. no. 1903.8.8.493. From Chutung; w. Yunnan; China. Date: 21/ 3/1902. Source: Col. Rippon. Material: F; whole/good skin.

Reg. no. 1905.11.11.26. From ; ; [China], [Captivity]. Date: //. Source: Lord Tweedmouth. Material: adult M; whole/good skin. Probably from captivity.

Reg. no. 1905.12.24.996. From Kuatun; Fujian (Fukien); China. Date: / 8/1901. Source: C.B. Rickett. Material: adult M; whole/good skin.

Reg. no. 1907.12.31.35. From Kuatun; Fujian (Fukien); China. Date: /11/1901. Source: Styan coll.. Material: adult M; whole/good skin.

Reg. no. 1907.12.31.67. From ; ; China. Date: //. Source: Styan coll.. Material: adult M; whole/good skin.

Reg. no. 1907.12.31.68. From ; ; China. Date: //. Source: Styan coll.. Material: adult F; whole/good skin.

Reg. no. 1916.9.20.18. From ; ; Captivity. Date: 10/ 3/1890. Source: E.A. Butler; A.L. Butler. Material: adult M; whole/good skin. Bred in captivity. From Herringfleet Hall, Suffolk.

Reg. no. 1918.6.25.241. From ; Fujian (Fukien); China. Date: /12/. Source: W.R. Ogilvie Grant. Material: adult F; whole/good skin.

Reg. no. 1940.7.2.1. From ; ; Captivity. Date: 16/ 7/1913. Source: I.C.I. Material: juv ; whole/good skin. Bred in captivity. From I.C.I. Game Research Estate, Knebworth Park, Herts.. Two and a half months old.

Reg. no. 1940.7.2.2. From ; ; Captivity. Date: 16/ 7/1913. Source: I.C.I. Material: chick ; whole/good skin. Bred in captivity. From I.C.I. Game Research Estate, Knebworth Park, Herts.. 24 hours old.

Reg. no. 1941.5.30.9469. From ; ; Captivity. Date: //1923. Source: Zoological Society; Ticehurst coll.. Material: juv ; whole/good skin. Bred in captivity.

Reg. no. 1955.3.110. From ; ; Captivity. Date: 7/12/1931. Source: G.E. Lodge coll.. Material: adult M; whole/good skin. Bred in captivity. From Ballywalter, Co.Down, N.Ireland.

Reg. no. 1955.3.111. From ; ; Captivity. Date: 7/12/1931. Source: G.E. Lodge coll.. Material: imm F; whole/good skin. Bred in captivity. From Ballywalter, Co.Down, N.Ireland.

Reg. no. 1969.49.9. From ; ; China. Date: //. Source: Hewitt coll.. Material: adult F; whole/good skin.

Reg. no. 1987.24.75. From ; ; [China]. Date: //. Source: Rothschild coll.. Material: adult F; whole/good skin. Mount. Moved from Tring annex store to main collection in 1980s.

Reg. no. 1988.19.1. From ; ; Captivity. Date: //. Source: Poyser; W.M. Crowfoot coll.. Material: adult M; whole/good skin. From Mr Poyser's aviary.

Reg. no. 1988.21.33. From ; ; [China]. Date: //. Source: Rothschild coll.. Material: adult M; whole/good skin. Mount in Tring gallery.

Reg. no. 1891.6.22.27–30. From ; ; [Captivity]. Date: //. Source: Zoological Society. Material: 4 ; whole/good eggs.

Reg. no. E1962.11.1. From ; ; Captivity. Date: / 5/1962. Source: P. Wayre, Ornamental Pheasant Trust. Material: 5 ; whole/good eggs (clutch). Laid in captivity at Gt. Witchingham, Norfolk. Five eggs from a clutch of nine.

Reg. no. E1936.6.3.1. From ; ; Captivity. Date: / 6/. Source: Lt. Col. J.D. Scale. Material: 1 ; whole/good egg.

Reg. no. E1941.3.1.87. From ; ; Captivity. Date: 25/ 4/1908. Source: Munt coll.. Material: 1 ; whole/good egg. Laid in captivity.

Reg. no. E1941.3.1.88. From ; ; Captivity. Date: 30/ 3/1901. Source: Munt coll.. Material: 1 ; whole/good egg. Laid in captivity.

Reg. no. E1941.3.1.89. From ; ; Captivity. Date: 28/ 5/1913. Source: Munt coll.. Material: 1 ; whole/good egg. Laid in captivity.

Argusianus bipunctatus

Double-banded Argus

Known only from a portion of a single feather of unknown origin, although Tioman Island, eastern Malaya has been suggested. A doubtfully valid species.

Reg. no. 1891.2.9.1. From ; ; . Date: //. Source: T.W. Wood; E. Bartlett. Material: adult M; part/good skin. Holotype. Part of a primary feather. Wood, *Field* 37 (1871): 281. See also *Ann. Mag. nat. Hist.*, series 4, 8 (1871): 67–8 and *Bull. Brit. Orn. Club* 103 (1983): 86–8.

Agelastes meleagrides

White-breasted Guineafowl

Occurs only in primary or mature secondary forest in West Africa, and extremely susceptible to disturbance. It now survives only in isolated pockets in Liberia, Sierra Leone, Ivory Coast and Ghana.

Reg. no. 1862.2.11.6. From Elminia, Fantee; ; [Ghana]. Date: //. Source: Brandt. Material: adult M; whole/good skin.

Reg. no. 1873.5.24.36. From Kazarakoo, Denkera; Ashanti; Ghana. Date: / 9/1872. Source: Mr Aubrim; Ussher; R.B. Sharpe coll.. Material: adult F; whole/good skin. The collector is given as Mr Aubrim, 'a native', in the register, where Ussher's name is not mentioned.

Reg. no. 1891.5.1.127. From Denkera; Ashanti; [Ghana]. Date: //. Source: Ussher; Shelley coll.. Material: adult F; whole/good skin.

Reg. no. 1905.5.28.3. From ; ; Liberia. Date: //. Source: F. & H. Wheldon; H. Reynolds. Material: adult ; whole/good skin.

Reg. no. 1930.4.7.1. From Enchi; ; Ghana (Gold Coast Colony). Date: 27/ 2/1930. Source: Mr Guthrie Hall; Junior Army & Navy Club. Material: adult ; whole/good skin. From the Enchi rain forest.

Reg. no. 1977.20.3094. From Cavally River; ; e. Liberia. Date: 4/12/1978. Source: A.D. Forbes Watson, Mount Nimba Expedition. Material: adult M; whole/good skin. Iris ox–brown, bill ivory.

Grus americana

Whooping Crane

Probably formerly occurred widely through North America but now restricted to breeding in Idaho and Northwest Territories. Winters in Texas and New Mexico. The greatest decline occurred in the 1890s. In 1941 the total population was thought to be as low as 15 birds. Following a recovery, there were 159 birds in the two wintering areas in 1989, with a further 55 in captivity.

Reg. no. 1863.1.28.135. From Fort Simpson; Northwest Territories; Canada. Date: 10/ 6/. Source: B.R. Ross. Material: adult ; whole/good skin.

Reg. no. 1880.2.28.2. From Marias River; ; North America. Date: / 6/1878. Source: G. Dawson; North American Boundary Commission. Material: adult ; whole/good skin. Locality uncertain. From along the forty–ninth parallel.

Reg. no. 1893.2.1.23. From Brownsville; Texas; United States. Date: 2/ 4/1890. Source: F.B. Armstrong; Salvin–Godman coll.. Material: adult M; whole/good skin.

Reg. no. 1913.12.26.19. From Padry Island; Texas; United States. Date: 20/ 2/1896. Source: W.P. Lowe. Material: adult M; whole/good skin.

Reg. no. 1926.12.22.6. From ; ; Canada. Date: //. Source: Imperial Institute. Material: adult ; whole/good skin.

Reg. no. 1856.3.26.1. From ; ; . Date: //. Source: Zoological Society. Material: ; whole/good skeleton. Identity uncertain; may be too small for *americana*, probably *canadensis* (C.J.O. Harrison). Previously registered as Skeleton Vellum Cat. 1180a.

Reg. no. 1901.7.10.583–4. From ; Iowa; United States. Date: //. Source: Foottit coll.; Crowley. Material: 2 ; whole/good eggs.

Reg. no. 1902.10.15.730–1. From ; ; North America. Date: //. Source: Radcliffe–Saunders coll.. Material: 2 ; whole/good eggs.

Reg. no. E1941.4.6.884. From nr. Nazareth; Iowa; United States. Date: 3/ 1/1872. Source: E. Ricksecker; Buckley coll.. Material: 2 ; whole/good eggs. 'From E. Ricksecker, Nazareth. Jan 3.1872.' The date is also written on one of the eggs, but they were probably collected before then.

Reg. no. E1930.10.22.1. From ; ; Captivity. Date: //1923. Source: Lord Lilford. Material: 1 ; whole/good egg. Laid in captivity.

Reg. no. E1941.4.6.882. From ; ; [America]. Date: //1863. Source: Buckley(?); Rothschild. Material: 1 ; whole/good egg. No data; registered as part of the Buckley coll. but with no positive identification, although the egg bears *Grus americana* in pencil. Register gives the date 1863, but it is not in the Buckley cat..

Reg. no. E1941.4.6.883. From Ackley, Sullivan Co.; Iowa; United States. Date: //1870. Source: W. Rice; J. Wallace; Buckley coll.. Material: 1 ; whole/good egg. Laid prior to 1870.

Grus leucogeranus

Siberian Crane

Formerly occurred widely throughout Siberia, but now reduced to two isolated breeding populations, one wintering in India and Iran, and the other in China. The western group has been declining in size since 1984–1985 and numbered only 14 in 1990–1991. There are about 2500 birds in the eastern population.

Reg. no. . From ; ; Russia. Date: //. Source: J. Gould coll.. Material: adult ; skin. Mount. Details taken from *Cat. Bds.*. Not found 1988.

Reg. no. 1878.12.31.92. From Barnaul; Siberia; Russia. Date: 22/ 6/1876. Source: O. Finsch; Westsibirische Expedition; Geographical Soc. of Bremen. Material: imm M; whole/good skin.

Reg. no. 1890.2.10.10. From Toman Jhil (or Jheel) or Jemanjhul, Etawah; ; India. Date: 4/ 2/1871. Source: W.E. Brooks; Hume coll.. Material: adult ; whole/good skin. Register gives date as 24 February 1871.

Reg. no. 1890.2.10.11. From Toman Jhil (or Jheel) or Jemanjhul, Etawah; ; India. Date: 4/ 2/1871. Source: W.E. Brooks; Hume coll.. Material: adult F; whole/good skin. Register gives date as 22 February 1871.

Reg. no. 1890.2.10.12. From Toman Jhil (or Jheel) or Jemanjhul, Etawah; ; India. Date: 22/ 2/1871. Source: W.E. Brooks; Hume coll.. Material: adult F; whole/good skin. Register gives date as 4 February 1871.

Reg. no. 1890.2.10.13. From Etawah; ; India. Date: //1865. Source: Hume coll.. Material: adult ; whole/good skin.

Reg. no. 1890.2.10.14. From Etawah; ; India. Date: //1865. Source: Hume coll.. Material: adult ; whole/good skin.

Reg. no. 1890.2.10.14b. From Etawah; ; India. Date: / 7/1865. Source: Hume coll.. Material: adult ; whole/good skin. This reg. no. does not apparently exist. Another specimen of this species bears the reg. no. 1890.2.10.14.

Reg. no. 1890.2.10.15. From Futehgurh; ; India. Date: 6/12/1874. Source: A. Anderson; Hume coll.. Material: adult F; whole/good skin. Date from collector's label; Museum label presumably incorrect (2/1875).

Reg. no. 1890.2.10.16. From Futehgurh; ; India. Date: 31/12/1876. Source: A.

Anderson; Hume coll.. Material: adult ; whole/good skin.

Reg. no. 1890.2.10.17. From Palee, Futehgurh; ; India. Date: 15/ 2/1873. Source: A. Anderson; Hume coll.. Material: adult M; whole/good skin. The data given is taken from the label. An apparently older piece of paper which had been wrapped round the left leg says 'two shot in company same date, *Mot*[? *acilla*] *Dukhunensis* male, Ahraura [Mirzapur, India], 17 November 1874'. This appears to relate to different birds.

Reg. no. 1891.4.11.20. From ; [Siberia]; [Russia]. Date: //. Source: Lord Lilford. Material: adult ; whole/good skin.

Reg. no. 1908.1.5.41. From Wuhu; ; China. Date: / 1/1895. Source: Styan coll.. Material: adult ; part/good skin. Head, neck and leg only. Specimen purchased in Shanghai street market. Skull in skeleton collection with same reg. no..

Reg. no. 1869.3.5.17. From ; ; . Date: //. Source: Material: ; whole/good skeleton. Previously registered as Skeleton Vellum Cat. 1376a, but one bone marked 1376c. Probably misidentified as this species.

Reg. no. 1908.1.5.41. From Wuhu; ; China. Date: / 1/1895. Source: F.W. Styan. Material: ; part/good skeleton. Specimen purchased in Shanghai street market. Skull only; taken from skin in collection with same reg. no..

Reg. no. S1952.2.150. From ; ; . Date: //. Source: Material: adult ; whole/good skeleton. The identity is queried in the register.

Reg. no. E1963.9.97. From ; ; . Date: 10/ 5/1919. Source: W. Rothschild. Material: 2 ; whole/good eggs. One egg inscribed 10 May 1919.

Reg. no. E1973.32.294. From ; n.e. Siberia; Russia. Date: / 6/1904. Source: M. Kuschel; Stuart Baker coll.. Material: 2 ; whole/good eggs (clutch). These eggs probably came to Kuschel from Bianchi. The eggs were taken by a Russian collector from a nest of rushes and reeds on the ground by a lake.

Reg. no. E1925.12.25.4251. From ; ; Captivity. Date: 10/ 6/1900. Source: Lord Lilford; Davidson coll.. Material: 1 ; whole/good egg. Laid in captivity at Lord Lilford's aviaries.

Reg. no. E1941.3.1.243. From ; ; Captivity. Date: //1907. Source: Munt coll.. Material: 1 ; whole/good egg. Laid in captivity.

Reg. no. E1941.3.1.244. From ; ; Captivity. Date: 17/ 5/1909. Source: Munt coll.. Material: 1 ; whole/good egg. Laid in captivity.

Reg. no. E1941.4.6.885. From Dardsun, Dauria; Chitinskaya; Russia. Date: / 5/1868. Source: A. Verdey; Buckley coll.. Material: 1 ; whole/good egg.

Rallus elegans levipes

Clapper Rail

Restricted to coastal salt marshes of southern California and Mexico. Pressures to drain these marshes are intense and the rail's habitat is rapidly becoming scarce.

Reg. no. 1965.M.2349. From San Diego; California; United States. Date: 4/ 3/1939. Source: Meinertzhagen coll.. Material: adult F; whole/good skin.

Rallus antarcticus

Austral Rail

Occurs in southern Chile and Argentina, where it is now very rare and may be extinct, not having been recorded since 1959. Draining and grazing of its wet grassland habitat is thought to be responsible.

Reg. no. 1844.1.8.15. From ; ; Chile. Date: //. Source: C. Bridges; Cuming. Material: adult ; whole/good skin.

Reg. no. 1880.11.18.569. From Sandy Point, Straits of Magellan; ; [Chile]. Date: //. Source: Challenger Expedition. Material: imm F; whole/good skin.

Reg. no. 1889.11.20.28. From ; Prov. of Santiago; Chile. Date: //1866. Source: [Leybold]; Dr Philippi; Gould; Salvin–Godman coll.. Material: adult ; whole/good skin. See *Proc. zool. Soc. Lond.* 1867: 333; 1868: 445. In the register, this specimen is said to have been from Leybold.

Reg. no. 1889.11.20.29. From ; ; Chile. Date: //1875. Source: Leybold; H.B. James; Salvin–Godman coll.. Material: adult ; whole/good skin. Ex–mount. In *Cat. Bds.*, one of the four H.B. James skins is said to be an immature.

Reg. no. 1892.2.10.33. From ; ; c. Chile. Date: //. Source: H.B. James coll.. Material: adult ; whole/good skin. In *Cat. Bds.*, one of the four H.B. James skins is said to be an immature.

Reg. no. 1892.2.10.34. From ; ; c. Chile. Date: //. Source: H.B. James coll.. Material: adult ; whole/good skin. In *Cat. Bds.*, one of the four H.B. James skins is said to be an immature.

Reg. no. 1892.2.10.35. From ; ; c. Chile. Date: //. Source: H.B. James coll.. Material: adult ; whole/good skin. In *Cat. Bds.*, one of the four H.B. James skins is said to be an immature.

Reg. no. 1892.6.29.36. From Lomas de Zamora, Buenos Ayres; ; [Argentina]. Date: 19/ 6/1884. Source: F. Withington; Sclater coll.. Material: adult F; whole/good skin. Iris reddish–brown, bill– upper mandible dark red, lower mandible bright red, legs and toes dark purple.

Reg. no. 1903.12.8.23. From Valle de Lago Blanco; Chubut, Patagonia; [Argentina]. Date: 27/11/1901. Source: J. Koslowsky; F.W.H. Rosenberg. Material: adult ; whole/good skin.

Reg. no. 1931.7.10.19. From Viamonte; Tierra del Fuego; . Date: 15/ 2/1931. Source: P.W. Reynolds. Material: adult F; whole/good skin. Iris brown, bill dusky–horn, lower mandible (except tip) pinkish–drab, feet drab–grey, front of tarsus and upper surface of toes pale.

Rallus philippensis macquariensis

Macquarie Island Banded Rail

Confined to Macquarie Island. First discovered in 1879 when the race was quite common. It became extinct some time between 1880 and 1894, believed killed off by rats.

Reg. no. 1881.12.27.1. From ; ; Macquarie Island. Date: //. Source: G. Grey. Material: adult F; whole/good skin.

Reg. no. 1881.12.27.2. From ; ; Macquarie Island. Date: //. Source: G. Grey. Material: adult M; whole/good skin.

Rallus wakensis

Wake Island Rail

Endemic to Wake Island in the Pacific Ocean. It disappeared during World War II, and is believed to have been exterminated for food by Japanese forces occupying the island.

Reg. no. 1908.12.21.7. From ; Wake Island; Pacific Ocean. Date: //. Source: Rothschild. Material: ; whole/good skin.

Rallus pacificus

Tahiti Rail

Endemic to Tahiti. Known only from a plate by George Forster. The species is presumed to have become extinct in the 18th century.

Reg. no. . From ; Tahiti (Taheitee); Society Islands. Date: //. Source: G. Forster. Material: ; painting. A drawing in the NHM library, by Georg Forster, (old) no. 128. See *Bull. Br. Mus. nat. Hist. (Hist.)* 1 (1959): 302.

Nesolimnas dieffenbachii

Dieffenbach's Rail

Endemic to the Chatham Islands, and known only from the type collected in 1840. The species is thought to have been exterminated by rats and cats.

Reg. no. 1842.9.29.12. From ; ; Chatham Islands. Date: //1840. Source: Dieffenbach; New Zealand Company. Material: adult ; whole/good skin. Holotype. Ex-mount. G.R. Gray, 1843, in Dieffenbach, *Travels to New Zealand*, 2: 197.

Cabalus modestus

Chatham Islands Rail

Endemic to the Chatham Islands, mainly from Mangare, where it became extinct between 1895 and 1900 as a result of habitat destruction and predation by introduced cats.

Reg. no. . From ; ; Chatham Islands. Date: //. Source: W. Hawkins; H.O. Forbes. Material: juv ; skin. Details taken from *Cat. Bds.*. Not found 1988.

Reg. no. 1892.10.31.9. From Mangere Island; ; Chatham Islands. Date: 1/ 6/1892. Source: W. Hawkins; H.O. Forbes. Material: M; whole/good skin.

Reg. no. 1893.6.24.5. From Mangere Island; ; Chatham Islands. Date: //. Source: W. Hawkins; H.O. Forbes. Material: F; whole/good skin. Eyes bluish-black with reddish-brown ring.

Reg. no. 1893.6.24.6. From Mangere Island; ; Chatham Islands. Date: //. Source: W. Hawkins; H.O. Forbes. Material: M; whole/good skin. Eyes bluish-black with reddish-brown ring.

Reg. no. E1941.2.2.5. From Little Mangare Island; ; Chatham Islands. Date: / 1/1892. Source: Hawkins; Dr H.O. Forbes. Material: 1 ; whole/good egg.

Atlantisia podarces

Great St. Helena Rail

Known only from bones of uncertain age from St. Helena.

Reg. no. S1963.1.1. From Prosperous Bay; ; St. Helena. Date: 24/ 4/1959. Source: N.P. Ashmole & D.F. Dorwood, British Ornithologists' Union. Material: ; part/good skeleton. Holotype. Part of the left tarso-metatarsus. Wetmore, *Ibis* 103b (1963): 379–81. There are six other bones in the collection, with the same reg. no.. Date of excavation given.

Tricholimnas lafresnayanus

New Caledonian Rail

Endemic to New Caledonia, mainly in lowland forest. It is believed to have become extinct around 1904, but there have been unconfirmed reports of sightings as recently as 1984.

Reg. no. 1870.12.3.3. From ; ; New Caledonia. Date: //. Source: A. Verdey; Verreaux coll.. Material: adult M; whole/good skin.

Reg. no. 1870.12.3.4. From ; ; New Caledonia. Date: //. Source: A. Verdey; Verreaux coll.. Material: adult F; whole/good skin.

Reg. no. 1889.11.1.375. From Boulari; ; New Caledonia. Date: 16/12/1876. Source: E.L. Layard; Tweeddale coll.. Material: juv F; whole/good skin. Bill and legs dark horn, iris crimson.

Reg. no. 1898.12.2.540. From Mount D'Or; ; New Caledonia. Date: 2/ 7/1877. Source: E.L. Layard; Seebohm coll.. Material: adult F; whole/good skin. Bill dark horn, legs brown, iris dark crimson.

Reg. no. 1901.11.20.356. From ; ; Captivity. Date: //1871. Source: London Zoo; Bartlett; Tristram coll.; Crowley coll.. Material: 1 ; whole/good egg. 'Laid in the gardens of the Zoological Society of London, 1871. Obtained from Bartlett June 1871'.

Tricholimnas sylvestris

Lord Howe Island Rail

Lord Howe Island. Probably occurred formerly in the lowland forests, but now confined to damp mountain forests in the southern part of the island. Captive breeding in recent years has increased the population from 30 to over 200 birds, and a programme to eradicate feral pigs from parts of the island has been largely successful. The species is now thought to be secure.

Reg. no. 1881.5.1.5699. From ; ; Lord Howe Island. Date: / 6/1869. Source: R.D. Fitzgerald, H.M.S. Thetis; Dr G. Bennett;

Gould coll.. Material: imm F; whole/good skin. Holotype. Ex-mount; the specimen was previously preserved in spirits. P.L. Sclater, *Proc. zool. Soc. Lond.* 1869: 471–3, pl. 35. Previously thought to be an adult, but now believed to be a first-year bird, probably a female.

Reg. no. 1898.12.2.538. From ; ; Lord Howe Island. Date: / 9/1887. Source: Sydney Museum; Seebohm coll.. Material: adult M; whole/good skin.

Reg. no. 1898.12.2.539. From ; ; Lord Howe Island. Date: / 9/1887. Source: Seebohm coll.. Material: adult ; whole/good skin.

Reg. no. 1919.7.15.274. From Little Slope; ; Lord Howe Island. Date: 11/ 2/1914. Source: G.M. Mathews. Material: adult F; whole/good skin. Bill dark horn, eye bright reddish-brown, feet dark flesh-grey.

Reg. no. 1919.7.15.275. From Little Slope; ; Lord Howe Island. Date: 12/ 2/1914. Source: G.M. Mathews. Material: adult F; whole/good skin. Bill dark horn, eye brownish-red, feet dark flesh-grey.

Reg. no. 1919.7.15.276. From Little Slope; ; Lord Howe Island. Date: 12/ 2/1914. Source: G.M. Mathews. Material: adult F; whole/good skin. Bill horn, eye red-brown, feet and legs bluish-green.

Reg. no. 1919.7.15.277. From Little Slope; ; Lord Howe Island. Date: 12/ 2/1914. Source: G.M. Mathews. Material: adult M; whole/good skin. Bill horn, eye red, feet dark flesh-grey.

Reg. no. 1919.7.15.278. From Little Slope; ; Lord Howe Island. Date: 8/ 5/1914. Source: G.M. Mathews. Material: adult M; whole/good skin. Bill dark horn-grey, eye bright orange-red, feet dark flesh-grey.

Reg. no. 1891.7.20.283. From ; ; Lord Howe Island. Date: //. Source: Gerrard; Salvin; Salvin-Godman coll.. Material: adult ; whole/good skeleton.

Reg. no. 1939.12.9.3705. From ; ; Lord Howe Island. Date: //. Source: Von Hugel coll.; Rothschild coll.. Material: ; whole/good skeleton. Ex-mount, from Tring gallery.

Dryolimnas cuvieri abbotti

Abbott's White-throated Rail

Endemic to Assumption Island but extinct before 1937.

Reg. no. 1906.12.21.138. From ; ; Assumption Island. Date: 12/ 3/1906. Source: M.J. Nicoll; Earl of Crawford, voyage of R.Y.S. Valhalla. Material: adult M; whole/good skin. Iris red, bill black, base of lower mandible to angle bright red, tarsi and toes dark brown. In heavy wing moult.

Reg. no. 1906.12.21.139. From ; ; Assumption Island. Date: 12/ 3/1906. Source: M.J. Nicoll; Earl of Crawford, voyage of R.Y.S. Valhalla. Material: adult M; whole/good skin. Iris red, bill black, except from base to angle of lower mandible which is bright red, tarsi and toes dark brown.

Reg. no. 1906.12.21.140. From ; ; Assumption Island. Date: 12/ 3/1906. Source: M.J. Nicoll; Earl of Crawford, voyage of R.Y.S. Valhalla. Material: adult M; whole/good skin. Iris red, bill red at base of lower mandible to angle, tip black, upper mandible almost entirely black, tarsi and toes dark brown.

Reg. no. 1906.12.21.141. From ; ; Assumption Island. Date: 12/ 3/1906. Source: M.J. Nicoll; Earl of Crawford, voyage of R.Y.S. Valhalla. Material: adult F; whole/good skin. Iris red, bill black, base of lower mandible to angle bright red, tarsi and toes dark brown.

Reg. no. 1906.12.21.142. From ; ; Assumption Island. Date: 13/ 3/1906. Source: M.J. Nicoll; Earl of Crawford, voyage of R.Y.S. Valhalla. Material: imm M; whole/good skin. Iris brownish-grey, bill, tarsi and toes brown.

Reg. no. 1910.4.8.1. From ; ; Assumption Island. Date: //. Source: Zoological Society coll.. Material: adult ; part/good skeleton. Incomplete skeleton. Died in London zoo.

Amaurolimnas concolor concolor

Uniform Crake

Formerly occurred in fresh-water marshes and by streams in Jamaica. It became extinct about 1890, and is believed to have been exterminated by the introduced mongoose.

Reg. no. 1846.5.26.13. From ; ; Jamaica. Date: //. Source: P. Gosse; H. Cuming. Material: adult ; whole/good skin. Syntype.

Reg. no. 1846.5.26.14. From ; ; Jamaica. Date: //. Source: P. Gosse; H. Cuming. Material: adult ; whole/good skin. Syntype.

Reg. no. 1847.6.8.5. From ; ; Jamaica. Date: //. Source: P. Gosse. Material: ; whole/good skin. Syntype. Ex-mount. Gosse, 1847, *Birds of Jamaica*, p. 369. There are another two syntypes in the collection. In *Hist. Colls.* this bird is regarded as the type, not 1846.5.26.13, as selected by Warren in the catalogue of type-specimens.

Reg. no. 1847.6.16.86. From ; ; Jamaica. Date: //. Source: P. Gosse. Material: ; part/good skeleton. Sternum only. Previously registered as Skeleton Vellum Cat. 749a.

Aramides gutturalis

Red-throated Wood Rail

Known only from the type, supposed to have come from western Peru in or prior to 1843.

Reg. no. 1843.5.24.134. From ; ; South America. Date: //. Source: Leadbeater. Material: adult; whole/good skin. Holotype. Sharpe, *Cat. Bds.* 23 (1894): 54, 57, pl. 5. The specimen is said to have come from Lima, Peru.

Nesoclopeus poeciloptera

Bar-wing Rail

Occurs on Viti Levu and Ovalau in Fiji. Believed to have become extinct in 1890 as a result of persecution by the mongoose, but rediscovered in 1973. If it still survives, it will be very rare.

Reg. no. 1864.12.1.6. From ; ; Fiji. Date: //. Source: Mr Cutter. Material: adult ; whole/good skin. Ex-mount.

Reg. no. 1889.11.3.68. From Korotumba; Viti Levu; Fiji. Date: 2/11/1875. Source: E.L. Layard; Tweeddale coll.. Material: adult M; whole/good skin. Bill orange and yellow, legs yellow, iris light brown. Insects in stomach.

Reg. no. 1889.11.3.69. From Korotumba; Viti Levu; Fiji. Date: 2/11/1875. Source: E.L. Layard; Tweeddale coll.. Material: adult F; whole/good skin. Bill orange and yellow, legs yellow, iris light brown. Insects in stomach.

Reg. no. 1889.11.3.70. From ; Ovalau; Fiji. Date: 16/ 1/1875. Source: E.L. Layard; Tweeddale coll.. Material: adult F; whole/good skin. Bill rich salmon, legs darker rich salmon, iris hair-brown. Stomach empty.

Reg. no. 1898.12.2.531. From Rewa; [Viti Levu]; Fiji. Date: 24/ 9/1875. Source: E.L. Layard. Material: adult F; whole/good skin. Bill and legs deep orange, iris red. Crustacea in stomach. The date is incorrectly given as 26 April 1875 in the register.

Reg. no. 1898.12.2.532. From Rewa; [Viti Levu]; Fiji. Date: 12/10/1875. Source: E.L. Layard; Seebohm coll.. Material: adult F; whole/good skin. Female with eggs. Food, insects, grass. Beak orange-red and yellow, legs orange-red, iris crimson.

Reg. no. 1940.12.8.87. From ; Viti Levu; Fiji. Date: //. Source: Rothschild coll.. Material: adult ; whole/good alc.

Gymnocrex intactus

Intact Rail

Known only from the type collected by L. Brazier in the Solomon Islands. This specimen has been consistently ignored, mainly because a number of birds allegedly collected by Brazier in the Solomons were actually collected on New Ireland, and it was considered to be a synonym of *G. plumbeiventris*. However, *intactus* differs in that the brown of the neck is darker and less chestnut with a slight maroon "bloom", the wings and back are slightly darker, and the belly very slightly paler. The locality is probably correct, and the specimen is apparently an extinct representative of *plumbeiventris* from the Solomons. It may be only subspecifically distinct.

Reg. no. 1887.1.20.12. From ; ; Solomon Islands. Date: //. Source: J. Buttray or L. Brazier; G. Kreft, Australian Mus.; P.L. Sclater. Material: adult ; whole/good skin. Holotype. P.L. Sclater, *Proc. zool. Soc. Lond.* 1869: 123, pl. 10. The locality may be incorrect.

Porzanula palmeri

Laysan Crake

Formerly occurred on Laysan Island, and introduced to other islands nearby. All populations became extinct by 1944, probably as a result of habitat destruction and predation by introduced rats.

Reg. no. 1898.4.29.22. From ; Laysan Island; Hawaiian Islands. Date: 22/ 6/1891. Source: H.L. Palmer; Rothschild. Material: M; whole/good skin.

Reg. no. 1898.4.29.23. From ; Laysan Island; Hawaiian Islands. Date: 19/ 6/1891. Source: H.L. Palmer; Rothschild. Material: F; skin. Iris reddish. Details taken from a label still extant, but the specimen was reported missing in 1968, and not found 1988.

Reg. no. 1898.4.29.29. From ; [Laysan Island]; [Hawaiian Islands]. Date: //. Source: H.L. Palmer; Rothschild. Material: M; whole/good skin.

Reg. no. 1939.12.9.21. From ; Laysan Island; Hawaiian Islands. Date: 2/ 9/1897. Source: Schauinsland; Rothschild coll.. Material: adult M; whole/good skin.

Reg. no. 1939.12.9.22. From ; Laysan Island; Hawaiian Islands. Date: 22/ 7/1896. Source: Schauinsland; Rothschild coll.. Material: adult F; whole/good skin.

Reg. no. 1939.12.9.543. From ; Laysan Island; Hawaiian Islands. Date: //. Source: Rothschild coll.. Material: adult ; whole/good skin. Mount.

Reg. no. 1939.12.9.544. From ; Laysan Island; Hawaiian Islands. Date: //. Source: Rothschild coll.. Material: adult ; whole/good skin. Mount.

Reg. no. 1939.12.9.545. From ; Laysan Island; Hawaiian Islands. Date: //. Source: Rothschild coll.. Material: adult ; whole/good skin. Mount.

Reg. no. 1939.12.9.546. From ; Laysan Island; Hawaiian Islands. Date: //. Source: Rothschild coll.. Material: adult ; whole/good skin. Mount.

Reg. no. 1927.2.3.2. From ; Laysan Island; Hawaiian Islands. Date: //. Source: Rothschild. Material: M; part/good skeleton. Skull only, from spirit specimen with same reg. no..

Reg. no. S1972.1.63. From ; ; . Date: //. Source: Rothschild coll.. Material: ; whole/good skeleton.

Reg. no. E1941.2.2.6. From ; Laysan Island; Hawaiian Islands. Date: 24/ 6/1891. Source: Palmer coll; Rothschild coll.. Material: 1 ; whole/good egg. In Rothschild, 1893, 1900, *The Avifauna of Laysan*, p. 9–10, 306: the nest is described, and it is stated that more than one egg was in the nest when found. Other eggs existed at the time in the collection of Dr Studer (*op. cit.*, p. 306).

Reg. no. 1927.2.3.1. From ; Laysan Island; Hawaiian Islands. Date: //. Source: H.L. Palmer; Rothschild. Material: adult ; whole/good alc. Specimen apparently located in 1987, but not found in 1988 or 1993.

Reg. no. 1927.2.3.2. From ; Laysan Island; Hawaiian Islands. Date: //. Source: H.L. Palmer; Rothschild. Material: adult ; part/good alc. The registration numbers of this bird and 1927.2.3.3 had been transposed in the register. The skull is in the skeleton collection with the same reg. no..

Reg. no. 1927.2.3.3. From ; Laysan Island; Hawaiian Islands. Date: //. Source: H.L. Palmer; Rothschild. Material: chick ; whole/good alc. The registration numbers of this bird and 1927.2.3.2 had been transposed in the register.

Reg. no. 1940.12.8.11. From ; Laysan Island; Hawaiian Islands. Date: //. Source: H.L. Palmer; Rothschild. Material: ; alc. Previously identified as *Pennula sandwichensis*. Not in the collection, exchanged with the Delaware Museum in 1975.

Reg. no. 1940.12.8.12–17. From ; Laysan Island; Hawaiian Islands. Date: //. Source: H.L. Palmer; Rothschild. Material: ; whole/good alc. Six spirit specimens (2 adults, 4 chicks), previously identified as *Pennula sandwichensis*.

Pennula millsi

Mills' Crake

Endemic to the island of Hawaii, becoming extinct between 1884 and 1893. Possibly only a colour morph of *P. sandwichensis*.

Reg. no. 1939.12.9.553. From ; Hawaii; Hawaiian Islands. Date: //. Source: Rothschild coll.. Material: adult ; whole/good skin. Mount.

Seven spirit specimens are listed in *Bull. Br. Mus. nat. Hist. (Zool.)* 34 (1979): 138, but this is believed to be a typographic error.

Pennula sandwichensis

Hawaiian Crake

Endemic to the island of Hawaii, becoming extinct between 1884 and 1893.

Reg. no. . From ; ; Hawaiian Islands (Sandwich Isles). Date: //1779. Source: Material: ; painting. A drawing in the NHM library, by William Ellis, no. 70. See *Bull. Br. Mus. nat. Hist. (Hist.)* 1 (1959): 334, and *Auk* 67 (1950): 80–1.

Nesophylax niger

Miller's Rail

Tahiti. Known only from plates by Forster and Miller, the species presumably became extinct in the 18th century. See Walters, *Notornis* 35: 265–269 and *Bulletin of Zoological Nomenclature* 46: 50–52.

Reg. no. . From ; Tahiti (Taheitee); Society Islands. Date: //. Source: J.R. Forster. Material: ; painting. Holotype. J.F. Miller, 1784, *Icones Animalium*, pt. 9, pl. 50b, apparently based on a drawing in the NHM library, by Georg Forster, (old) no. 130. See *Bull. Br. Mus. nat. Hist. (Hist.)* 1 (1959): 302–3.

Sarothrura watersi

Slender-billed Flufftail

Known only from a few areas of marsh in the rainforests of eastern Madagascar.

Reg. no. 1880.5.6.1. From s.e. Betsileo; ; Madagascar. Date: /12/1875. Source: T. Waters; E. Bartlett. Material: adult M; whole/good skin. Syntype. Bartlett, *Proc. zool. Soc. Lond.* 1879: 772–3, pl. 63. There is a female syntype in the collection, and a third syntype in the University Museum of Zoology, Cambridge, England.

Reg. no. 1880.5.6.2. From Fangulathova, s.e. Betsileo; ; Madagascar. Date: /12/1875. Source: [Rev. W. Deans Cowan]; T. Waters; E. Bartlett. Material: adult F; whole/good skin. Syntype. The label bears no details of the collector/ provenance: Waters and Bartlett are taken from the register, but Cowan is given in *Cat. Bds.*.

Reg. no. 1891.8.1.54. From ; ; [Madagascar]. Date: //. Source: Shelley coll.. Material: adult ; whole/good skin.

Reg. no. 1931.8.18.1263. From one day w. of Andapa; ; n. Madagascar. Date: 29/ 8/1930. Source: Archbold Vernay Expedition. Material: adult M; whole/good skin. Iris brown, bill brownish–black, feet pinkish–brown.

Reg. no. 1931.8.18.1264. From one day w. of Andapa; ; n. Madagascar. Date: 2/ 9/1930. Source: Archbold Vernay Expedition. Material: adult M; whole/good skin.

Reg. no. 1931.8.18.1265. From one day w. of Andapa; ; n. Madagascar. Date: 3/ 9/1930. Source: Archbold Vernay Expedition. Material: adult F; whole/good skin.

Reg. no. 1931.8.18.1266. From one day w. of Andapa; ; n. Madagascar. Date: 23/ 8/1930. Source: Archbold Vernay Expedition. Material: adult F; whole/good skin. Iris brown, bill blackish, feet pinkish–brown.

Poliolimnas cinereus brevipes

Iwo Jima Crake

Formerly occurred on Sulphur and Augustino Islands in the Iwo Jima group of the Volcano Islands. Already almost extinct when it was discovered in 1911, the race was last claimed to have been seen in 1925.

Reg. no. 1898.12.2.501. From Iwojima; Volcano Islands (Bonin Islands); Japan. Date: 7/ 6/1890. Source: P.A. Holst; Seebohm coll.. Material: F; whole/good skin. Irides and edge of eyelids bright red, tarsus, feet and claws greenish–grey, somewhat yellow at front part, e.g. below joint of feet, bill darker than feet (more illegible).

Reg. no. 1912.12.21.1. From Iwojima; Volcano Islands (Bonin Islands); Japan. Date: 31/ 5/1911. Source: A. Owston; Collingwood Ingram. Material: adult M; whole/good skin. Syntype. Ingram, *Bull. Brit. Orn. Club* 29 (1911): 21–2. A female syntype is also in the collection.

Reg. no. 1912.12.21.2. From Iwojima; Volcano Islands (Bonin Islands); Japan. Date: 31/ 5/1911. Source: A. Owston; Collingwood Ingram. Material: adult F; whole/good skin. Syntype.

Reg. no. E1941.2.1.24. From Dionisio; Volcano Islands (Vulcan Islands); [Japan]. Date: 25/ 5/1904. Source: ? Owston; Mathews coll.. Material: 3 ; whole/good eggs. From 24°N 141°E.

Reg. no. E1941.2.1.25. From Dionisio; Volcano Islands (Vulcan Islands); [Japan]. Date: 28/ 5/1904. Source: ? Owston;

Mathews coll.. Material: 2 ; whole/good eggs. From 24°N 141°E.

Reg. no. E1988.4.24. From s. Dionisio; Volcano Islands (Vulcan Islands); [Japan]. Date: 20/ 5/1904. Source: A. Owston; Collingwood Ingram; Jourdain coll.. Material: 2 ; whole/good eggs (clutch). From 24°N 141°E. Eggs exhibited, *Bull. Brit. Orn. Club* 37 (1916): 19–20.

Gallinula chloropus sandvicensis

Hawaiian Moorhen

Formerly occurred on all the main Hawaiian Islands, but now confined to Kauai, Oahu and Molokai, where very few birds survive. Marsh drainage is considered to be the main cause of decline, perhaps also predation by introduced mammals.

Reg. no. E1963.9.237. From Oahu; ; Hawaiian Islands. Date: //. Source: Rothschild. Material: 4 ; whole/good eggs.

Reg. no. E1988.4.25. From ; ; Hawaiian Islands. Date: //. Source: F. Gay; Rothschild. Material: 1 ; whole/good egg.

Porphyriornis nesiotis

Tristan Moorhen

Endemic to Tristan da Cunha. The species was common in 1818 but had disappeared by the end of the century, though the date of its extinction is not precisely known. The closely related *P. comeri* of Gough Island has recently been introduced to Tristan.

Reg. no. 1861.9.16.1. From ; ; Tristan da Cunha. Date: //. Source: G. Grey; P.L. Sclater. Material: adult ; whole/good skin. Syntype. Ex-mount. P.L. Sclater, *Proc. zool. Soc. Lond.* 1861: 209, 261, pl. 30. See also Buller, 1888, *History of the Birds of New Zealand*, 2nd. ed., 2: 104.

Reg. no. 1864.7.30.1. From ; ; Tristan da Cunha. Date: //. Source: G. Grey; Zoological Society. Material: ; whole/good skin. Probable syntype. See *Proc. zool. Soc. Lond.* 1861: 260–3. This is probably the bird which lived for some time at the Zoo.

Reg. no. S1952.3.8. From ; ; Tristan da Cunha. Date: //. Source: Sir G. Grey; Zoological Society. Material: ; part/good skeleton. Syntype. Sternum, pelvis, vertebrae, both coracoids and scapulae, furcula and both femurs, from a specimen preserved in spirit. See *Proc. zool. Soc. Lond.* 1861: 260–3, where the sternum and pelvis are figured.

Pareudiastes pacificus

Samoan Moorhen

Endemic to Samoa, where it was first discovered in 1869 and was extinct by 1873. The cause of extinction is not known.

Reg. no. . From ; ; Samoa. Date: //. Source: Material: adult ; skin. Said to have been from the Old Collection. Details taken from *Cat. Bds.*. Not in the collection; exchanged with the AMNH in 1930.

Reg. no. 1878.1.19.97. From ; ; Samoa. Date: //. Source: S.J. Whitmee. Material: ; whole/good skin.

Reg. no. 1923.9.7.10. From ; ; Samoa. Date: //. Source: G. Brown; Sir J. Murray.

Material: ; whole/good skin. This may be the specimen collected off a nest with two eggs (one of which was later given to Whitmee, thence to Sclater) on Savaii in October 1873 (see *Proc. zool. Soc. Lond.* 1874: 185).

Reg. no. 1874.11.18.36. From ; Savaii; Samoa. Date: /10/1873. Source: [Rev. G. Brown]; S.J. Whitmee; P.L. Sclater. Material: 1 ; whole/good egg. See Greenway, 1967, *Extinct and Vanishing Birds of the World*, 2nd edn., p. 241. This appears to be the only egg of the species. See also *Proc. zool. Soc. Lond.* 1874: 184–6. The specimen has been cracked for some time. The nest was on the ground and composed of a few twigs and a little grass.

Reg. no. 1874.2.20.38. From ; ; [Samoa]. Date: //. Source: [Whitmee]; [Tristram]. Material: ; whole/good alc. Said to have been from the old spirit collection, but this is unlikely as it is thought to be the specimen which Whitmee obtained in September 1872 and sent to Tristram; see *Proc. zool. Soc. Lond.* 1874: 184–5.

Notornis mantelli

Takahe

New Zealand. Formerly widespread in South Island, a similar or related form is known from bones on North Island. Now confined to a few small valleys in the Murchison Mountains. The bird was long thought to be extinct until rediscovered in 1948. The small population is now increasing following captive breeding and release onto predator-free islands.

Reg. no. . From Deas Cove, Thompson Sound; South Island; New Zealand. Date: //1851. Source: Capt. Howell; W. Mantell. Material: adult F; skin. Mount; mounted by A.D. Bartlett. Presumed female on plumage. Not in the collection. Presented to the Dominion Museum in 1953. *Trans. R. Soc. N.Z.* 88 (1960): 235–236.

Reg. no. 1856.5.28.1. From Dusky Bay, behind Resolution Island; South Island; New Zealand. Date: //1849. Source: W. Mantell. Material: adult M; whole/good skin. Mount. Presumed to be a male on plumage. Mounted by A.D. Bartlett. Caught by sealers in a gully of a sound behind Resolution Island, during the austral winter (said to have been snow on the ground). Earliest known skin. See *Proc. zool. Soc. Lond.* 1850: 209–214.

Reg. no. S1961.17.1. From Valley of Murchison Mountains; [South Island]; New Zealand. Date: //. Source: Dominion Museum. Material: ; whole/good skeleton.

Reg. no. S1974.9.3. From Fiordland; South Island; New Zealand. Date: /12/1967. Source: National Museum of New Zealand. Material: ; whole/good skeleton. Specimen excavated December 1967.

Rhynochetos jubatus

Kagu

New Caledonia. Formerly widespread, but now restricted to the most inaccessible valleys in the central mountain range.

Reg. no. . From ; ; New Caledonia. Date: //. Source: Material: adult ; skin. Mount. Details taken from *Cat. Bds.*. Not found

1988. Unlikely to be the mounted specimen in Tring gallery.

Reg. no. 1897.10.30.85. From Mount D'Or; ; New Caledonia. Date: 2/11/1876. Source: E.L. Layard; Seebohm coll.. Material: adult M; whole/good skin. Bill and legs orange, iris crimson.

Reg. no. 1897.10.30.86. From Mount D'Or; ; New Caledonia. Date: 12/11/1879. Source: E.L. Layard; Seebohm coll.. Material: adult F; whole/good skin. Bill and legs red–orange, iris crimson.

Reg. no. 1900.1.10.115. From Noumea; ; New Caledonia. Date: 25/ 8/1899. Source: A.M. Farquhar. Material: adult M; whole/good skin. Legs, irides and bill dark red.

Reg. no. 1925.11.1.582. From ; ; New Caledonia. Date: / 7/1924. Source: J.H. Wright; British Museum Expedition. Material: adult ; whole/good skin.

Reg. no. 1926.4.27.2. From ; ; Captivity. Date: 27/ 4/1926. Source: H. Whitley. Material: chick ; whole/good skin. Died in captivity.

Reg. no. 1931.1.15.2. From ; ; Captivity. Date: 28/ 1/1931. Source: H. Whitley. Material: adult ; part/good skin. Died in captivity. Legs and wings only, head in spirit collection with same reg. no..

Reg. no. 1936.11.9.1. From ; ; New Caledonia. Date: //1865. Source: Gould coll.. Material: imm ; whole/good skin. Ex–mount.

Reg. no. 1939.12.9.3712. From ; ; Captivity. Date: //. Source: Finckh; Rothschild coll.. Material: chick ; whole/good skin.

Reg. no. 1988.21.37. From ; ; [New Caledonia]. Date: //. Source: Rothschild coll.. Material: adult ; whole/good skin. Mount in Tring gallery. No details with the specimen.

Reg. no. 1896.2.16.26. From ; ; New Caledonia. Date: //. Source: Seebohm coll.. Material: ; part/good skeleton. Consists of parts of 2 specimens.

Reg. no. S1952.2.66. From ; ; [New Caledonia] Captivity. Date: //. Source: Zoological Society. Material: ; whole/good skeleton.

Reg. no. S1956.21.1. From ; ; [New Caledonia]. Date: //. Source: Material: ; whole/good skeleton. Earlier investigation suggested that this was the one which was purchased in 1894, and is listed as specimen c. in the *Cat. Bds.*. Taken from the gallery.

Reg. no. S1958.3.1. From ; ; New Caledonia, Captivity. Date: / 1/1958. Source: Zoological Society. Material: F; whole/good skeleton.

Reg. no. S1987.24.1. From ; ; [New Caledonia]. Date: //. Source: Abbe R. de Naurois. Material: ; whole/good skeleton.

Reg. no. E1941.4.6.1261. From ; ; Captivity. Date: //. Source: Buckley coll.. Material: 2 ; whole/good eggs. Laid in captivity.

Reg. no. 1904.7.25.1–2. From ; ; Captivity. Date: //. Source: H.E. Finckh. Material: 2 ; whole/good eggs (clutch). Laid in captivity.

Reg. no. 1901.7.10.573. From ; ; Captivity. Date: //1869. Source: London Zoo; Tristram; P. Crowley. Material: 1 ; whole/good egg. Egg laid in 1869, prior to August.

Reg. no. E1941.9.4.2767. From ; ; Captivity. Date: / 4/1931. Source: F. Shaw Meyer. Material: 1 ; whole/good egg. Laid in captivity in Sydney, New South Wales, Australia.

Reg. no. 1926.3.18.1. From ; ; Captivity. Date: //. Source: H. Whitley. Material: chick ; whole/good alc. Died in captivity in Devonshire. Both the label and the register state 'chick with egg', but the shell is not present.

Reg. no. 1931.1.15.2. From ; ; Captivity. Date: 28/ 1/1931. Source: H. Whitley. Material: adult ; part/good alc. Head only; the legs and wings are in the skin collection with the same reg. no..

Reg. no. 1933.4.3.1. From ; ; Captivity. Date: 7/ 4/1933. Source: Zoological Society. Material: adult F; whole/good alc.

Reg. no. 1934.12.1.2. From ; ; Captivity. Date: 14/12/1934. Source: H. Whitley. Material: adult ; whole/good alc.

Choriotis nigriceps

Indian Bustard

Formerly widespread through the Indian sub-continent, now confined to grassland areas of western India where it occurs in small numbers. A recent survey suggested a total population of 500–2000 individuals, depending on whether the birds were sedentary or moved long distances and could have been re-counted.

Reg. no. . From ; ; [India]. Date: //. Source: India Museum. Material: ; skin. One of six birds listed in the *Cat. Bds.* as having come from the India Museum collection, not found 1987.

Reg. no. . From ; ; [India]. Date: //. Source: India Museum. Material: ; skin. One of six birds listed in the *Cat. Bds.* as having come from the India Museum collection, not found 1987.

Reg. no. . From ; ; [India]. Date: //. Source: India Museum. Material: ; skin. One of six birds listed in the *Cat. Bds.* as having come from the India Museum collection, not found 1987.

Reg. no. . From ; ; [India]. Date: //. Source: India Museum. Material: ; skin. One of six birds listed in the *Cat. Bds.* as having come from the India Museum collection, not found 1987.

Reg. no. . From ; ; [India]. Date: //. Source: India Museum. Material: ; skin. One of six birds listed in the *Cat. Bds.* as having come from the India Museum collection, not found 1987.

Reg. no. . From ; ; [India]. Date: //. Source: India Museum. Material: ; skin. One of six birds listed in the *Cat. Bds.* as having come from the India Museum collection, not found 1987.

Reg. no. 1846.5.29.3. From ; ; . Date: //. Source: Argent. Material: ; whole/good skin.

Reg. no. 1881.7.30.128. From ; ; . Date: //. Source: Capt. S. Pinwill. Material: adult ; whole/good skin. Damaged: tail loose.

Reg. no. 1881.7.30.129. From ; ; . Date: //. Source: Capt. S. Pinwill. Material: adult ; whole/good skin. Damaged: neck loose.

Reg. no. 1890.2.20.1. From ; Punjab; India. Date: /11/1874. Source: Hume coll.. Material: adult ; whole/good skin.

Reg. no. 1890.2.20.10. From Sirsa Dist.; Haryana; India. Date: //1870. Source: Hume coll.. Material: adult M; whole/good skin. Collected during the cold season, 1870–1871. Damaged: foot detached.

Reg. no. 1890.2.20.11. From Sirsa; Haryana; India. Date: //. Source: Hume coll.. Material: adult F; whole/good skin.

Reg. no. 1890.2.20.12. From Sirsa; Haryana; India. Date: //. Source: Hume coll.. Material: adult M; whole/good skin.

Reg. no. 1890.2.20.13. From Sirsa; Haryana; India. Date: //1870. Source: Hume coll.. Material: adult M; whole/good skin. Collected during the cold season, 1870–1871.

Reg. no. 1890.2.20.14. From Sirsa; Haryana; India. Date: //. Source: Hume coll.. Material: adult M; skin. Details taken from the register, specimen not found 1987. This may be one of the birds marked '836'.

Reg. no. 1890.2.20.15. From Sirsa; Haryana; India. Date: //1870. Source: Hume coll.. Material: adult M; whole/good skin. Collected during the cold season, 1870–1871.

Reg. no. 1890.2.20.16. From Sirsa; Haryana; India. Date: //. Source: Hume coll.. Material: adult F; skin. Details taken from the register, specimen not found 1987. This may be one of the birds marked '836'.

Reg. no. 1890.2.20.17. From Sirsa; Haryana; India. Date: //1870. Source: Hume coll.. Material: adult M; whole/good skin. Collected during the cold season, 1870–1871.

Reg. no. 1890.2.20.18. From Sirsa; Haryana; India. Date: //1870. Source: Hume coll.. Material: adult ; whole/good skin. Damaged: tail detached.

Reg. no. 1890.2.20.19. From Sirsa; Haryana; India. Date: //1870. Source: Hume coll.. Material: adult M; whole/good skin. Collected during the cold season, 1870–1871.

Reg. no. 1890.2.20.2. From ; Punjab; India. Date: //. Source: Hume coll.. Material: adult F; skin. Details taken from the register, specimen not found 1987. This may be one of the birds marked '836'.

Reg. no. 1890.2.20.20. From Sirsa; Haryana; India. Date: //. Source: Hume coll.. Material: adult ; whole/good skin.

Reg. no. 1890.2.20.21. From Sirsa; Haryana; India. Date: //. Source: Hume coll.. Material: adult ; skin. Details taken from the register, specimen not found 1987. This may be one of the birds marked '836'.

Reg. no. 1890.2.20.22. From Sirsa; Haryana; India. Date: //. Source: Hume coll.. Material: adult M; whole/good skin.

Reg. no. 1890.2.20.23. From Sirsa; Haryana; India. Date: //. Source: Hume coll.. Material: adult ; skin. Details taken from the register, specimen not found 1987. This may be one of the birds marked '836'.

Reg. no. 1890.2.20.24. From Sirsa; Haryana; India. Date: //1870. Source: Hume coll.. Material: adult M; whole/good skin. Collected during the cold season, 1870–1871.

Reg. no. 1890.2.20.25. From Sirsa; Haryana; India. Date: /11/1871. Source: Hume coll.. Material: F; whole/good skin.

Reg. no. 1890.2.20.26. From Sirsa; Haryana; India. Date: 5/12/1869. Source: Hume coll.. Material: adult M; whole/good skin.

Reg. no. 1890.2.20.27. From Sirsa; Haryana; India. Date: //. Source: Hume coll.. Material: adult M; skin. Details taken from the register, specimen not found 1987. This may be one of the birds marked '836'.

Reg. no. 1890.2.20.28. From Sirsa; Haryana; India. Date: //1871. Source: Hume coll.. Material: adult M; whole/good skin.

Reg. no. 1890.2.20.29. From Sirsa; Haryana; India. Date: //1870. Source: Hume coll.. Material: adult M; whole/good skin. Collected during the cold season, 1870–1871.

Reg. no. 1890.2.20.3. From Hissar; Punjab; India. Date: /10/1875. Source: Hume coll.. Material: adult ; whole/good skin.

Reg. no. 1890.2.20.30. From Sirsa; Haryana; India. Date: //. Source: Hume coll.. Material: adult ; skin. Details taken from the register, specimen not found 1987. This may be one of the birds marked '836'.

Reg. no. 1890.2.20.31. From Sirsa; Haryana; India. Date: //. Source: Hume coll.. Material: adult ; whole/good skin.

Reg. no. 1890.2.20.32. From Sirsa; Haryana; India. Date: //. Source: Hume coll.. Material: adult ; whole/good skin.

Reg. no. 1890.2.20.33. From Sirsa; Haryana; India. Date: //. Source: Hume coll.. Material: adult ; skin. Details taken from the register, specimen not found 1987. This may be one of the birds marked '836'.

Reg. no. 1890.2.20.34. From Sirsa; Haryana; India. Date: //. Source: Hume coll.. Material: adult ; skin. Details taken from the register, specimen not found 1987. This may be one of the birds marked '836'.

Reg. no. 1890.2.20.35. From Sirsa; Haryana; India. Date: //. Source: Hume coll.. Material: adult ; whole/good skin.

Reg. no. 1890.2.20.36. From Sirsa; Haryana; India. Date: //. Source: Hume coll.. Material: adult ; skin. Details taken from the register, specimen not found 1987. This may be one of the birds marked '836'.

Reg. no. 1890.2.20.37. From Sirsa; Haryana; India. Date: 20/10/1867. Source: A.O. Hume; Hume coll.. Material: adult M; whole/good skin.

Reg. no. 1890.2.20.4. From Hissar; Punjab; India. Date: /10/1875. Source: Hume coll.. Material: adult M; whole/good skin.

Reg. no. 1890.2.20.5. From Jhansi; Jalaun; India. Date: / 8/1868. Source: Hume coll.. Material: adult ; whole/good skin.

Reg. no. 1890.2.20.6. From Gwalior; Gird Gwalior; India. Date: / 7/1866. Source: Hume coll.. Material: adult ; whole/good skin.

Reg. no. 1890.2.20.7. From Gwalior; ; India. Date: / 7/1866. Source: Hume coll.. Material: adult ; skin. Not found 1988, although this may be the bird which was recently registered as 1988.9.12.

Reg. no. 1890.2.20.8. From Gwalior; Gird Gwalior; India. Date: / 7/1866. Source: Hume coll.. Material: adult ; whole/good skin.

Reg. no. 1890.2.20.9. From Petasa, Kotri (Kotree); ; Pakistan. Date: //. Source: E.A. Butler; Hume coll.. Material: adult F; whole/good skin. Petasa, 14 miles n. of Kotree on the Indus River.

Reg. no. 1898.12.12.115. From nr. Chanda C.P., Chandrapur; Bharat; India. Date: /11/1870. Source: W.T. Blanford. Material: adult M; whole/good skin.

Reg. no. 1949.WHI.1.6655. From Abohar, Ferozepore Dist.; Punjab; India. Date: 7/ 3/1912. Source: Whistler coll.. Material: adult M; whole/good skin.

Reg. no. 1949.WHI.1.6656. From nr. Maunola, Bhandara Dist.; Madhya Pradesh; India. Date: 4/10/1933. Source: Whistler coll.. Material: adult F; whole/good skin.

Reg. no. 1988.9.1. From ; ; [India]. Date: //. Source: Material: ; whole/good skin. No data. The style of preparation of this skin suggests it may be one of the missing Hume specimens.

Reg. no. 1988.9.10. From ; ; [India]. Date: //. Source: Material: ; whole/good skin. No data. The style of preparation of this skin suggests it may be one of the missing Hume specimens.

Reg. no. 1988.9.11. From ; ; [India]. Date: //. Source: Material: ; whole/good skin. No data. The style of preparation of this skin suggests it may be one of the missing Hume specimens.

Reg. no. 1988.9.12. From Gwalior; ; India. Date: 6/ 7/1866. Source: Hume coll.. Material: adult ; whole/good skin. This may be the bird which was previously registered as 1890.2.20.7.

Reg. no. 1988.9.13. From Deccan; ; [India]. Date: //. Source: Sykes; India Museum. Material: adult ; whole/good skin. The reverse of the Museum label bears the stamp 'India Museum 1879'.

Reg. no. 1988.9.14. From ; ; . Date: //. Source: Material: ; part/good skin. Wing only.

Reg. no. 1988.9.15. From Sirsa; Punjab; [India]. Date: //. Source: Material: juv ; whole/good skin.

Reg. no. 1988.9.16. From Sirsa; [Punjab]; [India]. Date: //. Source: Material: juv ; whole/good skin.

Reg. no. 1988.9.17. From Urneewalla, Sirsa; [Punjab]; [India]. Date: /11/1871. Source: Material: juv ; whole/good skin.

Reg. no. 1988.9.18. From Urneewalla, Sirsa; [Punjab]; [India]. Date: //1872. Source: Material: juv M; whole/good skin.

Reg. no. 1988.9.19. From [Khandeish]; ; [India]. Date: //. Source: Material: ; whole/good skin. This bird has a label on its leg which reads 'Khandeish, to be placed at the close of the exhibition at the disposal of Lord Waldon, 15, Cromwell Rd., London W.', there is no other information.

Reg. no. 1988.9.2. From Sirsa; ; [India]. Date: //1867. Source: Material: ; whole/good skin. This bird has the number 836 on the label, but little else. As some of the Hume collection bears this number, this specimen may also be from that collection.

Reg. no. 1988.9.3. From Urneewalla; Punjab; [India]. Date: /11/1871. Source: Material: adult F; whole/good skin. This bird has the number 836 on the label, but little else. As some of the Hume collection bears this number, this specimen may also be from that collection.

Reg. no. 1988.9.4. From ; ; [India]. Date: //. Source: Material: adult ; whole/poor skin. Damaged: skin in pieces.

Reg. no. 1988.9.5. From Urnewalla; [Punjab]; [India]. Date: //1870. Source: Material: adult M; whole/good skin. Collected during the cold season, 1870–1871. This bird has the number 836 on the label, but little else. As some of the Hume collection bears this number, this specimen may also be from that collection.

Reg. no. 1988.9.6. From ; ; [India]. Date: //. Source: Material: ; whole/good skin. No data. The style of preparation of this skin suggests it may be one of the missing Hume specimens.

Reg. no. 1988.9.7. From ; ; [India]. Date: //. Source: Material: ; whole/good skin. No data. The style of preparation of this skin suggests it may be one of the missing Hume specimens.

Reg. no. 1988.9.8. From ; ; [India]. Date: //. Source: Material: ; whole/good skin. No data. The style of preparation of this skin suggests it may be one of the missing Hume specimens.

Reg. no. 1988.9.9. From ; ; [India]. Date: //. Source: Material: ; whole/good skin. No data. The style of preparation of this skin suggests it may be one of the missing Hume specimens.

Reg. no. . From Kharaghoda; [Gujarat]; [India]. Date: 6/ 8/1896. Source: Col. K.J. Buchanan coll.; Jourdain coll.. Material: 2 ; eggs. Two single egg clutches with the same data. Not found <1988; probably not received with the Jourdain collection.

Reg. no. . From ; ; . Date: //. Source: Jourdain coll.. Material: 2 ; eggs. Two single eggs from different clutches. Not found <1988; probably not received with the Jourdain collection.

Reg. no. 1891.3.2.3003–7. From Urneewala, Sirsa; Punjab; India. Date: / 7/1867. Source: Hume coll.. Material: 5 ; whole/good eggs.

Reg. no. 1891.3.2.3008–12. From Sirsa; Punjab; India. Date: / 7/1868. Source: Hume coll.. Material: 5 ; whole/good eggs.

Reg. no. 1891.3.2.3022+. From Sirsa; Punjab; India. Date: //. Source: Hume coll.. Material: 8 ; whole/good eggs. Eight eggs with reg. nos. 1891.3.2.3022, 3023, 3024, 3027, 3030, 3034, 3035, 3038. Collected in 1869–70.

Reg. no. 1891.3.2.3025+. From Sirsa; Punjab; India. Date: //. Source: Hume coll.. Material: 4 ; whole/good eggs. Four eggs with reg. nos. 1891.3.2.3025, 3036, 3063, 3079. 3036 is deformed. Collected about 1869–70. The Hume eggs of this species seem to have been collected by Khan Nizam-oo-deen Khan. See Hume, 1873–4, *Nests and Eggs of Indian birds*, rough draft (i.e. 1st edn.), p. 557–9.

Reg. no. 1891.3.2.3026+. From Sirsa; Punjab; India. Date: //. Source: Hume coll.. Material: 8 ; whole/good eggs. Eight eggs with reg. nos. 1891.3.2.3026, 3042, 3043, 3047, 3052, 3053, 3062, 3076. Collected about 1869–70.

Reg. no. 1891.3.2.3028+. From Sirsa; Punjab; India. Date: //. Source: Hume coll.. Material: 8 ; whole/good eggs. Eight eggs with reg. nos. 1891.3.2.3028, 3032, 3044, 3046, 3057, 3068, 3075, 3077. Collected about 1869–70.

Reg. no. 1891.3.2.3029+. From Sirsa; Punjab; India. Date: //. Source: Hume coll.. Material: 8 ; whole/good eggs. Eight eggs with reg. nos. 1891.3.2.3029, 3033, 3037, 3048, 3058, 3060, 3067, 3073. Collected in 1869–70.

Reg. no. 1891.3.2.3031+. From Sirsa; Punjab; India. Date: //. Source: Hume coll.. Material: 6 ; whole/good eggs. Six eggs with reg. nos. 1891.3.2.3031, 3039, 3040, 3065, 3071, 3074. Collected in 1869–70.

Reg. no. 1891.3.2.3041+. From Sirsa; Punjab; India. Date: //. Source: Hume coll.. Material: 8 ; whole/good eggs. Eight eggs with reg. nos. 1891.3.2.3041, 3055, 3056, 3061, 3064, 3066, 3069, 3072. Collected in 1869–70.

Reg. no. 1891.3.2.3045+. From Sirsa; Punjab; India. Date: //. Source: Hume coll.. Material: 8 ; whole/good eggs. Eight eggs with reg. nos. 1891.3.2.3045, 3049, 3050, 3051, 3059, 3064, 3070, 3078. Collected in 1869–70.

Reg. no. 1901.7.10.543–4. From Sirsa; Punjab; India. Date: //1869. Source: [? Hume coll.]; Foottit coll.; Crowley coll.. Material: 2 ; whole/good eggs.

Reg. no. E1966.9.34–5. From Sholapur, Bombay; ; India. Date: //1874. Source: L.S. Fenton coll.. Material: 2 ; whole/good eggs.

Reg. no. E1973.32.619. From Kharaghoda, Rann of Cutch; [Gujarat]; India. Date: 12/

8/1901. Source: Harrington Bulkeley; Stuart Baker coll.. Material: 2 ; whole/good eggs.

Reg. no. . From Kharaghoda; [Gujarat]; [India]. Date: 6/ 8/1896. Source: Harrington Bulkeley; Col. K. Buchanan coll.; Jourdain coll.. Material: ; egg. One or two eggs. Not found <1988; probably not received with the Jourdain collection.

Reg. no. . From Kharaghoda; [Gujarat]; [India]. Date: 7/ 9/1896. Source: Col. K.J. Buchanan coll.; Jourdain coll.. Material: 1 ; egg. Not found <1988; probably not received with the Jourdain collection.

Reg. no. . From ; ; . Date: //. Source: H. Pearson; Jourdain coll.. Material: 1 ; egg. Not found <1988; probably not received with the Jourdain collection.

Reg. no. 1859.6.5.79. From ; ; India. Date: //. Source: Gould coll.. Material: 1 ; whole/good egg. Badly damaged.

Reg. no. 1865.2.3.1. From ; ; India. Date: //. Source: Gould coll.. Material: 1 ; whole/good egg.

Reg. no. 1891.3.2.3002. From Sirsa; Punjab; India. Date: 30/ 6/1869. Source: Khan Nizam-oo-deen; Hume coll.. Material: 1 ; whole/good egg.

Reg. no. 1891.3.2.3013. From Sirsa; Punjab; India. Date: 3/ 7/1869. Source: Hume coll.. Material: 1 ; whole/good egg.

Reg. no. 1891.3.2.3014. From Sirsa; Punjab; India. Date: 1/ 7/1869. Source: Hume coll.. Material: 1 ; whole/good egg.

Reg. no. 1891.3.2.3015. From Sirsa; Punjab; India. Date: 10/ 7/1869. Source: Hume coll.. Material: 1 ; whole/good egg.

Reg. no. 1891.3.2.3016. From Sirsa; Punjab; India. Date: 6/ 7/1869. Source: Hume coll.. Material: 1 ; whole/good egg.

Reg. no. 1891.3.2.3017. From Sirsa; Punjab; India. Date: 11/ 7/1867. Source: Hume coll.. Material: 1 ; whole/good egg.

Reg. no. 1891.3.2.3018. From Sirsa; Punjab; India. Date: 10/ 8/1869. Source: Hume coll.. Material: 1 ; whole/good egg.

Reg. no. 1891.3.2.3019. From Sirsa; Punjab; India. Date: 23/ 8/1869. Source: Hume coll.. Material: 1 ; whole/good egg. The day is given as 13th in Seebohm's ms catalogue.

Reg. no. 1891.3.2.3020. From Sirsa; Punjab; India. Date: 27/ 8/1869. Source: Hume coll.. Material: 1 ; whole/good egg.

Reg. no. 1891.3.2.3021. From Sirsa; Punjab; India. Date: / 8/1870. Source: Hume coll.. Material: 1 ; whole/good egg.

Reg. no. E1922.11.9.169. From Urneswala, Sirsa; Punjab; [India]. Date: / 7/1867. Source: A.O. Hume coll.; Miss Cockburn coll.. Material: 1 ; whole/good egg.

Reg. no. E1925.12.25.2554. From Sholapur; ; [India]. Date: 2/ 9/1874. Source: Davidson coll.. Material: 1 ; whole/good egg. Taken by Davidson.

Reg. no. E1925.12.25.3352. From Sholapur; ; [India]. Date: / 9/1874. Source:

Davidson coll.. Material: 1 ; whole/good egg.

Reg. no. E1925.12.25.3353. From Sholapur; ; [India]. Date: 20/ 7/1874. Source: Davidson coll.. Material: 1 ; whole/good egg.

Reg. no. E1925.12.25.3355. From Kharaghoda; Gujarat (Guzerat); [India]. Date: 1/ 8/1893. Source: Harrington Bulkeley; Davidson coll.. Material: 1 ; whole/good egg.

Reg. no. E1925.12.25.3356. From Kharaghoda; Gujarat (Guzerat); [India]. Date: / 9/1892. Source: Harrington Bulkeley; Davidson coll.. Material: 1 ; whole/good egg.

Reg. no. E1925.12.25.3357. From Kharaghoda; Gujarat (Guzerat); [India]. Date: 5/ 8/1892. Source: Mrs Pearson; Barnes coll.; Davidson coll.. Material: 1 ; whole/good egg.

Reg. no. E1925.12.27.139. From ; ; [India]. Date: 7/ 9/1892. Source: Buckley coll.; Barnes coll.; Davidson coll.. Material: 1 ; whole/good egg. Taken by Buckley. This egg was found among the 'duplicates' in 1977.

Reg. no. E1941.1.7.1026. From Sholapore, Deccan; ; [India]. Date: 29/ 8/1874. Source: J. Davidson; Butler coll.. Material: 1 ; whole/good egg.

Reg. no. E1947.16.146. From Sholapore, Deccan; ; India. Date: 8/ 7/1894. Source: H.K.P. Carter; J.D. Salmon coll.. Material: 1 ; whole/good egg.

Reg. no. E1966.9.36. From Rajkot, Kathiwar; ; India. Date: //. Source: L.S. Fenton coll.. Material: 1 ; whole/good egg.

Reg. no. E1972.13.258. From Karagora; ; . Date: //. Source: General Jackson; Rothschild coll.. Material: 1 ; whole/good egg.

Reg. no. E1972.13.259. From Dengaum, Kharaghara; ; . Date: 23/ 8/1891. Source: Rothschild coll.. Material: 1 ; whole/good egg.

Reg. no. E1973.32.617. From Kharaghoda; Gujarat (Guzerat); India. Date: 19/10/1902. Source: Harrington Bulkeley; Stuart Baker coll.. Material: 1 ; whole/good egg.

Reg. no. E1973.32.618. From Kharaghoda; Gujarat (Guzerat); India. Date: 7/ 9/1901. Source: Harrington Bulkeley; Stuart Baker coll.. Material: 1 ; whole/good egg.

Reg. no. E1973.32.620. From Kharaghoda; Gujarat (Guzerat); India. Date: 15/ 9/1894. Source: Harrington Bulkeley; Stuart Baker coll.. Material: 1 ; whole/good egg.

Reg. no. E1973.32.621. From Kharaghoda; Gujarat (Guzerat); India. Date: 15/ 9/1894. Source: Harrington Bulkeley; Stuart Baker coll.. Material: 1 ; whole/good egg.

Reg. no. E1973.32.622. From Kharaghoda; Gujarat (Guzerat); India. Date: 12/ 8/1901. Source: Harrington Bulkeley; Stuart Baker coll.. Material: 1 ; whole/good egg.

Reg. no. E1973.32.623. From Sholapur; ; [India]. Date: 4/ 4/1898. Source: Vidal coll.; Stuart Baker coll.. Material: 1 ; whole/good egg.

Reg. no. E1973.32.624. From Sholapur; ; [India]. Date: 7/ 5/1898. Source: Vidal coll.; Stuart Baker coll.. Material: 1 ; whole/good egg.

Reg. no. E1973.32.625. From Kharaghoda; Gujarat (Guzerat); India. Date: 27/ 9/1902. Source: Harrington Bulkeley; Stuart Baker coll.. Material: 1 ; whole/good egg.

Reg. no. E1973.32.626. From Kharaghoda; Gujarat (Guzerat); India. Date: 18/ 8/1901. Source: Harrington Bulkeley; Stuart Baker coll.. Material: 1 ; whole/good egg.

Reg. no. E1973.32.627. From Cutch; ; w. India. Date: 27/ 1/1903. Source: F.E. Kemp; Stuart Baker coll.. Material: 1 ; whole/good egg.

Reg. no. E1973.32.628. From Deccan; ; India. Date: 3/ 8/1908. Source: Whymper coll.; Stuart Baker coll.. Material: 1 ; whole/good egg.

Reg. no. E1976.1.809. From ; ; . Date: 16/ 8/1897. Source: Bell coll.. Material: 1 ; whole/good egg.

Reg. no. E1968.1.56. From Jhinjiwada, Rann of Cutch; [Gujarat]; [India]. Date: 11/ 8/1947. Source: Pir Taj Mahomed; K.R. Eates. Material: 1 ; whole/good egg (clutch). Incubation, fresh. Egg laid in scrape in ground in shade of clump of coarse grass and a few scrub bushes.

Reg. no. E1968.1.57. From Mawasiri, Rann of Cutch, s.e. Naga Parkar; [Sind]; [Pakistan]. Date: 22/ 8/1948. Source: Pir Taj Mahomed; K.R. Eates. Material: 1 ; whole/good egg (clutch). Incubation, fresh. Egg laid in bare scrape in ground in shade of clump of coarse grass in open grass–land and some scrub.

Reg. no. E1968.1.58. From Poonj Wagar, nr. Diplo; Sind; Pakistan. Date: 27/ 9/1948. Source: Baloch Khan; K.R. Eates. Material: 1 ; whole/good egg (clutch). Incubation, fresh. Egg laid in scrape in ground in patchy grass–land growing stunted scrub in parts.

Reg. no. E1968.1.59. From Vamri village, O.P. Haro, Naga Parkar; [Sind]; [Pakistan]. Date: 28/ 8/1949. Source: Pir Taj Mahomed; K.R. Eates. Material: 1 ; whole/good egg (clutch). Incubation, fresh. Egg laid in scrape in bare ground in coarse grass land bordering fallow fields.

Haematopus ostralegus chathamensis **Chatham Islands Oystercatcher**

Confined to the Chatham Islands, where it is now reduced to about 80 birds. This form is now generally regarded as a separate species.

Reg. no. 1896.7.1.147. From Wellington; ; New Zealand. Date: / 6/1873. Source: Travers; Seebohm coll.. Material: adult F; whole/good skin. Listed in register and on label as *H. longirostris*, with Wellington stated as the locality. The bird has been identified as *H.o. chathamensis* by R.A. Falla, who questions the locality.

Reg. no. 1896.7.1.150. From ; ; Chatham Islands. Date: 5/ 8/1871. Source: Sir J.

Hector; Seebohm coll.. Material: adult F; whole/good skin. Listed in the register and on the label as *H. longirostris*, but the label also carries the comment from R.A. Falla, 'This is *chathamensis*'.

Reg. no. 1896.7.1.151. From ; ; Chatham Islands. Date: 5/ 8/1871. Source: Sir J. Hector; Seebohm coll.. Material: adult M; whole/good skin. Subspecies probably incorrect. The register and label give the locality of this bird as Chatham Islands, but R.A. Falla has commented on the label, 'This is not *chathamensis*, but *finschi*, which is not found on Chatham Island'.

Haematopus moquini meadewaldoi
Canary Islands Oystercatcher

Endemic to the Canary Islands, where it was always rare and is thought to have become extinct sometime during the 20th century. There have been some recent alleged sightings, and it is just possible that a few survive. Habitat loss is believed to have been responsible for the decline. This form is now considered to be a separate species.

Reg. no. 1905.12.22.322. From ; Graciosa; Canary Islands. Date: 6/ 4/1890. Source: E.G.B. Meade Waldo. Material: adult F; whole/good skin. One of a pair. The other (the male) was in the possession of Canon Tristram. See *Ibis* 1890: 437.

Reg. no. 1905.12.22.323. From Jandea; Fuerteventura; Canary Islands. Date: 7/ 4/1888. Source: E.G.B. Meade Waldo. Material: adult F; whole/good skin. Holotype. Bannerman, *Bull. Brit. Orn. Club* 31 (1913): 33–4. Said to have been a breeding bird, containing well-developed eggs. See also *Ibis* 1889: 13.

Reg. no. 1939.12.9.25. From ; Graciosa; Canary Islands. Date: 3/ 6/1913. Source: D.A. Bannerman. Material: adult M; whole/good skin. The date is incorrectly given as 13 June 1913 in the register. Bill orange-vermillion, feet pale pink, iris crimson, eyelids orange-vermillion.

Reg. no. . From ; ; . Date: //. Source: Material: 1 ; egg. Not in the collection. Bannerman believed there were no extant eggs, but there is said to be one in a private collection. The correspondence is on file at the NHM.

Himantopus novaezelandiae
Black Stilt

Now restricted to a confined area in New Zealand, where it hybridizes with the Pied Stilt *H. himantopus leucocephalus*. The Black Stilt suffers heavy predation from cats and ferrets. The population is believed to be about 30 birds, and is the subject of intense conservation efforts.

Reg. no. . From ; ; New Zealand. Date: //. Source: Capt. Stokes. Material: adult ; skin. Mount. Details taken from *Cat. Bds.*. Not found 1988. There is no reason to believe that 1987.24.169 is the missing bird.

Reg. no. . From ; ; New Zealand. Date: //. Source: New Zealand Company. Material: imm ; skin. Details taken from *Cat. Bds.*. Not found 1988.

Reg. no. . From Otago; ; New Zealand. Date: / 5/. Source: Hume coll.. Material: adult M; skin. Details taken from *Cat. Bds.*. Not found 1988.

Reg. no. . From Otago; ; New Zealand. Date: / 9/. Source: Hume coll.. Material: imm F; skin. Details taken from *Cat. Bds.*. Not found 1988.

Reg. no. 1843.1.17.6. From Port Nicholson; North Island; New Zealand. Date: //. Source: New Zealand Company. Material: adult ; whole/good skin. Holotype or syntype. Ex–mount. Gould, *Proc. zool. Soc. Lond.* 1841: 8. The pied specimen 1843.1.17.7 may be a syntype. Subsequent usage has restricted the name to the black form. See Oliver, 1955, *New Zealand Birds*, 2nd. ed., p. 304–6.

Reg. no. 1845.1.13.11. From ; ; New Zealand. Date: //. Source: P. Earl. Material: ; whole/good skin. Ex–mount.

Reg. no. 1847.1.8.22. From ; ; New Zealand. Date: //. Source: G. Grey. Material: imm ; whole/good skin.

Reg. no. 1852.1.17.27. From Massacre Bay; ; New Zealand. Date: //. Source: Lyall; Stokes. Material: adult ; whole/good skin. Iris scarlet.

Reg. no. 1892.8.3.258. From Salt Water Creek; ; New Zealand. Date: / 5/1873. Source: Seebohm coll.. Material: ; whole/good skin.

Reg. no. 1892.8.3.259. From Wellington; ; New Zealand. Date: //. Source: Gould; Seebohm coll.. Material: M; whole/good skin.

Reg. no. 1892.8.3.260. From Selwyn; ; New Zealand. Date: / 9/1872. Source: Seebohm coll.. Material: adult M; skin. Not found 1988.

Reg. no. 1892.8.3.261. From Selwyn; ; New Zealand. Date: 29/ 3/. Source: Seebohm coll.. Material: juv ; skin. Not found 1988.

Reg. no. 1895.3.15.34. From ; ; New Zealand. Date: //. Source: Sir S. Scott. Material: ; whole/good skin.

Reg. no. 1895.3.15.35. From ; ; New Zealand. Date: //. Source: Sir S. Scott. Material: ; whole/good skin.

Reg. no. 1896.7.1.719. From ; ; New Zealand. Date: //. Source: Seebohm coll.. Material: ; part/good skin.

Reg. no. 1896.7.1.720. From Selwyn; ; New Zealand. Date: / 9/1872. Source: Haast; Seebohm coll.. Material: F; whole/good skin.

Reg. no. 1896.7.1.721. From Dunedin, Otago; ; New Zealand. Date: //. Source: Frank; Seebohm coll.. Material: adult F; whole/good skin.

Reg. no. 1939.12.9.29. From ; South Island; New Zealand. Date: / 9/1870. Source: G.M. Mathews; Rothschild coll.. Material: adult M; whole/good skin.

Reg. no. 1939.12.9.30. From ; ; New Zealand. Date: //. Source: Rothschild coll.. Material: adult ; whole/good skin.

Reg. no. 1949.64.145. From ; ; New Zealand. Date: //. Source: G.M. Mathews. Material: ; whole/good skin.

Reg. no. 1987.24.169. From ; ; [New Zealand]. Date: //. Source: Material: adult F; whole/good skin. Ex-mount. Moved from Tring annex store to main collection in 1980s.

Reg. no. E1901.1.1.4338-40. From Otago River; South Island; New Zealand. Date: 14/10/1872. Source: J.R. Cook; Harting coll.; Seebohm coll.. Material: 3 ; whole/good eggs.

Reg. no. E1941.4.6.950. From ; ; New Zealand. Date: //1875. Source: H. Travers; Dr Hector; Buckley coll.. Material: 1 ; whole/good egg. Apparently collected prior to 1875: 'From H. Travers, per Dr Hector, Oct 20th 1875'. There is a label on the egg: '80, *Himantopus melas*', and in pencil 'Black Stilt'.

Rhinoptilus bitorquatus

Jerdon's Courser

Only known from a very restricted area of eastern India, but probably more widespread in prehistoric times. Had last been seen in 1900 and presumed extinct until rediscovered in 1986.

Reg. no. 1891.10.1.1036. From nr. Bhadrachallam, n. of Godavari River; Andhra Pradesh; India. Date: 9/ 3/1871. Source: W.T. Blanford; Hume coll.. Material: adult M; whole/good skin. This bird is described in the *Cat. Bds.* 24: 50–51, where the day is incorrectly given as 7th.

Reg. no. 1896.7.1.51. From nr. Bhadrachallam, n. of Godavari River; Andhra Pradesh; India. Date: 5/ 3/1871. Source: W.T. Blanford; Seebohm coll.. Material: M; whole/good skin.

***Vanellus macropterus* (*Rogibyx tricolor*)**

Javanese Lapwing

Formerly occurred in Java and perhaps also in Sumatra and Timor. Survived on Java until about 1940, but has not been seen since, and the species is presumed to be extinct.

Reg. no. 1880.1.1.2963. From ; Java; Indonesia. Date: //. Source: India Museum. Material: adult ; whole/good skin. Holotype of *Vanellus tricolor*. Horsfield, *Trans. Linn. Soc.* 13 (1821): 186.

Reg. no. 1987.24.41. From ; [Java]; [Indonesia]. Date: //. Source: [Leiden Museum]. Material: adult ; whole/good skin. Mount, with no data, moved from Tring annex store to main collection in 1980s. This may be specimen c., in *Cat. Bds.*, a mounted bird collected in Java, and acquired from the Leiden Museum.

Reg. no. VEL.39.114a. From ; Java; Indonesia. Date: //. Source: [Horsfield]; East India Company. Material: adult ; whole/poor skin. Damaged: skin brittle and in pieces. This specimen is similar to the type, and was probably collected by

Horsfield, even though they arrived at the Museum separately (see *Cat. Bds.*).

Charadrius novaeseelandiae

Thinornis (New Zealand Shore Plover)

Formerly occurred widely in the New Zealand area, but now restricted to South East Island in the Chatham group. The population is estimated to be about 100 birds. *Thinornis rossi*, known only from the type, is usually regarded as a synonym of *novaeseelandiae*, but may represent a valid (extinct) form. The type is apparently in immature plumage, and differs from immature plumage of *novaeseelandiae*.

Reg. no. 1842.12.16.78. From ; Auckland Islands; New Zealand. Date: //. Source: Voyage of H.M.S. Erebus & Terror; the Admiralty. Material: imm F; whole/good skin. Holotype and only specimen of *Thinornis rossi*. Ex-mount. G.R. Gray, 1845, *The Zoology of the Voyage of H.M.S. Erebus and Terror*, Birds, p. 12, pl. 11. Described as an adult, but apparently immature. Listed as a female in the register. Irides a bluish-grey.

Reg. no. 1845.1.13.10. From ; ; New Zealand. Date: //. Source: P. Earl. Material: adult ; whole/good skin. Ex-mount.

Reg. no. 1845.1.13.9. From ; ; New Zealand. Date: //. Source: P. Earl. Material: adult ; whole/good skin. Mount.

Reg. no. 1892.10.31.5. From ; ; Chatham Islands. Date: 12/ 5/1892. Source: H.O. Forbes. Material: adult M; whole/good skin.

Reg. no. 1892.10.31.6. From ; ; Chatham Islands. Date: 8/ 5/1892. Source: H.O. Forbes. Material: adult M; whole/good skin.

Reg. no. 1892.10.31.7. From ; ; Chatham Islands. Date: 8/ 5/1892. Source: H.O. Forbes. Material: imm F; whole/good skin.

Reg. no. 1892.10.31.8. From ; ; Chatham Islands. Date: 12/ 5/1892. Source: H.O. Forbes. Material: juv M; whole/good skin.

Reg. no. 1893.6.24.11. From ; ; Chatham Islands. Date: //. Source: H.O. Forbes. Material: chick ; whole/good skin.

Reg. no. 1896.7.1.684. From ; ; Chatham Islands. Date: / 7/1874. Source: M. Milne Edwards; Seebohm coll.. Material: juv ; whole/good skin.

Reg. no. 1896.7.1.685. From ; ; Chatham Islands. Date: //. Source: W. Hector; J.E. Harting; Seebohm coll.. Material: adult F; whole/good skin. Locality given as Otago, New Zealand, in the register.

Reg. no. 1896.7.1.686. From Pitt Island; ; Chatham Islands. Date: /12/1871. Source: Hector; Seebohm coll.. Material: adult F; whole/good skin.

Reg. no. 1896.7.1.687. From Pitt Island; ; Chatham Islands. Date: //. Source: Hector; Seebohm coll.. Material: adult M; skin. Details taken from the register. Not found 1988.

Reg. no. 1939.12.9.36. From ; ; Chatham Islands. Date: //. Source: Rothschild coll.. Material: adult M; whole/good skin.

Reg. no. 1939.12.9.37. From ; ; Chatham Islands. Date: //1895. Source: Rothschild coll.. Material: adult F; whole/good skin.

Reg. no. 1939.12.9.38. From ; ; Chatham Islands. Date: //. Source: Palmer coll.; Rothschild coll.. Material: ; whole/good skin. Albino.

Numenius borealis

Eskimo Curlew

Formerly bred widely and commonly on the Canadian tundras and wintered on grasslands from southern Brazil to Argentina. Now very rare and once thought to be extinct. A few have been seen on migration and on former breeding grounds though no breeding or wintering areas are currently known.

Reg. no. . From ; ; Paraguay. Date: //. Source: E. Wilson. Material: adult ; skin. Mount. Details taken from *Cat. Bds.*. Not found <1986.

Reg. no. . From ; ; Paraguay. Date: 9/10/. Source: A. Peel. Material: adult ; skin. Details taken from *Cat. Bds.*. Not found 1988.

Reg. no. 1847.8.10.20. From ; ; Barbados. Date: //. Source: R. Schomburgk. Material: adult ; whole/good skin.

Reg. no. 1863.1.28.127. From Fort Good Hope; [Northwest Territories]; [Canada]. Date: //. Source: B.R. Ross. Material: adult ; whole/good skin.

Reg. no. 1878.1.25.21. From Buenos Aires; ; Argentina. Date: 9/10/1878. Source: A. Peel. Material: ; whole/good skin. Bought in the street. Probably the specimen that is listed in the *Cat. Bds.* from Paraguay.

Reg. no. 1885.11.20.182. From ; Chubut (Chufat), Patagonia; Argentina. Date: 8/11/1877. Source: H. Durnford. Material: ; whole/good skin. Beak black, legs and feet dark slate, iris wood–brown.

Reg. no. 1885.11.20.183. From ; Chubut (Chufat), Patagonia; Argentina. Date: 8/10/1877. Source: H. Durnford. Material: ; whole/good skin.

Reg. no. 1888.10.10.2651. From St. Michaels; Alaska; United States. Date: 25/5/1880. Source: E.W. Nelson; H.W. Henshaw; Salvin–Godman coll.. Material: M; whole/good skin.

Reg. no. 1888.10.10.2652. From St. Michaels; Alaska; United States. Date: 14/5/1881. Source: E.W. Nelson; H.W. Henshaw; Salvin–Godman coll.. Material: adult M; whole/good skin.

Reg. no. 1888.10.10.2653. From St. Michaels; Alaska; United States. Date: 4/7/1877. Source: E.W. Nelson. Material: adult F; whole/good skin. US National Explorations in Alaska. Locality given as Massachusetts in the register, but presumably in error.

Reg. no. 1888.10.10.2654. From ; Massachussetts; United States. Date: 10/10/1871. Source: H.W. Henshaw; Salvin–Godman coll.. Material: M; whole/good skin.

Reg. no. 1891.10.20.1027. From Fort Simpson; [Northwest Territories]; [Canada]. Date: //. Source: B. McDonald; Salvin–Godman coll.. Material: adult ; whole/good skin.

Reg. no. 1891.10.20.1028. From Henley Harbour; Labrador; Canada. Date: 23/ 8/. Source: E. Coues; Salvin–Godman coll.. Material: adult F; whole/good skin.

Reg. no. 1891.10.20.1029. From Brownsville; Texas; United States. Date: 28/ 3/1889. Source: F.B. Armstrong; Salvin–Godman coll.. Material: adult M; whole/good skin.

Reg. no. 1891.10.20.1030. From Brownsville; Texas; United States. Date: //. Source: F.B. Armstrong; Salvin–Godman coll.. Material: adult ; skin. Details taken from *Cat. Bds.*. Not found <1986.

Reg. no. 1891.10.20.1031. From Brownsville; Texas; United States. Date: 28/ 3/1889. Source: F.B. Armstrong; Salvin–Godman coll.. Material: adult F; whole/good skin.

Reg. no. 1891.10.20.1032. From Brownsville; Texas; United States. Date: 2/ 4/1889. Source: F.B. Armstrong; Salvin–Godman coll.. Material: adult M; whole/good skin.

Reg. no. 1891.10.20.1033. From Brownsville; Texas; United States. Date: 28/ 3/1889. Source: F.B. Armstrong; Salvin–Godman coll.. Material: adult F; whole/good skin.

Reg. no. 1891.10.20.1034. From Brownsville; Texas; United States. Date: 28/ 3/1890. Source: F.B. Armstrong; Salvin–Godman coll.. Material: M; whole/good skin. The year of collection is given as 1889 in the register.

Reg. no. 1891.10.20.1035. From Brownsville; Texas; United States. Date: 25/ 3/1889. Source: F.B. Armstrong; Salvin–Godman coll.. Material: adult M; skin. Details taken from the register. Not found <1986.

Reg. no. 1891.10.20.1036. From Brownsville; Texas; United States. Date: 22/ 3/1889. Source: F.B. Armstrong; Salvin–Godman coll.. Material: adult M; skin. Details taken from the register. Not found <1986.

Reg. no. 1891.10.20.1037. From Brownsville; Texas; United States. Date: 27/ 3/1889. Source: F.B. Armstrong; Salvin–Godman coll.. Material: adult M; whole/good skin.

Reg. no. 1891.10.20.1038. From Brownsville; Texas; United States. Date: 26/ 3/1889. Source: F.B. Armstrong; Salvin–Godman coll.. Material: F; skin. Not found 1988.

Reg. no. 1891.10.20.1039. From San Geronimo, Vera Paz; ; Guatemala. Date: / 4/1860. Source: R. Owen; Salvin–Godman coll.. Material: adult M; whole/good skin.

Reg. no. 1891.10.20.1040. From Xavier; ; Brazil. Date: 7/10/1826. Source: J. Natterer; Salvin–Godman coll.. Material: adult M; whole/good skin.

Reg. no. 1891.10.20.1464. From ; ; Grenada. Date: 22/ 9/1891. Source: D.W. Smith; Salvin–Godman coll.. Material: adult ; whole/good skin.

Reg. no. 1894.2.19.70. From Henley Harbour; Labrador; Canada. Date: 27/ 8/1887. Source: Coues; Seebohm coll.. Material: adult ; whole/good skin.

Reg. no. 1894.2.19.71. From Henley Harbour; Labrador; Canada. Date: 20/ 8/1861. Source: A.L. Heermann; Seebohm coll.. Material: M; whole/good skin.

Reg. no. 1894.2.19.72. From Colonia; ; Uruguay. Date: 5/12/1883. Source: H.W. Harrison; Seebohm coll.. Material: M; whole/good skin.

Reg. no. 1896.7.1.779. From ; Labrador; Canada. Date: //1869. Source: Seebohm coll.. Material: juv ; whole/good skin. On long term loan to Newfoundland Mus., St. Johns, Canada. The date is from Hahn, 1963, *Where is that Vanished Bird?*, Toronto.

Reg. no. 1896.7.1.780. From Ipswich; Massachussetts; United States. Date: / 9/1872. Source: C.J. Maynard; Seebohm coll.. Material: juv F; whole/good skin.

Reg. no. 1896.7.1.781. From ; ; Mexico. Date: //. Source: Seebohm coll.. Material: adult ; whole/good skin.

Reg. no. 1897.11.10.418. From Wolf Island, Kingston; Ontario; Canada. Date: 10/10/1873. Source: A.R. Dresser; E. Hargitt. Material: adult F; whole/good skin.

Reg. no. 1897.11.10.419. From ; ; Greenland. Date: 13/ 8/1880. Source: R. Miller. Material: adult F; skin. Not found 1988.

Reg. no. 1906.12.7.267. From North Truro; Massachussetts; United States. Date: 1/ 9/1889. Source: G.S. Miller coll.. Material: adult M; skin. Details taken from the register. Not found 1988, but see the entries for the other North Truro birds.

Reg. no. 1906.12.7.268. From North Truro; Massachussetts; United States. Date: 24/ 9/1889. Source: G.S. Miller coll.. Material: adult M; skin. Details taken from the register. Not in the collection. 1906.12.7.267, 268 and 270 are missing. One was exchanged with Maj. Allan Brooks in 1919, probably this one.

Reg. no. 1906.12.7.269. From North Truro; Massachussetts; United States. Date: 24/ 9/1889. Source: G.S. Miller coll.. Material: M; whole/good skin.

Reg. no. 1906.12.7.270. From North Truro; Massachussetts; United States. Date: 24/ 9/1889. Source: G.S. Miller coll.. Material: adult F; skin. Details taken from the register. Not found 1988, but see the entries for the other North Truro birds.

Reg. no. 1906.12.7.271. From North Truro; Massachussetts; United States. Date: 10/ 9/1890. Source: G.S. Miller coll.. Material: adult F; whole/good skin.

Reg. no. 1922.3.5.558. From ; ; Guyana (British Guiana). Date: //. Source: McConnell coll.. Material: F; whole/good skin. Described in Chubb, 1916, *The Birds of British Guiana*, 1: 114–5; presumably female, but not noted as such on the label. Damaged: head detached.

Reg. no. 1934.1.1.901. From Cook County; Illinois; United States. Date: 10/ 4/. Source: Witherby coll.. Material: M; whole/good skin.

Reg. no. 1950.24.2. From ; ; North America. Date: //. Source: J.E. Harting; A. Lister. Material: F; whole/good skin.

Reg. no. 1965.M.2960. From Brownsville; Texas; United States. Date: 26/ 3/1902. Source: J. Dwight; R. Meinertzhagen coll.. Material: M; whole/good skin.

Reg. no. 1965.M.2961. From Brownsville; Texas; United States. Date: 26/ 3/1902. Source: J. Dwight; R. Meinertzhagen coll.. Material: F; whole/good skin.

Reg. no. 1979.10.1. From Ungava Bay (Ungarva Bay); Quebec; Canada. Date: / 8/1886. Source: D. Johnston; Blackpool College of Tech. & Art. Material: adult ; whole/good skin.

Reg. no. 1988.21.10. From ; Massachussetts; United States. Date: 10/ 9/1872. Source: Tweeddale coll.. Material: F; whole/good skin.

Reg. no. 1988.21.11. From North Truro; Massachussetts; United States. Date: 1/ 9/1888. Source: Material: adult M; whole/good skin.

Reg. no. 1988.21.12. From North Truro; Massachussetts; United States. Date: //. Source: G.S. Miller coll.. Material: adult ; whole/good skin. Label broken; the fragment remaining lacks the reg. no. and the date, but this may be the bird previously registered as 1906.12.7.270, a female collected on 24 September 1889.

Reg. no. 1988.8.19. From ; ; North America. Date: //. Source: J.J. Audubon. Material: ; whole/good skin. Previously registered as VEL.38.50b.

Reg. no. 1988.8.20. From ; ; North America. Date: //. Source: J.J. Audubon. Material: ; whole/good skin. Previously registered as VEL.38.50a.

Reg. no. 1896.2.16.179. From ; ; [America]. Date: //. Source: Seebohm coll.. Material: ; part/poor skeleton. Damaged; part of skull only.

Reg. no. E1941.4.6.947. From Anderson River Fort, lr. Anderson River; Northwest Territories; Canada. Date: 13/ 6/1863. Source: R. McFarlane; Smithsonian; Buckley. Material: 4 ; whole/good eggs. Female parent snared on the nest. See *Proc. U.S. natn. Mus.* 14 (1891): 429 for descriptions of the nests.

Reg. no. E1973.GS.95. From the Barren Gound, e. of Anderson River; Northwest Territories; Canada. Date: 25/ 6/1864. Source: R. McFarlane; Smithsonian Inst.; Godman Salvin coll.. Material: 2 ; damaged eggs. From the n.e. arctic coast. The nest was a few withered leaves placed in a depression in the ground. It is uncertain if the data–slip refers to both eggs.

Reg. no. 1902.3.10.54. From Anderson River; Northwest Territories; Canada. Date:

//. Source: R. McFarlane; Crowley. Material: 1 ; damaged egg. Specimen badly damaged. Figured, see Seebohm, 1896, *Coloured Figures of the Eggs of British Birds*, p. 133–4, pl. 45.

Prosobonia leucoptera

White-winged Sandpiper

Endemic to Tahiti and Moorea. Known only from three specimens, two of which are now lost, and several original paintings. The evidence suggests that two species were involved (Walters, *Boll. Mus. reg. Sci. nat. Torino* 9: 217–226), one on Tahiti (*leucoptera*) and one on Moorea (*ellisi*). Both have been extinct since about 1777.

Reg. no. . From ; Moorea; Society Islands. Date: //. Source: Material: ; painting. Holotype of *P. ellisi*. Sharpe, *Bull. Brit. Orn. Club* 16 (1906): 86. Based on a drawing in the NHM library, by William Ellis, no. 65. See *Bull. Br. Mus. nat. Hist. (Hist.)* 1 (1959): 333.

Reg. no. . From ; Tahiti (Taheitee); Society Islands. Date: //. Source: Material: ; painting. A drawing in the NHM library, by Georg Forster, (old) no. 120, based on the type-specimen now in Leiden Museum. See *Bull. Br. Mus. nat. Hist. (Hist.)* 1 (1959): 300.

Coenocorypha aucklandica iredalei

Iredale's Snipe

Endemic to South Cape Island off Stewart Island where it became extinct in 1964 as a result of the depredations of rats.

Reg. no. 1932.10.25.1. From Big South Cape Island; ; New Zealand. Date: /12/1932. Source: E.F. Stead. Material: adult M; whole/good skin.

Reg. no. 1939.12.9.41. From ; ; [New Zealand]. Date: 24/ 4/1901. Source: Rothschild coll.. Material: adult F; whole/good skin. Damaged: head detached. The reg. no. is untraced and may be incorrect; this may be 1939.12.9.50, given in the register as this subspecies, but without sex or date. 1939.12.9.41 was allocated to a pigeon.

Pinguinus impennis

Great Auk

Formerly bred on low rocky shores and islets in the North Atlantic. In historic times its range was greatly reduced, and its final demise was expedited by human persecution. The last proven occurrence was 1844, but there were unconfirmed sightings for several years thereafter.

Reg. no. 1858.4.27.1. From ; ; Iceland. Date: //1833. Source: Prof Lidth de Jeude; Warwick. Material: ; whole/good skin. Mount. Labrador in *Cat. Bds.*. ?Labrador in register. See *Nat. Hist. Rev.* 12 (1865): 473, where it is noted that the BM obtained a Great Auk from de Jeude in 1856. Though marked 'Labrador' on its stand, it was procured from the R. Mus. Copenhagen where it had been sent from Iceland after 1830.

Reg. no. 1939.12.9.537. From ; ; [Atlantic Ocean]. Date: //. Source: E. Delegorgue; Boucard; Leopold Field; R. Ward; Rothschild coll.. Material: ; whole/good skin. Said to have come from a collection in Brunswick or Mainz. See *Trans. Edinb. nat. Fld Club* 3 (1897): 248–9, pl. 3; 3 (1898): 328–30. Figured in Naumann, 1903, *Naturgeschichte der Vogel Mitteleuropas*, 12: 169–208, plate 17a, no. 2.

Reg. no. 1949.17.1. From ; ; Iceland. Date: //1833. Source: Lilford. Material: ; whole/good skin. Mount. Collected about 1833. See Grieve, 1885, *The Great Auk*, appendix, p. 15, no. 44.

Reg. no. 1988.21.13. From Fowl's Craig, Papa Westray, Orkney Islands; Scotland; United Kingdom. Date: / 5/1813. Source: W. Foulis; W. Bullock. Material: adult M; whole/good skin. No. B277a. Mount. Probably collected in May or summer 1813. See MacGillivray, 1852, *A History of British Birds*, 5: 361; *Nat. Hist. Rev.* 12 (1865): 467–88; Buckley & Harvie-Brown, 1891, *A Vertebrate Fauna of Orkney*, p. 245–257.

Reg. no. . From ; Newfoundland; Canada. Date: //. Source: Material: ; skeleton. Various bones. Details taken from *Cat. Bds.*. Not found 1988.

Reg. no. 1864.12.10.1. From Funk Island; Newfoundland; Canada. Date: //. Source: J. Matthew Jones, or Dr Kirk. Material: ; whole/good skeleton. Ex-mount. The label and the *Cat. Bds.* say the specimen was presented by J. Matthew Jones, but the register says it was presented by Dr Kirk. See *Nat. Hist. Rev.* 12 (1865): 484–5.

Reg. no. 1881.5.1.6238–9. From Funk Island; Newfoundland; Canada. Date: //. Source: Gould coll.. Material: ; part skeleton. 2 skulls. Details taken from the register. Not found 1988 or 1992. *Cat. Bds.* lists "Bones, various. Newfoundland. Purchased" which could include these.

Reg. no. 1939.12.9.538. From Funk Island; Newfoundland; Canada. Date: //. Source: Rothschild coll.. Material: ; whole/good skeleton. Ex-mount.

Reg. no. S1952.1.195. From White Park Bay, County Antrim; Northern Ireland; United Kingdom. Date: 18/ 4/1893. Source: W.J. Knowles. Material: ; part/good skeleton. Humerus only. Discovered in kitchen midden on date shown.

Reg. no. S1972.1.156. From ; ; [North Atlantic Ocean]. Date: //. Source: H. Shaw; Rothschild coll.. Material: ; part/good skeleton. Two each: humerus, ulna, radius, tibio-tarsus; from skin in Hawthorne Museum. The old label appears to read 'Lord Hill's bird. H. Shaw 1862'. There is also a letter from Shaw to Rothschild dated 28 August 1896 offering him the bones for 4 pounds (sterling).

Reg. no. 1901.11.15.814. From ; ; [Iceland]. Date: //1844. Source: Kjarbolling; Williams; J.de C. Wise; Tristram; Crowley. Material: 1 ; whole/good egg. *Ann. Mag. nat. Hist.* 14 (1864): 236. Grieve, 1885, *The Great Auk*, appendix, p. 27, no. 19. *Trans. Edinb. nat. Fld Club* 3 (1897): 260. One from the last batch of eggs sent to Denmark from Iceland. Date may not be correct.

Reg. no. E1941.1.1.1. From ; ; [Iceland]. Date: //. Source: Brandt; Fr Schultz;

Huehnel; Count Rodern; Rothschild. Material: 1 ; whole/good egg. Grieve, 1885, *The Great Auk*, appendix, p. 25–6, no. 8. *Trans. Edinb. nat. Fld Club* 3 (1897): 263. Naumann, 1903, *Naturgeschichte der Vogel Mitteleuropas*, 12: pl. 17b, no. 1.

Reg. no. E1941.1.1.2. From ; ; [North Atlantic Ocean]. Date: //1761. Source: Prof. Spallanzani; Mus. of Anatomy, Pavia; R. Champley; Rothschild. Material: 1 ; whole/good egg. See *Trans. Edinb. nat. Fld Club* 2 (1888): 114–6. The egg was discovered at Pavia by Champley in 1861. It was said that it had been deposited there a century before, by Spallanzani.

Reg. no. E1949.7.1. From ; ; [North Atlantic Ocean]. Date: //. Source: Alfred Newton, College of Surgeons; A.W. Crichton; Lord Lilford. Material: 1 ; whole/good egg. Found with nine others by Alfred Newton at the College of Surgeons in 1861. *Trans. Edinb. nat. Fld Club* 3 (1897): 259–60.

Reg. no. E1962.1.5. From ; ; . Date: //. Source: Bullock coll.. Material: 1 ; damaged egg. Broken at the larger end; inscribed '*Alca impennis*'. Previously glued to a board in the gallery, probably bleached. Bought for the Museum by W.E. Leach in 1819.

Reg. no. E1962.1.6. From ; ; [North Atlantic Ocean]. Date: //. Source: Bullock coll.. Material: 1 ; whole/good egg. Inscribed '139' at smaller end. Previously glued to a board in the gallery, probably bleached. The board has 'Sloane Museum?' on back. Bought for the Museum by W.E. Leach in 1819.

Raphus cucullatus

Dodo

Previously confined to the island of Mauritius where it was exterminated about 1662.

Reg. no. . From ; ; Mauritius. Date: //. Source: Material: ; skin. Two plaster casts each of the head and the foot (one cast broken), from the Ashmolean, Oxford. Stored in the skeleton collection.

Reg. no. . From ; ; [Mauritius]. Date: //. Source: Material: ; part skin. Foot only. Not found for many years. Listed in *Cat. Bds.* and figured by Mitchell, 1844–1849, *Genera of Birds*, pl. 120 and Strickland & Melville, 1848, *The Dodo and Kindred Birds*, pl. vi.

Reg. no. S1988.50.1. From Mare aux Songes; ; Mauritius. Date: //. Source: Rothschild coll.. Material: ; whole/good skeleton. Mount, formerly in the Tring gallery. A composite skeleton with bones from more than one bird. There are more specimens in the Palaeontology Department.

Pezophaps solitaria

Rodriguez Solitaire

Previously confined to the island of Rodriguez where it was exterminated around 1760. Storer has recommended that this bird be placed in a separate family.

Reg. no. S1988.49.1. From ; ; Rodriguez. Date: //. Source: H.H. Slater, Transit of Venus Expedition; Royal Society. Material: ; part/poor skeleton. Several bone fragments, together with a label reading 'Rodrigues 1874'.

Columba vitiensis godmanae

Godman's Pigeon

Occurred on Lord Howe Island. Known only from traveller's accounts and paintings.

Reg. no. . From ; ; Lord Howe Island. Date: //1790. Source: G. Raper, voyage of H.M.S. Sirius. Material: ; painting. Holotype. Mathews, *Austral avian Rec.* 3 (1915): 21–4. A drawing in the NHM library, by George Raper, no. 72, dating about 1790. See also Mathews, 1928, *The Birds of Norfolk & Lord Howe Islands*, p. 2, pl. 2.

Columba versicolor

Bonin Wood–Pigeon

Formerly occurred on several islands in the Bonin group, but believed extinct since 1889. The reason for its extinction is unknown.

Reg. no. 1892.11.23.3. From Nakondo Shima; Bonin Islands; Japan. Date: 15/ 9/1889. Source: P.A. Holst; Seebohm coll.. Material: adult M; whole/good skin. Beak greenish–yellow, iris dark blue, tarsus dark red.

Columba jouyi

Ryukyu Pigeon

Ryukyu Islands. Last recorded in 1936 and certainly extinct on Okinawa, but it might possibly survive on a few of the smaller islets.

Reg. no. 1887.10.1.3. From ; Ryukyu Islands (Liukiu Islands); [Japan]. Date: //. Source: Pryer; Seebohm coll.. Material: adult ; whole/good skin. See 1897.10.30.432, 433. Collected in December–January 1887, but noted as December in *Cat. Bds.*, where the collector is identified as Pryer.

Reg. no. 1897.10.30.432. From ; Ryukyu Islands (Loo Choo Islands); [Japan]. Date: / 1/1887. Source: [Pryer]; Seebohm coll.. Material: adult ; whole/good skin. 1897.10.30.432, 433 & 1887.10.1.3 (this species) all have similar collector's labels. This was Pryer, according to *Cat. Bds.*, referring to the specimen registered in 1887.

Reg. no. 1897.10.30.433. From ; Ryukyu Islands (Loo Choo Islands); [Japan]. Date: / 1/1887. Source: [Pryer]; Seebohm coll.. Material: adult ; whole/good skin. 1897.10.30.432, 433 & 1887.10.1.3 (this species) all have similar collector's labels. This was Pryer, according to *Cat. Bds.*, referring to the specimen registered in 1887.

Columba mayeri

Pink Pigeon

Mauritius. At one time considered extinct, but small numbers survived, and captive breeding has 'saved' the species, though it still remains endangered.

Reg. no. 1861.12.10.10. From ; ; Mauritius. Date: //1857. Source: Verreaux. Material: adult M; whole/good skin.

Reg. no. 1873.10.25.29. From ; ; Mauritius. Date: //. Source: R.B. Sharpe. Material: adult ; whole/good skin.

Reg. no. 1892.8.2.47. From ; ; Mauritius. Date: //. Source: Van Dam; Bewsher; Shelley coll.. Material: ; whole/good skin.

Reg. no. 1909.3.2.1. From ; ; Mauritius. Date: //. Source: Desjardins; M. Ulcoy. Material: adult M; whole/good skin.

Reg. no. 1909.3.2.2. From ; ; Mauritius. Date: //. Source: Desjardins; M. Ulcoy. Material: adult F; whole/good skin.

Reg. no. 1914.10.6.1. From ; ; Mauritius. Date: //1910. Source: Aylesbury Museum. Material: ; whole/good skin.

Reg. no. E1941.3.1.508a–b. From ; ; Captivity. Date: / 9/1909. Source: Munt coll.. Material: 2 ; whole/good eggs. Laid in captivity. All the eggs in the Munt coll. were laid by the same bird.

Reg. no. E1984.4.27–8. From ; ; Captivity. Date: //1979. Source: Jersey Wildlife Preservation Trust. Material: 2 ; whole/good eggs. Laid in captivity.

Reg. no. E1984.4.29–33. From ; ; Captivity. Date: //1980. Source: Jersey Wildlife Preservation Trust. Material: 5 ; whole/good eggs. Two eggs are small, and were without yolks.

Reg. no. E1941.3.1.503. From ; ; Captivity. Date: 11/ 6/1909. Source: H. Munt coll.. Material: 1 ; whole/good egg. Laid in captivity. All the eggs in the Munt coll. were laid by the same bird.

Reg. no. E1941.3.1.504. From ; ; Captivity. Date: 13/ 6/1909. Source: Munt coll.. Material: 1 ; whole/good egg. Laid in captivity. All the eggs in the Munt coll. were laid by the same bird.

Reg. no. E1941.3.1.505. From ; ; Captivity. Date: 27/ 6/1909. Source: Munt coll.. Material: 1 ; whole/good egg. Laid in captivity. All the eggs in the Munt coll. were laid by the same bird.

Reg. no. E1941.3.1.506. From ; ; Captivity. Date: 7/ 7/1909. Source: Munt coll.. Material: 1 ; whole/good egg. Laid in captivity. All the eggs in the Munt coll. were laid by the same bird.

Reg. no. E1941.3.1.507. From ; ; Captivity. Date: 18/ 7/1909. Source: Munt coll.. Material: 1 ; whole/good egg. Laid in captivity. All the eggs in the Munt coll. were laid by the same bird.

Reg. no. E1941.3.1.509. From ; ; Captivity. Date: 14/10/1909. Source: Munt coll.. Material: 1 ; whole/good egg. Laid in captivity. All the eggs in the Munt coll. were laid by the same bird.

Reg. no. E1941.9.4.2574. From ; ; Captivity. Date: 17/ 9/1907. Source: Rothschild. Material: 1 ; whole/good egg. Laid in captivity, in a zoo (? London Zoo).

Reg. no. E1984.4.26. From ; ; Captivity. Date: //1983. Source: Jersey Wildlife Preservation Trust. Material: 1 ; damaged egg. Cracked by parents.

Reg. no. E1984.4.34. From ; ; Captivity. Date: 29/ 6/. Source: Jersey Wildlife

Preservation Trust. Material: 1 ; whole/good egg. Laid in captivity.

Reg. no. A1976.18.2. From ; ; Mauritius. Date: / 9/1976. Source: J.E. Cooper. Material: juv ; alc.

Reg. no. A1976.18.3. From ; ; Mauritius. Date: / 9/1976. Source: J.E. Cooper. Material: juv ; alc.

Streptopelia picturata coppingeri
Glorioso Turtle-Dove

Island of Glorioso, north of Madagascar. It became extinct between 1884 and 1906, but the reasons for this are unknown as the island seemed relatively undisturbed.

Reg. no. 1882.10.25.1. From ; Glorioso Islands; [Madagascar]. Date: / 5/1882. Source: Dr R.W. Coppinger, voyage of H.M.S. Alert. Material: adult F; whole/good skin. Holotype. Sharpe, 1884, *Report on the Zoological Collections made ... during the Voyage of H.M.S. Alert 1881-2*, p. 484. Iris orange-red, bill dark, legs and feet dark red.

Streptopelia picturata rostrata
Seychelles Turtle-Dove

Seychelles. Exterminated as a result of hybridisation with introduced birds of the nominate race. By 1975 no pure *rostrata* were thought to remain.

Reg. no. 1889.2.10.718. From ; Marie Anne Island (Marianne Island); Seychelles. Date: //. Source: H.B. Tristram; Tweeddale coll.. Material: adult ; whole/good skin.

Reg. no. 1889.2.12.230. From ; Praslin Island; Seychelles. Date: //. Source: H.B. Tristram; Salvin-Godman coll.. Material: adult ; whole/good skin.

Reg. no. 1889.2.12.231. From ; Praslin Island; Seychelles. Date: //. Source: H.B. Tristram; Salvin-Godman coll.. Material: adult ; whole/good skin.

Reg. no. 1892.8.2.131. From ; Marie Anne Island (Marianne Island); Seychelles. Date: //. Source: Shelley coll.. Material: M; whole/good skin.

Reg. no. 1927.12.18.88. From ; Cousin Island; Seychelles. Date: 6/ 2/1888. Source: J.J. Lister coll.. Material: adult M; whole/good skin. Iris orange, stomach contained castor oil seeds and some flowers.

Reg. no. 1927.12.18.89. From ; Cousin Island; Seychelles. Date: 6/ 2/1888. Source: J.J. Lister coll.. Material: adult M; whole/good skin.

Reg. no. 1927.12.18.90. From ; Cousin Island; Seychelles. Date: 6/ 2/1888. Source: J.J. Lister coll.. Material: ; whole/good skin.

Reg. no. 1946.75.16. From ; Bird Island; Seychelles. Date: / 5/1940. Source: Sapsworth and Goodfellow. Material: adult F; whole/good skin. Iris brown, eyelids blood-red, bill horn-grey, feet blood-red.

Reg. no. 1946.75.17. From ; La Digue Island; Seychelles. Date: / 4/1940. Source: A.D. Sapsworth. Material: adult M; skin.

Details taken from the register. Not found 1988.

Reg. no. 1955.63.1. From ; Cousin Island; Seychelles. Date: 26/ 8/1955. Source: R. Percy & M. Ridley. Material: adult F; whole/good skin. Bill blue, cere and feet crimson, iris chestnut.

Ectopistes migratorius

Passenger Pigeon

North America. Once occurred in huge numbers, but large-scale persecution reduced these until the populations were no longer viable. No wild birds remained after 1899, and the last individual died on 1 September 1914 in Cincinnati Zoo.

Reg. no. . From ; ; North America. Date: //. Source: Material: adult M; whole/good skin. Mount in South Kensington gallery.

Reg. no. 1855.10.4.2. From ; California; United States. Date: //. Source: Warwick. Material: imm ; whole/good skin. Ex-mount. Locality is questioned in the *Cat. Bds.*.

Reg. no. 1860.2.23.6. From [Saumas Plains]; [British Columbia]; [Canada]. Date: /10/. Source: J.K. Lord; North American Boundary Commission. Material: imm F; whole/good skin. Female, October & Saumas Plains given in the register, but not on the label. Lord collected mainly in British Columbia (see *Hist. Colls.*).

Reg. no. 1863.1.7.292. From the w. side of the Rocky Mountains; [British Columbia]; [Canada]. Date: //. Source: J.K. Lord. Material: adult M; whole/good skin. Lord collected mainly in British Columbia (see *Hist. Colls.*).

Reg. no. 1863.1.7.293. From the w. side of the Rocky Mountains; [British Columbia]; [Canada]. Date: //. Source: J.K. Lord. Material: adult F; whole/good skin. Lord collected mainly in British Columbia (see *Hist. Colls.*).

Reg. no. 1876.4.15.298. From Fort Dufferin; ; North America. Date: 23/ 6/1873. Source: Dawson; North American Boundary Commission; Foreign Office. Material: F; whole/good skin. Two specimens bear this reg. no., the second being a heron.

Reg. no. 1876.4.15.45. From Fort Dufferin; ; North America. Date: //. Source: Dawson; North American Boundary Commission; Foreign Office. Material: M; whole/good skin. Eyes red. Noted as the mate of 1876.4.15.46.

Reg. no. 1876.4.15.46. From Fort Dufferin; ; North America. Date: //. Source: Dawson; North American Boundary Commission; Foreign Office. Material: M; whole/good skin. Registered as female. Noted as the mate of 1876.4.15.45.

Reg. no. 1876.4.15.47. From ; ; North America. Date: 17/ 5/1874. Source: Dawson; North American Boundary Commission; Foreign Office. Material: adult M; whole/good skin. From along the 49th parallel.

Reg. no. 1888.10.10.2884. From Big Otter Lake, Herk Co.; New York; United States. Date: 17/ 4/1878. Source: C. Hart Merriam;

Salvin–Godman coll.. Material: adult M; whole/good skin.

Reg. no. 1888.10.10.2885. From Big Otter Lake, Herk Co.; New York; United States. Date: 17/ 4/1878. Source: C. Hart Merriam; Salvin–Godman coll.. Material: adult F; whole/good skin.

Reg. no. 1888.10.10.2886. From Adirondacks; New York; United States. Date: 11/ 6/1878. Source: C. Hart Merriam; Salvin–Godman coll.. Material: imm ; whole/good skin.

Reg. no. 1888.10.10.2887. From Adirondacks; New York; United States. Date: 4/ 6/1878. Source: C. Hart Merriam; Salvin–Godman coll.. Material: juv F; whole/good skin.

Reg. no. 1888.10.10.2888. From Lyons Falls, New York; New York; United States. Date: 26/ 4/1879. Source: A. Jenings Dayan; Salvin–Godman coll.. Material: adult F; whole/good skin. Collection date given as September in the register.

Reg. no. 1888.10.10.2889. From Gropitville; Massachussetts; United States. Date: 31/ 5/1872. Source: H.W. Henshaw; Salvin–Godman coll.. Material: adult M; whole/good skin. Locality given as Grantville in the register.

Reg. no. 1888.10.10.2890. From Ontonagon; Michigan; United States. Date: 2/10/1881. Source: A. Jenings Dayan; Salvin–Godman coll.. Material: juv M; whole/good skin. Collection date 10.2.1881 = 10 February or 2 October: plumage condition suggests 2 October.

Reg. no. 1889.2.10.311. From ; ; Canada. Date: //. Source: G.E. Shelley; Tweeddale coll.. Material: adult F; whole/good skin.

Reg. no. 1889.4.20.164. From Market; District of Columbia; United States. Date: 16/ 2/1860. Source: E. Coues; Salvin–Godman coll.. Material: adult M; whole/good skin.

Reg. no. 1906.12.7.36. From ; ; North America. Date: //. Source: G.S. Miller coll.. Material: adult ; whole/good skin.

Reg. no. 1906.12.7.37. From ; ; United States. Date: 24/ 3/1875. Source: G.S. Miller coll.. Material: adult F; skin. Details taken from the register. Not found <1986.

Reg. no. 1906.12.7.38. From Boston; Massachussetts; United States. Date: /11/1886. Source: G.S. Miller coll.. Material: adult M; whole/good skin. Bought in a market.

Reg. no. 1906.12.7.39. From Boston; Massachussetts; United States. Date: /11/1886. Source: G.S. Miller coll.. Material: adult M; skin. Bought in a market. Details copied from the register. Not found <1986.

Reg. no. 1911.11.6.1. From Kingston; [Ontario]; Canada. Date: //1879. Source: Maj. Gen. Oliver. Material: M; whole/good skin.

Reg. no. 1918.12.3.1. From ; ; [North America]. Date: //. Source: LCC Art School. Material: adult ; whole/good skin.

Reg. no. 1918.12.3.2. From ; ; [North America]. Date: //. Source: LCC Art School. Material: adult ; whole/good skin.

Reg. no. 1921.8.3.1. From ; ; North America. Date: //. Source: A. Beverley Welford. Material: ; whole/good skin. Mount.

Reg. no. 1926.12.22.46. From ; Quebec; Canada. Date: //. Source: E. Dionne, Tax. Univ. Lav. Quebec; Imperial Institute. Material: M; whole/good skin.

Reg. no. 1932.9.19.89. From ; ; [North America]. Date: //. Source: C.H.B. Grant coll.. Material: M; whole/good skin.

Reg. no. 1933.6.18.53. From ; ; Canada. Date: //. Source: H.E. Dresser; Miss Wright. Material: adult M; whole/good skin.

Reg. no. 1939.12.9.4071. From ; ; [North America]. Date: //. Source: Rothschild coll.. Material: adult ; whole/good skin. Mount. Abnormal plumage.

Reg. no. 1939.12.9.41. From Moose River; New York; United States. Date: 6/11/1878. Source: Rothschild coll.. Material: adult M; whole/good skin.

Reg. no. 1939.12.9.540. From ; s. Indiana; United States. Date: 1/ 4/1887. Source: Rothschild coll.. Material: adult M; whole/good skin. Mount.

Reg. no. 1939.12.9.541. From ; ; North America. Date: //. Source: Rothschild coll.. Material: adult F; skin. Mount in Tring gallery. Specimen labelled as male.

Reg. no. 1939.12.9.542. From ; ; [North America]. Date: //. Source: Rothschild coll.. Material: ; whole/good skin. Mount.

Reg. no. 1949.64.142. From ; ; [North America]. Date: //. Source: G.M. Mathews. Material: adult F; whole/good skin.

Reg. no. 1988.21.14. From ; ; North America. Date: //. Source: Hudson's Bay Company. Material: adult M; whole/good skin. Apparently unregistered until 1988, but data matches VEL.32.57a listed here.

Reg. no. VEL.32.57a. From ; ; North America. Date: //. Source: Hudson's Bay Company. Material: adult M; skin. Not found 1986, but the data matches a specimen which was apparently not registered until 1988.

Reg. no. VEL.32.57b. From ; ; North America. Date: //. Source: J.J. Audubon. Material: ; skin. Not found <1986, not listed in *Cat. Bds.*.

Reg. no. VEL.32.57c. From ; ; North America. Date: //. Source: Hardwicke coll.. Material: adult M; whole/good skin. Mount in South Kensington gallery.

Reg. no. G1862.5.20.14. From ; ; [North America] Captivity. Date: //. Source: Zoological Society. Material: ; whole/good skeleton.

Reg. no. S1959.5.1. From Raddatz / Rockshelter, Sauk Co.; Wisconsin; United States. Date: //1957. Source: W. Wittrey; P. Hahn; Royal Ontario Museum. Material: ; part/good skeleton. Humerus, ulna and carpo-metacarpus only. Sub-fossil; recovered from midden debris.

Reg. no. . From Chatham; Georgia (GA); United States. Date: 14/ 5/1887. Source: G. Fairchild; Arderna coll.; C.H. Gowland; Lt.Col. R.F. Meiklejohn. Material: 2 ; eggs. Partly incubated. Eggs not found 1982.

Reg. no. 1891.3.1.514-6. From ; ; North America. Date: //. Source: Henshaw coll.; Salvin-Godman. Material: 3 ; whole/good eggs.

Reg. no. E1901.1.1.769-70. From ; ; Canada. Date: //. Source: Gale; Stevens; Hargitt coll.; Seebohm coll.. Material: 2 ; whole/good eggs.

Reg. no. E1941.3.1.513a-c. From ; ; [North America]. Date: //. Source: Rothschild. Material: 3 ; whole/good eggs.

Reg. no. E1941.9.4.1182. From ; ; North America. Date: //. Source: John Krider coll.; Buckley coll.. Material: 4 ; whole/good eggs.

Reg. no. E1941.9.4.1184. From ; ; [North America]. Date: //. Source: M.P. Kennard; Buckley coll.. Material: 5 ; whole/good eggs. Taken out of birds in the market and received by Buckley on 2 May 1871.

Reg. no. E1961.15.105. From ; ; United States. Date: / 6/1880. Source: F.F. Freeth. Material: 2 ; whole/good eggs.

Reg. no. . From ; Texas; United States. Date: //. Source: H.E. Dresser; Seebohm coll.. Material: 1 ; egg. Egg not found. Listed in Seebohm ms cat., and in Oates' *Cat. Eggs*, but apparently not found when the eggs of this species were revised in the 1960s.

Reg. no. . From ; Michigan; United States. Date: 12/ 6/1883. Source: F. Wood-Jones coll.; Stevens'; Meiklejohn. Material: 1 ; egg. Egg not found <1982. The nest was a scanty collection of sticks and twigs placed high up in a maple.

Reg. no. . From Lake City; ; United States. Date: 13/ 5/1881. Source: J. Sharpe; R.F. Meiklejohn. Material: 1 ; egg. Egg not found <1982.

Reg. no. 1852.3.20.38. From ; ; Captivity. Date: //. Source: H.F. Walter. Material: 1 ; whole/good egg. Laid in captivity in England.

Reg. no. 1885.4.20.3094. From East Windsor, Hill; Connecticut; United States. Date: //. Source: W. Wood; Smithsonian; Salvin-Godman. Material: 1 ; whole/good egg.

Reg. no. 1885.4.20.3095. From Rockford; Illinois; United States. Date: //. Source: Blackman; Smithsonian; Salvin-Godman. Material: 1 ; whole/good egg.

Reg. no. 1901.11.20.90. From ; ; Captivity. Date: //. Source: Tristram; Crowley. Material: 1 ; whole/good egg. Laid in captivity at London Zoo.

Reg. no. E1941.2.2.7. From Musquash; New Brunswick; Canada. Date: //1859. Source: Material: 1 ; whole/good egg. Oviduct egg from shot bird, taken in summer 1859.

Reg. no. E1941.2.2.8. From Sillson; Massachussetts; United States. Date: //. Source: Dr T.M. Brewer; A. Newton; Rothschild coll.. Material: 1 ; whole/good egg.

Reg. no. E1941.9.4.1183. From ; Illinois; United States. Date: //. Source: M.M. Miller; E.A. Samuels coll.; Buckley coll.. Material: 1 ; whole/good egg.

Reg. no. E1941.9.4.2564. From ; Michigan; United States. Date: //1850. Source: C.M. Phelps; Rothschild. Material: 1 ; whole/good egg. Bought at a sale on 15 May 1917.

Reg. no. E1941.9.4.2565. From ; ; . Date: 29/ 8/1868. Source: Rothschild. Material: 1 ; whole/good egg.

Reg. no. E1941.9.4.2566. From ; ; . Date: //. Source: Rothschild. Material: 1 ; whole/good egg.

Reg. no. E1941.9.4.2567. From ; Texas; United States. Date: 15/ 6/1890. Source: Rothschild. Material: 1 ; whole/good egg. Taken out of a dead bird.

Reg. no. E1941.9.4.2568. From ; ; . Date: //. Source: Rothschild. Material: 1 ; whole/good egg.

Reg. no. E1960.6.52. From ; ; . Date: //. Source: W.J. Horley; Stevens. Material: 1 ; whole/good egg. Bought at Stevens' sale on 15 May 1901.

Reg. no. E1973.1.1009. From ; ; North America. Date: //. Source: Jourdain coll.. Material: 1 ; whole/good egg. Found in a box of Jourdain 'duplicates' and identified only by pasted on paper label.

Reg. no. E1977.13.1. From ; ; United States. Date: //1879. Source: R.J. Connor. Material: 1 ; whole/good egg.

Reg. no. E1977.13.2. From ; ; North America. Date: //. Source: R.J. Connor. Material: 1 ; whole/good egg.

Zenaida macroura graysoni

Socorro Dove

Formerly endemic to Socorro Island in the Revillagigedos group. It is presumed to have become extinct on the island between 1958 and 1978, but still survives in captivity. Extinction is believed to have been the result of predation by feral cats introduced in 1957.

Reg. no. 1927.11.14.2. From Socorro Island (Soccoco Island); ; Mexico. Date: 21/ 6/1927. Source: H. Bright; Mus. Derbianum, Liverpool. Material: adult F; whole/good skin.

Leptotila wellsi

Grenada Dove

Restricted to scrubland in the southeastern corner of Grenada in the West Indies, where it is now very rare. In 1989/1990 there were 75–85 birds, a decrease from about 100 in 1987.

Reg. no. 1898.9.20.1. From ; ; Tobago. Date: //. Source: O.E. Creswell. Material: adult M; whole/good skeleton.

Reg. no. 1905.1.20.1–3. From ; ; Captivity. Date: 1/ 3/1905. Source: A.G. Butler. Material: 3 ; whole/good eggs. Laid in captivity in Tobago.

Reg. no. E1941.3.1.625. From ; ; Captivity. Date: / 4/1905. Source: A.G. Butler; H. Munt. Material: 2 ; whole/good eggs. Laid in captivity at Beckenham, Kent.

Leptotila ochraceiventris

Ochre-bellied Dove

Confined to a very restricted area of southwestern Ecuador and northern Peru where it inhabits deciduous and moist evergreen forest. It is much at risk from habitat destruction.

Reg. no. 1926.6.13.1. From Zaruma; del Oro; Ecuador. Date: 28/ 9/1913. Source: W.B. Richardson; American Museum of Natural History. Material: adult M; whole/good skin.

Caloenas nicobarica pelewensis

Palau Pigeon

A race of Nicobar Pigeon confined to the Palau Islands. It inhabits forest on four of the islands in the group. The Palau Pigeon has been very rare since it was first discovered in the 1870s and it was not recorded at all between 1880 and 1945.

Reg. no. 1878.10.29.36. From Peleliu; Palau Islands; Caroline Islands. Date: //. Source: Cohen. Material: adult ; whole/good skin.

Gallicolumba ferruginea

Tanna Ground-Dove

Known only from a painting by Georg Forster, and the description by J.R. Forster of a specimen (no longer extant) collected on Tanna in 1774 during Cook's voyages. No one has seen the dove since, and it must be presumed extinct.

Reg. no. . From ; Tanna; New Hebrides. Date: 17/ 8/1774. Source: G. Forster. Material: F; painting. A drawing in the NHM library, by Georg Forster, (old) no. 142. See *Bull. Br. Mus. nat. Hist. (Hist.)* 1 (1959): 305.

Microgoura meeki

Choiseul Pigeon

Choiseul, Solomon Islands. Known only from 6 skins and an egg collected by A.S. Meek in 1904. The species is now presumed extinct, although it may have survived as late as 1965.

Reg. no. 1939.12.9.40. From Choiseul; ; Solomon Islands. Date: 5/ 1/1904. Source: A.S. Meek; Rothschild coll.. Material: M; whole/good skin.

Reg. no. E1941.9.4.2469. From Choiseul; ; Solomon Islands. Date: 10/ 1/1904. Source: A.S. Meek; Rothschild. Material: 1 ; whole/good egg.

Ptilinopus mercierii tristrami

Hivaoa Fruit-Dove

Confined to Hivaoa in the Marquesas, and known only from a few specimens collected in the 19th century. It has been extinct since 1922.

Reg. no. 1925.12.22.174. From ; Hiva Oa; Marquesas Islands. Date: 31/12/1924. Source: H.J. Kelsall, St. George Expedition. Material: adult F; whole/good skin. Irides dull yellow (raw umber), bill olive-green horn, legs and feet purplish-grey.

Reg. no. 1928.10.27.42. From ; Hiva Oa; Marquesas Islands. Date: 6/12/1922. Source: R.H. Beck; American Museum of Natural History. Material: adult M; whole/good skin. Irides yellow, bill greenish, feet purplish.

Reg. no. 1928.10.27.43. From ; Hiva Oa; Marquesas Islands. Date: 5/11/1922. Source: E.H. Quayle; American Museum of Natural History. Material: juv F; whole/good skin. Iris light olive-brown, orbital ring drab, bill dull green tinged brown, feet dull dark purple.

Ducula galeata

Marquesan Imperial-Pigeon

Known only from Nukuhiva in the Marquesas, where it is now restricted to valleys at the western end of the island. About 400 birds were thought to survive in 1975.

Reg. no. 1861.12.10.2. From ; ; Marquesas Islands. Date: //. Source: Verreaux. Material: M; whole/poor skin. Ex-mount.

Reg. no. 1928.10.27.137. From ; Nukuhiva; Marquesas Islands. Date: //. Source: E.H. Quayle; American Museum. Material: adult F; whole/good skin. The reg. no. is untraced, but appears to be incorrect, as 1928.10.27.- only runs to 91 specimens. See 1928.10.27.17 below, but collector's number does not match this specimen.

Reg. no. 1928.10.27.17. From ; Nukuhiva; Marquesas Islands (Society Islands). Date: 11/ 6/1902. Source: E.H. Quayle; American Museum. Material: F; skin. Details taken from the register. Not found 1988. See 1928.10.27.137 above, although collector's number does not match that specimen.

Reg. no. 1928.10.27.18. From ; Nukuhiva; Marquesas Islands. Date: 17/10/1922. Source: E.H. Quayle; American Museum. Material: M; whole/good skin. Iris white, bill black, feet cherry-red.

Reg. no. 1938.11.5.1. From Hakapa Valley; Nukuhiva; Marquesas Islands. Date: //. Source: A.D. Sanson; A. Seligman. Material: adult ; whole/good skin.

Reg. no. S1975.9.5. From Tovii; Nukuhiva; Marquesas Islands. Date: 22/ 8/1972. Source: D.T. Holyoak. Material: adult F; whole/good skeleton.

Hemiphaga novaeseelandiae spadicea

Norfolk Island Pigeon

Almost nothing is known about this pigeon which became extinct in the 19th century, probably between 1833 and 1848. The date of extinction has sometimes been given incorrectly as 1801.

Reg. no. VEL.32.9a. From ; ; [Norfolk Island]. Date: //. Source: Material: ; whole/good skin. Mount. From the Old Collection. The locality is given as New Holland in the register.

Reg. no. VEL.32.9b. From ; ; Norfolk Island (New Holland). Date: //. Source: Lady Carrington. Material: ; skin. Details taken from the register. Not included in the *Cat. Bds.*, and not found 1988.

Hemiphaga novaeseelandiae chathamensis Chatham Island Pigeon

Now restricted to the two largest of the Chatham Islands where it is very rare. The woodlands in which it lives continue to deteriorate. In 1977 the population was believed to be below 25 birds.

Reg. no. 1892.10.22.1. From ; ; Chatham Islands. Date: 31/ 1/1892. Source: H.O. Forbes. Material: adult M; whole/good skin. Syntype of *Carpophaga chathamica*. Forbes, *Nature, Lond.* 46 (1892): 252–3. There is another syntype in the collection.

Reg. no. 1892.10.22.2. From ; ; Chatham Islands. Date: //. Source: H.O. Forbes. Material: adult ; whole/good skin. Syntype of *Carpophaga chathamica*. A humerus from this specimen is in the skeleton collection, S1973.76.1.

Reg. no. 1894.4.18.1. From ; ; Chatham Islands. Date: //. Source: H.O. Forbes. Material: ; whole/good skin.

Reg. no. 1894.4.18.2. From ; ; Chatham Islands. Date: //. Source: H.O. Forbes. Material: ; whole/good skin.

Reg. no. 1894.4.18.3. From ; ; Chatham Islands. Date: //. Source: H.O. Forbes. Material: ; whole/good skin.

Reg. no. 1894.4.18.4. From ; ; Chatham Islands. Date: //. Source: H.O. Forbes. Material: ; whole/good skin.

Reg. no. 1894.4.18.5. From ; ; Chatham Islands. Date: //. Source: H.O. Forbes. Material: adult ; whole/good skin.

Reg. no. 1939.12.9.3738. From Wellington; ; New Zealand. Date: //. Source: Rothschild coll.. Material: F; whole/good skin. Ex-mount. Partial albino. The Museum label says '*Hemiphaga chathamensis*'. Because of the nature of the plumage of this specimen, the subspecific identity may not be correct. If this is *chathamensis*, the locality is presumably erroneous.

Reg. no. 1939.12.9.39. From Chatham Island; ; Chatham Islands. Date: //. Source: Rothschild coll.. Material: M; whole/good skin.

Reg. no. S1973.76.1. From ; ; Chatham Islands. Date: //. Source: H.O. Forbes. Material: ; part/good skeleton. Humerus only. Taken from a skin in the collection, 1892.10.22.2.

Lorius tibialis Blue–thighed Lory

Known only from the type purchased in Calcutta Market and of unknown locality. Some authorities believe it to be an aberrant specimen of *L. domicellus*.

Reg. no. 1891.4.1.5. From ; [Moluccas]; [Indonesia], Captivity. Date: 10/10/1875. Source: W. Jamrach; Zoological Society. Material: adult F; whole/good skin. Holotype. P.L. Sclater, *Proc. zool. Soc.*

Lond. 1871: 499, pl. 40. Brought home by Jamrach from the markets of Calcutta; 'doubtless ... from one of the less explored islands of the Molucca group'. Described from life; died at London Zoo.

Nestor productus

Norfolk Island Kaka

Known only from Norfolk Island and nearby Philip Island. It became extinct in the 19th century, but the exact date is unknown. The last bird is said to have died in a cage in London some time after 1851.

Reg. no. . From ; ; Philip Island. Date: //. Source: Material: ; whole/good skin. Mount in South Kensington gallery.

Reg. no. 1837.9.26.12. From ; ; Philip Island. Date: //. Source: G. Bennett; Mus. Austral.. Material: ; whole/good skin. From the Old Collection. Previously registered as VEL.25.282b. The number 1837.9.26.12 is untraced in the General registers, but is listed in Gray's vellum volume for 1837–8.

Reg. no. 1955.6.N.19.3. From ; ; Philip Island. Date: //. Source: W. Wright. Material: ; whole/good skin.

Reg. no. VEL.25.282a. From ; ; Philip Island. Date: //. Source: Bell; Hardwicke coll.. Material: ; whole/good skin. Mount. From the Old Collection.

Psephotus pulcherrimus

Paradise Parrot

Occurred in a restricted area of southeastern Queensland and northern New South Wales, but was last reliably reported in 1927 and may be extinct in the wild. Hybrids between *P. chrysopterygius* and *P. varius*, which occur in captivity, are superficially very similar to *P. pulcherrimus* and have caused confusion.

Reg. no. . From ; ; Australia. Date: //. Source: T. Mitchell. Material: adult ; skin. Mount. Details taken from *Cat. Bds.*. Not found 1988.

Reg. no. 1846.10.5.24. From ; New South Wales; Australia. Date: //. Source: Leadbeater. Material: ; whole/good skin.

Reg. no. 1847.8.14.65. From ; New South Wales; Australia. Date: //. Source: Sir T.L. Mitchell. Material: M; whole/good skin.

Reg. no. 1847.8.14.66. From ; ; Australia. Date: 13/ 5/1846. Source: Sir T.L. Mitchell. Material: imm M; whole/good skin.

Reg. no. 1847.8.14.67. From ; ; Australia. Date: //. Source: Sir T.L. Mitchell. Material: F; whole/good skin.

Reg. no. 1858.12.2.6. From ; ; [Australia]. Date: //. Source: Gould. Material: ; whole/good skin. Hybrid.

Reg. no. 1878.7.30.2. From ; ; Australia. Date: //. Source: A. Wiener. Material: ; whole/good skin. Ex–mount, abnormal plumage. One foot damaged.

Reg. no. 1881.12.7. From ; Queensland; e. Australia. Date: //. Source: J.T. Cockerell; Godman coll.. Material: M; whole/good skin.

Reg. no. 1881.12.7.22. From ; ; e. Australia. Date: //. Source: J.T. Cockerell; Godman coll.. Material: F; whole/good skin.

Reg. no. 1881.5.1.4807. From ; ; Australia. Date: //. Source: Gould. Material: ; whole/good skin.

Reg. no. 1881.5.1.4808. From ; ; Australia. Date: //. Source: Gould. Material: ; whole/good skin.

Reg. no. 1889.1.20.40. From ; Queensland; Australia. Date: //. Source: Tweeddale coll.. Material: ; whole/good skin.

Reg. no. 1927.12.18.110. From Mount Hutton, West Grove; Queensland; Australia. Date: 3/ 7/1890. Source: J.J. Lister. Material: M; whole/good skin. Hybrid: *P. haematonotus* x *P. pulcherrimus* according to label.

Reg. no. 1938.4.8.1. From ; ; [Australia]. Date: //1850. Source: Miss Fitzgibbon. Material: ; whole/good skin. Ex-mount. Said to have been collected about 1850–1860.

Reg. no. G1868.4.15.4. From ; ; [Australia]. Date: 29/ 5/1867. Source: Zoological Society. Material: M; part/good skeleton. Sternum only.

Cyanoramphus novaezelandiae subflavescens

Lord Howe Island Parakeet

Formerly endemic to Lord Howe Island. The race became extinct in 1869, partly as a result of shooting by islanders because of its damage to crops.

Reg. no. 1851.5.1.4812. From ; ; Lord Howe Island. Date: 7/ 9/1853. Source: J. MacGillivray, voyage of H.M.S. Herald; Gould coll.. Material: adult F; whole/good skin. Syntype.

Reg. no. 1881.5.1.4813. From ; ; Lord Howe Island. Date: 7/ 9/1853. Source: J. MacGillivray, voyage of H.M.S. Herald; Gould coll.. Material: adult M; whole/good skin. Syntype. Salvadori, *Ann. Mag. nat. Hist.* (6) 7 (1891): 68. Locality given as New Zealand in the register. The female syntype is also in the collection.

Cyanoramphus novaezelandiae cookii

Norfolk Island Parakeet

Endemic to Norfolk Island, where it was once very common. Since 1908 it has been restricted to the small area of surviving indigenous forest. The race had declined to about 20 birds in 1969. It was shot in the 19th century because of damage to orchards and crops, and it is presently believed to be in competition for nest sites with the introduced Rosella *Platycercus elegans*. *Cooki* may warrant treatment as a full species.

Reg. no. 1859.1.10.5. From ; ; Norfolk Island. Date: / 6/1856. Source: T.M. Rayner, voyage of H.M.S. Herald; the Admiralty. Material: adult ; whole/good skin. Holotype of *Platycercus rayneri*. G.R. Gray, *Ibis* 1862: 228.

Reg. no. VEL.22.30a. From ; ; [Norfolk Island] (New Zealand). Date: //. Source:

Bullock coll.. Material: adult ; whole/good skin. Holotype. G.R. Gray, 1859, *List of the Specimens of Birds in the Collection of the British Museum*, part 3, section 2: 13. Previously said to be from New Zealand in error. Locality given as Pacific Islands in the register.

Reg. no. VEL.22.30b. From ; ; [Norfolk Island] (New Zealand). Date: //. Source: Bullock coll.. Material: ; skin. Details taken from the register, where the locality is given as the Pacific Islands. Not found 1988.

Reg. no. 1926.8.27.1. From ; ; Captivity. Date: 30/ 8/1926. Source: Lord Tavistock. Material: imm F; whole/good alc. Dissected specimen.

Cyanoramphus novaezelandiae erythrotis
Macquarie Island Parakeet
Formerly endemic to Macquarie Island where it was exterminated around 1914, apparently by feral cats.

Reg. no. 1855.12.19.317. From ; ; . Date: //. Source: Zoological Society. Material: ; whole/good skin.

Reg. no. 1859.11.22.18. From ; ; Macquarie Island. Date: //. Source: Massena coll.; Parzudaki. Material: ; whole/good skin.

Reg. no. 1859.11.22.19. From ; ; Macquarie Island. Date: //. Source: Massena coll.; Parzudaki. Material: ; whole/good skin.

Cyanoramphus auriceps forbesi
Forbes's Parakeet
Chatham Islands; formerly occurred on Mangere and Little Mangere but, following deforestation of Mangere, was restricted to the smaller island. Following revegetation the species began to recolonise Mangere but hybridization with *C. novaezelandiae chathamensis* has taken place.

Reg. no. 1893.6.24.7. From ; ; Chatham Islands. Date: //. Source: H.O. Forbes. Material: ; whole/good skin.

Cyanoramphus malherbi
Malherbe's Parakeet
Remote areas of South Island, New Zealand, where it was apparently always rare. The taxonomic status of this species has been questioned, although its recent treatment as a colour morph of *C. auriceps* seems to be based on weak evidence.

Reg. no. 1872.11.22.1. From ; Otago; New Zealand. Date: //. Source: W.L. Buller. Material: ; whole/good skin.

Reg. no. 1872.11.22.2. From ; Otago; New Zealand. Date: //. Source: W.L. Buller. Material: ; whole/good skin.

Reg. no. 1897.12.6.2. From ; Otago; New Zealand. Date: / 8/1887. Source: Otago University Museum. Material: F; whole/good skin. Year of collection is illegible on the label.

Reg. no. 1927.12.18.106. From w. coast sounds; ; New Zealand. Date: / 1/1889. Source: J.J. Lister coll.. Material: F; whole/good skin.

Reg. no. 1945.54.2. From ; ; New Zealand. Date: //. Source: W.J.P. Wardlaw. Material: F; whole/good skin. Eye yellow.

Reg. no. 1953.33.3. From Akaro, 120 miles from Dunedin; South Island; New Zealand. Date: 9/ 3/1872. Source: C.F. Tebbutt. Material: M; whole/good skin. Ex-mount.

Cyanoramphus zealandicus

Black-fronted Parakeet

Known only from Tahiti where it became extinct in the 19th century. The last known specimen was taken in 1844.

Reg. no. 1859.11.22.20. From ; Tahiti; Society Islands. Date: //. Source: Massena coll.; Parzudaki. Material: ; whole/good skin.

Cyanoramphus ulietanus

Raiatea Parakeet

Raiatea, Society Islands. Known only from two specimens thought to have been collected on Cook's voyage in 1773 or 1774. The type is in Vienna; the origin of the NHM specimen is uncertain, having been obtained from the Bullock Collection (not from the Massena Collection, *contra* Greenway (1967)).

Reg. no. VEL.22.31a. From ; [Raiatea]; Society Islands. Date: //. Source: Bullock coll.. Material: ; whole/good skin. The register gives the locality as New Holland, which was crossed out by G.R. Gray, and Tanna inserted.

Eunymphicus cornutus uvaeensis

Ouvea Parakeet

Restricted to Uvea (Ouvea) in the Loyalty Islands, where the population is now very low as a result of continuing deforestation. In 1974 there were about 200 individuals. Attempts to introduce birds to nearby islands have failed.

Reg. no. 1891.4.1.9. From ; Uvea; Loyalty Islands. Date: 13/ 6/1882. Source: Zoological Society. Material: adult F; whole/good skin. Received at the Zoo on 27 April 1882. Pupils black with orange ring, beak black, top of upper mandible bluish.

Reg. no. 1894.7.1.1. From ; ; Loyalty Islands. Date: //. Source: H.H. Sharland. Material: ; whole/good skin.

Reg. no. 1898.12.2.150. From ; Uvea; Loyalty Islands, Captivity. Date: 20/ 7/1880. Source: E.L. Layard; Seebohm coll.. Material: M; whole/good skin. Syntype. There are two other syntypes in the Merseyside County Museums, Liverpool. See Wagstaffe, 1978, *Type Specimens of Birds in the Merseyside County Museums*, p. 10.

Pezoporus wallicus flaviventris

Western Ground Parrot

Known only from a very limited area on the south coast of Western Australia although formerly more extensive. The race declined towards the end of the 19th century and by 1913 it was erroneously thought to be extinct.

Reg. no. 1888.6.18.17. From Swan River; Western Australia (s.w. Australia); Australia. Date: //. Source: R.B. Sharpe. Material: adult ; whole/good skin.

Lophopsittacus mauritianus

Broad-billed Parrot

Formerly occurred on Mauritius, but became extinct sometime after 1683. Known from bones and travellers' accounts.

Reg. no. . From ; ; Mauritius. Date: //. Source: Material: ; part/poor skeleton. Cranium only. See Cowles, in Diamond (ed.), 1987, *Studies of Mascarene Island Birds*, p. 97–8.

Coracopsis nigra barklyi

Seychelles Black Parrot

Now known mainly from the Vallee de Mai on Praslin in the Seychelles, though it was more widespread on the island during the last century. The population is low, but it is rigorously protected and considered to be relatively safe.9

Reg. no. 1890.10.10.33. From ; ; Seychelles. Date: //. Source: Shelley coll.. Material: adult M; whole/good skin. Listed in the register as 1890.10.10.33 and on the label as 1890.10.10.34.

Reg. no. 1890.10.10.34. From ; ; Seychelles. Date: //. Source: Shelley coll.. Material: adult F; whole/good skin. Listed in the register as 1890.10.10.34 and on the label as 1890.10.10.35.

Reg. no. 1890.10.10.35. From ; ; Seychelles. Date: //1879. Source: Tristram coll.; Shelley coll.. Material: adult ; whole/good skin. Listed in the register as 1890.10.10.35 and on the label as 1890.10.10.36.

Reg. no. 1870.4.19.1. From ; Praslin Island; Seychelles, Captivity. Date: //1867. Source: S. Ward; Zoological Society; E. Newton. Material: adult M; whole/good skin. Syntype. Newton, *Proc. zool. Soc. Lond.* 1867: 346, pl. 22. Sent home alive by Ward; died in captivity at London Zoo. From January or February 1867. Described from life. The other three syntypes are in the University Museum of Zoology, at Cambridge, England.

Eight further skins are listed on page 278.

Reg. no. E1962.1.640. From ; ; Captivity. Date: //. Source: Munt coll.. Material: 1 ; whole/good egg. Laid in captivity.

Loriculus philippensis chrysonotus

Cebu Hanging-Parrot

Formerly endemic to Cebu in the Philippines. Like nearly all the endemic Cebu forms it became extinct when the native forest was totally removed.

Reg. no. 1889.1.20.239. From ; Cebu; Philippines. Date: / 4/1872. Source: A.B. Meyer; Tweeddale coll.. Material: adult F; whole/good skin.

Reg. no. 1889.1.20.240. From ; Cebu; Philippines. Date: / 4/1872. Source: A.B. Meyer; Tweeddale coll.. Material: adult F; whole/good skin.

Reg. no. 1889.1.20.241. From ; Cebu; Philippines. Date: / 4/1872. Source: A.B. Meyer; Tweeddale coll.. Material: adult M; whole/good skin.

Reg. no. 1889.1.20.242. From ; Cebu; Philippines. Date: / 4/1877. Source: A.H. Everett; Tweeddale coll.. Material: adult M; whole/good skin.

Reg. no. 1896.6.6.1053. From ; Cebu; Philippines. Date: / 3/1888. Source: Steere coll.. Material: M; whole/good skin.

Reg. no. 1896.6.6.1054. From Carmen; Cebu; Philippines. Date: 19/ 3/1888. Source: F.S. Bourns; Steere coll.. Material: F; whole/good skin.

Reg. no. 1965.11.1. From ; Cebu; Philippines, Captivity. Date: //1871. Source: Zoological Society; E. Bartlett coll.. Material: adult M; whole/good skin. Syntype. Ex-mount. P.L. Sclater, *Ibis* 1872: 323–5, pl. 11. Acquired live by London Zoo from a dealer in Liverpool in March or April 1871. Died in captivity.

Loriculus philippensis siquijorensis

Siquijor Hanging-Parrot

Formerly occurred in coconut groves, fruit trees and bamboo clumps on the island of Siquijor in the Philippines. Believed extinct since 1908.

Reg. no. 1896.6.6.1055. From ; Siquijor; Philippines. Date: 14/ 2/1888. Source: J.B. Steere, Steere Expedition. Material: adult M; whole/good skin. Syntype. Steere, 1890, *A List of the Birds and Mammals collected by the Steere Expedition to the Philippines* ..., p. 6; see also *Ibis* 1891: 304.

Reg. no. 1896.6.6.1056. From ; Siquijor; Philippines. Date: 14/ 2/1888. Source: J.B. Steere, Steere Expedition. Material: adult F; whole/good skin. Syntype. Details taken from the register. Not found 1988.

Loriculus flosculus

Wallace's Hanging-Parrot

For a long time known only from the type specimen, collected on Flores Island, Indonesia, and three eggs. The species was rediscovered in 1978.

Reg. no. 1873.5.12.1555. From ; Flores Island; Indonesia. Date: //1862. Source: A.R. Wallace. Material: adult ; whole/good skin. Holotype. Wallace, *Proc. zool. Soc. Lond.* 1863 (publ. 1864): 488. Only known skin.

Loriculus salvadorii

Salvadori's Hanging-Parrot

Mindanao, Philippines. Known only from two specimens: the type collected in 1929, a male, formerly in Hachisuka's collection in Japan; and a female in the NHM collected between 1836 and 1839. This form may be an aberration of *L. philippensis apicalis*.

Reg. no. 1842.2.15.89. From ; Mindanao; Philippines. Date: //. Source: Cuming. Material: adult F; whole/good skin. Cuming was apparently active in the area in 1836–9.

Psittacula wardi

Seychelles Parakeet

Formerly occurred on Mahe and Silhouette Islands in the Seychelles, but became extinct sometime between 1870 and 1906.

Reg. no. 1880.6.3.1. From ; Mahe Island; Seychelles. Date: / 3/1880. Source: H.L. Warry; Tristram. Material: adult F; whole/good skin.

Reg. no. 1890.10.10.5. From ; ; Seychelles. Date: / 5/1880. Source: H.M. Warry; Canon Tristram; Shelley coll.. Material: ; whole/good skin. Month of collection from *Cat. Bds.*.

Reg. no. 1890.10.10.6. From ; ; Seychelles. Date: / 5/1881. Source: H.M. Warry; Canon Tristram; Shelley coll.. Material: ; whole/good skin.

Psittacula eques echo

Mauritius Parakeet

Now survives only in small numbers in the mountain forests of southwestern Mauritius. Only about 10 birds remain and they suffer from almost total breeding failure.

Reg. no. 1859.11.22.42. From ; ; Mauritius. Date: //. Source: Parzudaki. Material: adult ; whole/good skin. Listed in the register as *Palaeornis borbonica*.

Reg. no. 1861.12.10.16. From ; ; Mauritius (Maurice Is). Date: //. Source: Verreaux. Material: adult F; whole/good skin.

Reg. no. 1890.10.10.7. From ; ; Mauritius. Date: //. Source: Bewsher; Shelley coll.. Material: adult M; whole/good skin.

Reg. no. 1890.10.10.8. From ; ; Mauritius. Date: //. Source: Bewsher; Shelley coll.. Material: adult F; whole/good skin.

Reg. no. A1976.18.1. From ; ; Mauritius. Date: / 9/1976. Source: J.E. Cooper. Material: adult ; whole/good alc. The specimen was obtained through an ICBP project.

Psittacula intermedia

Intermediate Parakeet

Formerly known from 7 skins collected somewhere in the Indian region. It was presumed extinct but recently some specimens (reputedly from the plains of Uttar Pradesh) have appeared in the bird markets of India.

Reg. no. 1980.3.1. From ; ; India. Date: //. Source: Rothschild coll.. Material: adult ; whole/good skin. On extended loan from the American Museum of Natural History, which has all the other specimens of this species (ex Rothschild). See *J. Bombay nat. Hist. Soc.* 82 (1985): 197–9

Anodorhynchus glaucus

Glaucous Macaw

Formerly occurred in central South America (Paraguay and adjacent regions). It is now believed to be extinct, as none have been recorded for over 40 years.

Reg. no. 1859.11.22.1. From ; ; Paraguay. Date: //. Source: Massena coll.; Parzudaki. Material: adult ; whole/good skin.

Reg. no. 1889.1.30.2. From ; ; Paraguay. Date: //1883. Source: Boucard; Salvin–

Godman coll.. Material: adult M; whole/good skin.

Anodorhynchus leari

Lear's Macaw

Formerly known only from captive specimens. Its natural drylands habitat in a restricted area of Bahia, eastern Brazil, was discovered only in 1978. The total population is thought to be 60–100 birds in two groups.

Reg. no. 1890.6.1.1. From ; ; Brazil, Captivity. Date: 11/ 1/1883. Source: P.L. Sclater; Zoological Society. Material: adult M; whole/good skin.

Reg. no. E1962.1.628. From ; ; Captivity. Date: //. Source: Munt coll.. Material: 1 ; whole/good egg. Laid in captivity.

Cyanopsitta spixii

Spix's Macaw

The natural habitat of this species was only located in the late 1980s. The species was restricted to very small areas of gallery woodland along seasonal creeks of interior north–eastern Brazil. Only one bird remained in the wild in 1992, although another area held birds until 1988. A few survive in captivity in various parts of the world.

Reg. no. 1859.11.22.2. From ; ; Brazil. Date: //. Source: Massena coll.; Parzudaki. Material: adult ; whole/good skin.

Reg. no. 1890.6.1.2. From ; ; n. Brazil, Captivity. Date: 22/ 1/1884. Source: Zoological Society; Sclater coll.. Material: adult F; whole/good skin. Received at the Zoo 15 November 1878, died 22 January 1884. Eyes light yellow.

Reg. no. S1972.1.79. From ; ; [n. Brazil, Captivity]. Date: 8/ 8/1902. Source: Zoological Society; Rothschild coll.. Material: M; whole/good skeleton. Received at the Zoo 24 June 1901.

Ara caninde/A. glaucogularis

Blue–throated Macaw

Occurs only in a limited area of Bolivia where it numbers between 500 and 1000 birds. The species is still heavily exploited for the cage–bird trade. Ingels, Parkes & Farrand (1981, *Le Gerfaut* 71: 283–294) point out that the name *caninde* is a synonym of *A. ararauna*, and the former should be known as *A. glaucogularis*.

Reg. no. 1850.8.14.21. From River Amazon; ; [South America]. Date: //. Source: Stevens. Material: adult ; whole/good skin. Measurements given in Forshaw & Cooper, 1973, *Parrots of the World*, p. 360. See also *Le Gerfaut* 71 (1981): 283–294).

Reg. no. 1857.3.8.1. From ; ; Brazil. Date: //. Source: Warwick. Material: adult ; whole/good skin. From the interior of Brazil. Details taken from *Cat. Bds.* and *Le Gerfaut* 71 (1981): 283–294. Not found 1988, although apparently present to 1970s. Measurements given in Forshaw & Cooper, 1973, *Parrots of the World*, p. 360.

Ara tricolor/A. cubensis

Cuban Macaw

Formerly occurred on Cuba, the Isle of Pines and Hispaniola. The last specimen was collected in 1864 but it is believed that a few birds survived until 1885. Macaws formerly occurred on a number of other West Indian islands, but no specimens of these exist. Although at least some of them may represent valid species, they are known only from traveller's accounts, in some cases very old. It has been suggested that the name *A. tricolor* be restricted to birds from Hispaniola, last reported in 1820, and that birds from Cuba (and possibly the Isle of Pines) be called *A. cubensis*, last taken in 1864 (*Carib. J. Sci.* 21 (1985): 169–175).

Reg. no. 1851.7.21.5. From ; ; [Cuba]. Date: //. Source: [Lt. A. Smith]. Material: ; whole/good skin. Reg. no. untraced, but the specimen was received no later than 1859. This is probably 1851.7.24.5, '*Psittacus*', no locality, ? purchased of Lt. A. Smith.

Reg. no. 1858.5.13.1. From ; ; [Cuba]. Date: //. Source: Zoological Society. Material: ; whole/good skin. Ex-mount. Presumably a captive bird.

Ognorhynchus icterotis

Yellow-eared Parrot

Restricted to a small area of southwestern Colombia and adjacent northern Ecuador, where it has declined seriously due to habitat destruction.

Reg. no. 1854.1.25.6. From ; ; Colombia. Date: //. Source: Stevens. Material: adult ; whole/good skin.

Reg. no. 1859.11.22.62. From Ocana; ; Colombia (New Granada). Date: //. Source: Massena coll.; Parzudaki. Material: adult ; whole/good skin. Holotype. Massena & Souance, *Revue Mag. Zool.* 1854: 71.

Reg. no. 1889.1.30.55. From 'Bogota'; ; Colombia. Date: //. Source: Trubner; Salvin-Godman coll.. Material: adult ; whole/good skin.

Reg. no. 1889.1.30.56. From Intaj; ; Ecuador. Date: //. Source: C. Buckley; Salvin-Godman coll.. Material: adult ; whole/good skin. Received by Salvin & Godman in December 1877.

Reg. no. 1890.6.1.22. From 'Bogota'; ; Colombia. Date: //1868. Source: Trubner; Sclater coll.. Material: adult ; whole/good skin.

Reg. no. 1899.6.30.472. From Zitaguira, e. slope of e. Cordillera; ; Colombia. Date: //. Source: T.H. Wheeler; Salvin-Godman coll.. Material: adult ; whole/good skin.

Reg. no. 1921.12.29.8. From 'Bogota'; ; Colombia. Date: //. Source: Alexander; Lawrence coll.; Amer. Mus. Nat. Hist.; W.F.H. Rosenberg. Material: adult ; whole/good skin.

Conuropsis carolinensis carolinensis

Carolina Parakeet

The species formerly occurred widely over North America east of the Great Plains. It was persecuted by man as a pest of fruit orchards and declined drastically in the 19th century. The last definite record was in 1914, but there were unconfirmed reports of birds as late as 1938. This race was found mainly in the southeastern states, from Virginia to Florida.

Reg. no. 1890.6.1.43. From ; ; North America. Date: //. Source: Leadbeater; P.L. Sclater. Material: ; whole/good skin.

Reg. no. 1896.12.1.41. From Tarpon Springs; Florida; United States. Date: 22/ 1/1889. Source: W.E.D. Scott; Salvin–Godman coll.. Material: adult M; whole/good skin.

Reg. no. 1896.12.1.42. From Miami; Florida; United States. Date: 6/12/1871. Source: H.W. Henshaw (or C. Hart Merriam); Salvin–Godman coll.. Material: adult M; whole/good skin.

Reg. no. 1896.12.1.43. From Cedar Keys; Florida; United States. Date: 17/12/1870. Source: H.W. Henshaw (or C. Hart Merriam); Salvin–Godman coll.. Material: imm F; whole/good skin.

Reg. no. 1896.12.1.44. From Lake Okeechobee; Florida; United States. Date: / 4/1873. Source: C. Hart Merriam; Salvin–Godman coll.. Material: adult ; whole/good skin.

Reg. no. 1896.12.1.45. From Lake Okeechobee; Florida; United States. Date: 6/11/1888. Source: W. Hoxie; C. Hart Merriam; Salvin–Godman coll.. Material: adult F; whole/good skin.

Reg. no. 1896.12.1.46. From Lake Okeechobee; Florida; United States. Date: 6/11/1888. Source: W. Hoxie; C. Hart Merriam; Salvin–Godman coll.. Material: adult M; whole/good skin.

Reg. no. 1896.12.1.47. From Lake Okeechobee; Florida; United States. Date: 6/11/1888. Source: W. Hoxie; C. Hart Merriam; Salvin–Godman coll.. Material: F; whole/good skin.

Reg. no. 1896.12.1.48. From Fort Drum; Florida; United States. Date: 27/10/1888. Source: W. Hoxie; C. Hart Merriam; Salvin–Godman coll.. Material: adult M; whole/good skin.

Reg. no. 1896.12.1.49. From Fort Drum; Florida; United States. Date: 27/10/1888. Source: W. Hoxie; C. Hart Merriam; Salvin–Godman coll.. Material: adult M; whole/good skin.

Reg. no. 1896.12.1.50. From Fort Drum; Florida; United States. Date: 27/10/1888. Source: W. Hoxie; C. Hart Merriam; Salvin–Godman coll.. Material: adult M; whole/good skin.

Reg. no. 1896.12.1.51. From Fort Drum; Florida; United States. Date: 27/10/1888. Source: W. Hoxie; C. Hart Merriam; Salvin–Godman coll.. Material: adult M; whole/good skin.

Reg. no. 1896.12.1.52. From Fort Drum; Florida; United States. Date: 27/10/1888. Source: W. Hoxie; C. Hart Merriam; Salvin–Godman coll.. Material: adult F; whole/good skin.

Reg. no. 1896.12.1.53. From Linden; Florida; United States. Date: 15/ 2/1889.

Source: W.E.D. Scott; Salvin–Godman coll.. Material: adult M; whole/good skin.

Reg. no. 1896.12.1.54. From Linden; Florida; United States. Date: 15/ 2/1889. Source: W.E.D. Scott; Salvin–Godman coll.. Material: adult F; whole/good skin.

Reg. no. 1896.12.1.55. From Linden; Florida; United States. Date: 15/ 2/1889. Source: W.E.D. Scott; Salvin–Godman coll.. Material: adult M; whole/good skin.

Reg. no. 1896.12.1.56. From Linden; Florida; United States. Date: 15/ 2/1889. Source: W.E.D. Scott; Salvin–Godman coll.. Material: adult M; whole/good skin.

Reg. no. 1896.12.1.57. From Linden; Florida; United States. Date: 20/ 2/1889. Source: W.E.D. Scott; Salvin–Godman coll.. Material: adult F; whole/good skin.

Reg. no. 1896.12.1.58. From Linden; Florida; United States. Date: 20/ 2/1889. Source: W.E.D. Scott; Salvin–Godman coll.. Material: adult F; whole/good skin.

Reg. no. 1896.12.1.59. From Linden; Florida; United States. Date: 20/ 2/1889. Source: W.E.D. Scott; Salvin–Godman coll.. Material: adult F; whole/good skin.

Reg. no. 1896.12.1.60. From Linden; Florida; United States. Date: 20/ 2/1889. Source: W.E.D. Scott; Salvin–Godman coll.. Material: adult F; whole/good skin.

Reg. no. 1896.12.1.61. From Linden; Florida; United States. Date: 15/ 2/1889. Source: W.E.D. Scott; Salvin–Godman coll.. Material: adult F; whole/good skin.

Reg. no. 1896.12.1.62. From Linden; Florida; United States. Date: 20/ 2/1889. Source: W.E.D. Scott; Salvin–Godman coll.. Material: adult M; whole/good skin.

Reg. no. 1896.12.1.63. From Linden; Florida; United States. Date: 20/ 2/1889. Source: W.E.D. Scott; Salvin–Godman coll.. Material: adult F; whole/good skin.

Reg. no. 1896.12.1.64. From Linden; Florida; United States. Date: 15/ 2/1889. Source: W.E.D. Scott; Salvin–Godman coll.. Material: adult F; whole/good skin.

Reg. no. 1896.12.1.65. From Enterhuse or Enterprise; Florida; United States. Date: / 2/1869. Source: G. Boardman; H.W. Henshaw; Salvin–Godman coll.. Material: juv F; whole/good skin.

Reg. no. 1906.12.7.670. From Manatee County; Florida; United States. Date: 17/ 5/1887. Source: G.S. Miller coll.. Material: adult M; whole/good skin.

Reg. no. 1925.2.16.170. From ; ; Hawaiian Islands. Date: //. Source: S.B. Wilson. Material: adult ; whole/good skin. The locality is only given on the label printed by the Museum for the Wilson collection, and not in the register. It would seem to be erroneous.

Reg. no. 1988.21.38. From ; ; [United States]. Date: //. Source: Rothschild coll.. Material: adult M; whole/good skin. Mount in Tring gallery. There may be further data under a gummed label on the base of the mount.

Reg. no. 1853.7.12.11. From ; ; [United States]. Date: //. Source: Warwick. Material: ; whole/good skeleton. Previously registered as Skeleton Vellum Cat. 944a.

Reg. no. S1955.5.49. From North Fork, Sebastian River; Florida; United States. Date: / 4/1892. Source: J.F. Delaney; J. Ronley; Rothschild coll.. Material: F; whole/good skeleton.

Conuropsis carolinensis ludovicianus
Carolina Parakeet

See above. This race was found mainly in the interior states, along the Mississippi catchment area. Extinct.

Reg. no. 1988.21.39. From ; ; [United States]. Date: //. Source: Material: adult ; whole/good skin. Mount in South Kensington gallery. Claimed to be of this subspecies, but apparently a faded specimen.

Reg. no. 1988.8.37. From ; Louisiana; United States. Date: //. Source: J.J. Audubon. Material: adult ; whole/good skin. Ex-mount. Previously registered as VEL.22.33b. Subspecific identification not certain.

Reg. no. VEL.22.33a. From ; Louisiana; United States. Date: //. Source: J.J. Audubon. Material: ; skin. Details taken from *Cat. Bds.*. Not found 1988, although listed by Hahn, 1963, *Where is that Vanished Bird?*, Toronto.

***Conuropsis carolinensis* subspecies**

Reg. no. 1901.12.15.651. From ; ; [North America]. Date: //. Source: Nehrkorn; P. Crowley. Material: 1 ; whole/good egg.

Reg. no. E1962.1.607. From ; ; Captivity. Date: //. Source: Munt coll.. Material: 1 ; whole/good egg. Laid in captivity.

Reg. no. E1972.11.194. From ; ; . Date: //. Source: Rodern coll.; Rothschild. Material: 1 ; whole/good egg.

Amazona vittata vittata
Puerto Rican Amazon

Now restricted to portions of one forest at the eastern end of Puerto Rico; formerly occurred widely through the island. By 1975 the wild population had dropped to 13 birds, but intense conservation management led to an increase to about 50 in early 1989, with further birds in captivity. Only 23 birds were found after severe storms swept the island later that year, but a good breeding season in 1992 boosted the population to 39–40. The race *gracilipes* from nearby Culebra Island became extinct in 1899.

Reg. no. 1859.11.22.13. From ; ; [Puerto Rico]. Date: //. Source: Massena coll.; Parzudaki. Material: adult ; whole/good skin. Ex-mount.

Reg. no. 1870.4.13.11. From ; ; n. side of Puerto Rico. Date: //. Source: G. Latimer; Smithsonian Institute. Material: adult ; whole/good skin. Iris yellow.

Reg. no. 1881.1.30.8. From ; ; Puerto Rico. Date: //. Source: Zoological Society.

Material: adult ; whole/good skin. Ex-mount.

Reg. no. 1889.1.30.236. From ; ; Puerto Rico. Date: //1869. Source: G. Latimer; Salvin-Godman coll.. Material: adult ; whole/good skin.

Reg. no. 1890.6.1.155. From ; ; Puerto Rico. Date: //. Source: Taylor; P.L. Sclater. Material: adult ; whole/good skin.

Reg. no. 1891.4.1.30. From ; ; Puerto Rico. Date: 3/ 8/1884. Source: Zoological Society. Material: adult M; whole/good skin. Eyes yellow. The reg. no. was previously given as 1890.4.1.30 in error.

Reg. no. 1905.6.28.426. From ; ; Puerto Rico. Date: / 5/1863. Source: Cavendish Taylor coll.. Material: ; whole/good skin.

Amazona pretrei

Red-spectacled Amazon

Inhabits a very small area of southern Brazil and northeastern Argentina. It has declined seriously as a result of forest destruction and, formerly, hunting.

Reg. no. . From Rio de la Plata; ; [South America]. Date: //. Source: Massena coll.. Material: imm ; skin. Details taken from *Cat. Bds.*. Not found 1988.

Reg. no. . From ; ; Brazil. Date: //. Source: Sclater coll.. Material: adult M; skin. Details taken from *Cat. Bds.*. Not found 1988. Sex given in *Cat. Bds.* as female, subsequently altered to male in Museum copy.

Reg. no. 1859.11.22.75. From Rio de la Plata; ; [South America]. Date: //. Source: Parzudaki. Material: adult ; whole/good skin.

Reg. no. 1930.9.1.1. From ; ; Captivity. Date: 7/ 8/1930. Source: H. Whitley. Material: adult M; whole/good skin.

Amazona brasiliensis

Red-tailed Amazon

Confined to the coastal forests of southern Brazil where it is now rare. The population was believed to be about 3,000 birds in the late 1980s but there has been a severe decline since.

Reg. no. 1890.6.1.168. From ; ; South America. Date: //1879. Source: Westerman; P.L. Sclater. Material: adult ; whole/good skin.

Reg. no. 1891.4.1.36. From ; ; Brazil, Captivity. Date: 4/10/1885. Source: Zoological Society. Material: adult M; whole/good skin. The bird was received (? at the Zoo) on 21 April 1885.

Reg. no. 1891.4.1.37. From ; ; South America, Captivity. Date: 6/ 5/1883. Source: Zoological Society. Material: adult F; whole/good skin. The bird was received (? at the Zoo) on 18 December 1879.

Reg. no. 1988.21.67. From ; ; [South America]. Date: 3/ 9/1896. Source: Mr Jamrach; Brazenor; Rothschild coll.. Material: adult M; whole/good skin. Mount in Tring gallery. Brazenor Bros. were dealers in Brighton.

Reg. no. E1941.9.4.2445. From Rio Grande; ; [Brazil]. Date: //1886. Source: Rothschild. Material: 1 ; whole/good egg.

Amazona dufresniana rhodocorytha
Red-browed Amazon

Confined to eastern Brazil, where a major decline in numbers has occurred as a result of forest clearance and hunting. Now generally considered a separate species.

Reg. no. 1852.3.8.21. From ; ; Brazil. Date: //. Source: Verreaux. Material: adult ; whole/good skin.

Reg. no. 1889.1.30.303. From ; ; Brazil. Date: //. Source: Boucard; Salvin–Godman coll.. Material: juv ; whole/good skin.

Reg. no. 1889.1.30.304. From ; ; Brazil. Date: //. Source: T.C. Eyton; Salvin–Godman coll.. Material: adult ; whole/good skin. Received by Salvin & Godman in 1881.

Reg. no. 1890.6.1.175. From ; ; Brazil. Date: //. Source: Verreaux; Sclater coll.. Material: adult ; whole/good skin.

Reg. no. 1890.6.1.176. From ; ; s.e. Brazil, Captivity. Date: 26/ 6/1881. Source: Zoological Society. Material: adult M; whole/good skin. Received at the Zoo on 18 December 1879, died 26 June 1881. The *Cat. Bds.* erroneously ascribes this skin to the Sclater coll..

Reg. no. 1891.1.23.83. From 'Bogota'; ; [Colombia, ?= Brazil]. Date: //. Source: Gerrard. Material: juv ; whole/good skin.

Reg. no. 1891.4.1.40. From ; ; s.e. Brazil, Captivity. Date: 7/11/1879. Source: Zoological Society. Material: adult F; whole/good skin. Received at the Zoo on 20 October 1879, died 7 November 1879.

Reg. no. 1869.10.19.7. From ; ; Brazil. Date: //. Source: Zoological Society. Material: F; whole/good skeleton.

Amazona vinacea
Vinaceous Amazon

Restricted to southeastern Brazil, southeastern Paraguay and northeastern Argentina, where forest destruction has greatly reduced its numbers.

Reg. no. 1889.1.30.278. From Concepcion, Misiones; ; Argentina. Date: 27/ 6/1881. Source: E.W. White; Salvin–Godman coll.. Material: adult M; whole/good skin.

Reg. no. 1889.1.30.279. From Ytarare; ; Brazil. Date: 11/ 2/1821. Source: J. Natterer; Salvin–Godman coll.. Material: adult M; whole/good skin.

Reg. no. 1889.1.30.280. From ; Bahia; Brazil. Date: //. Source: Wucherer; Salvin–Godman coll.. Material: adult ; whole/good skin.

Reg. no. 1890.6.1.166. From ; ; Brazil. Date: //. Source: Parzudaki; Sclater coll.. Material: adult ; whole/good skin.

Reg. no. 1891.4.1.34. From ; ; Brazil, Captivity. Date: 17/ 1/1883. Source: Zoological Society. Material: adult F; whole/good skin. Received at the Zoo on

30 November 1881, died 17 January 1883. Eyes red.

Reg. no. 1908.12.14.100. From Villa Rica; ; Paraguay. Date: 22/ 6/1907. Source: Lord Brabourne. Material: adult F; whole/good skin. Bill red, tip whitish, iris red, feet slaty.

Reg. no. 1908.12.14.99. From Villa Rica; ; Paraguay. Date: 22/ 6/1907. Source: Lord Brabourne. Material: adult M; whole/good skin. Bill red, tip whitish, iris red, feet slaty.

Reg. no. VEL.22.48a. From ; ; Brazil. Date: //. Source: Hardwicke coll.. Material: adult ; whole/good skin. Not listed in *Cat. Bds.*.

Amazona versicolor

St. Lucia Amazon

Found only on the island of St. Lucia, where it is now restricted to the central mountain forests. Its population was estimated to be 100 in 1977 but, as a result of protection, had risen to 300–350 in 1990.

Reg. no. 1851.11.5.1. From ; ; West Indies (Antilles). Date: //. Source: Verreaux. Material: adult ; whole/good skin. Reg. no. untraced.

Reg. no. 1890.6.1.181. From ; ; St. Lucia. Date: //1881. Source: J. Semper; P.L. Sclater. Material: adult ; whole/good skin.

Reg. no. 1890.6.1.182. From ; ; St. Lucia. Date: 10/ 4/1881. Source: Semper; Zoological Society; P.L. Sclater. Material: adult F; whole/good skin.

Reg. no. 1890.6.1.183. From ; ; St. Lucia. Date: 15/ 5/1882. Source: Semper; Zoological Society; P.L. Sclater. Material: adult F; whole/good skin. Eyes yellow, beak black, feet greyish.

Reg. no. 1938.9.8.1. From ; ; St. Lucia. Date: //1899. Source: M.J.D. Cowie. Material: ; whole/good skin. Mount. Purchased from natives, about 1899.

Reg. no. E1984.4.41. From ; ; Captivity. Date: //1982. Source: Jersey Wildlife Preservation Trust. Material: 1 ; whole/good egg. Laid in captivity.

Reg. no. E1984.4.42. From ; ; Captivity. Date: //1982. Source: Jersey Wildlife Preservation Trust. Material: 1 ; whole/good egg. Laid in captivity.

Reg. no. E1984.4.43. From ; ; Captivity. Date: //1982. Source: Jersey Wildlife Preservation Trust. Material: 1 ; whole/good egg. Laid in captivity. Second clutch.

Reg. no. E1984.4.44. From ; ; Captivity. Date: //1983. Source: Jersey Wildlife Preservation Trust. Material: 1 ; whole/good egg. Laid in captivity.

Reg. no. E1984.4.45. From ; ; Captivity. Date: //1982. Source: Jersey Wildlife Preservation Trust. Material: 1 ; whole/good egg. Laid in captivity.

Amazona arausiaca

Red-necked Amazon

Restricted to the island of Dominica, mainly in the lowland forests in the northern part of the island. It is seriously threatened by clearance of the forests for banana plantations, and its population in 1992 was estimated at about 500 birds.

Reg. no. 1881.11.26.1. From ; ; Dominica. Date: //. Source: F. Ober; W.J. Williams. Material: ; whole/good skin.

Reg. no. 1889.1.30.308. From ; ; Dominica. Date: //1881. Source: F. Ober; Salvin–Godman coll.. Material: adult ; whole/good skin.

Reg. no. 1889.6.10.47. From ; ; Dominica. Date: 4/ 6/1888. Source: Ramage; British Association; West India Committee, Royal Society. Material: adult ; whole/good skin.

Reg. no. 1889.6.10.48. From ; ; Dominica. Date: / 6/. Source: Ramage; British Association; West India Committee, Royal Society. Material: adult M; whole/good skin.

Reg. no. 1890.6.1.179. From ; ; Dominica. Date: //1881. Source: F. Ober; P.L. Sclater. Material: adult ; whole/good skin.

Reg. no. 1890.6.1.180. From ; ; Dominica. Date: //. Source: Angas; P.L. Sclater. Material: adult ; whole/good skin.

Amazona guildingii

St. Vincent Amazon

Confined to several valleys along the mountain ridge of St. Vincent, where it was threatened with forest destruction and illegal hunting mainly for the pet trade. The population was estimated to be about 450–500 in the late 1980s. New protection laws, including a forest reserve and major education programme have gone far to assist the species.

Reg. no. 1855.12.19.219. From ; ; St. Vincent. Date: //. Source: Rev. L. Guilding; Zoological Society. Material: adult ; whole/good skin. Syntype. Vigors, *Proc. zool. Soc. Lond.* 1836 (publ. 1837): 80.

Reg. no. 1889.1.20.307. From ; ; St. Vincent. Date: 2/ 9/1875. Source: Zoological Society; Salvin–Godman coll.. Material: adult ; whole/good skin. The bird was received (? at the Zoo) on 29 May 1874. The reg. no. appears to be wrong. The correct no. may be 1889.1.30.306, although the locality for that is given as Dominica in the register.

Reg. no. 1890.6.1.185. From ; ; St. Vincent. Date: 23/11/1880. Source: Zoological Society; P.L. Sclater. Material: adult F; whole/good skin. The bird was received (? at the Zoo) on 9 March 1880.

Reg. no. 1898.2.8.289. From ; ; St. Vincent. Date: 8/ 1/1890. Source: D.W. Smith; Salvin–Godman coll.. Material: adult ; whole/good skin.

Reg. no. 1898.2.8.290. From ; ; St. Vincent. Date: 6/ 1/1890. Source: D.W. Smith; Salvin–Godman coll.. Material: adult ; whole/good skin.

Reg. no. 1898.2.8.291. From ; ; St. Vincent. Date: 16/ 1/1890. Source: D.W.

Smith; Salvin–Godman coll.. Material: adult M; whole/good skin.

Reg. no. 1898.2.8.292. From ; ; St. Vincent. Date: 15/ 2/1890. Source: D.W. Smith; Salvin–Godman coll.. Material: adult ; whole/good skin.

Reg. no. 1898.2.8.293. From ; ; St. Vincent. Date: 15/ 2/1890. Source: D.W. Smith; Salvin–Godman coll.. Material: adult F; whole/good skin.

Reg. no. 1898.2.8.294. From ; ; St. Vincent. Date: / 4/. Source: D.W. Smith; Salvin–Godman coll.. Material: adult ; whole/good skin.

Reg. no. 1898.2.8.295. From ; ; St. Vincent. Date: 12/ 8/1889. Source: D.W. Smith; Salvin–Godman coll.. Material: adult ; whole/good skin. The year is incorrectly given as 1890 in the register.

Reg. no. 1898.2.8.296. From ; ; St. Vincent. Date: 28/ 9/1889. Source: D.W. Smith; Salvin–Godman coll.. Material: adult M; whole/good skin. The year is incorrectly given as 1890 in the register.

Reg. no. 1898.2.8.297. From ; ; St. Vincent. Date: 14/11/1889. Source: D.W. Smith; Salvin–Godman coll.. Material: adult F; whole/good skin.

Reg. no. 1898.2.8.298. From ; ; St. Vincent. Date: 18/11/1889. Source: D.W. Smith; Salvin–Godman coll.. Material: adult ; whole/good skin.

Reg. no. 1932.3.17.1. From ; ; St. Vincent. Date: //. Source: E.O. Cochrane. Material: F; whole/good skin. The sternum of this bird is in the skeleton collection, 1932.3.17.2.

Reg. no. 1988.21.3. From ; ; . Date: //. Source: Rothschild coll.. Material: ; whole/good skin. Ex–mount.

Reg. no. 1988.21.65. From ; ; [South America]. Date: //. Source: Rothschild coll.. Material: adult ; whole/good skin. Mount in Tring gallery.

Reg. no. 1856.12.30.4. From ; ; West Indies. Date: //. Source: Zoological Society. Material: ; whole/good skeleton. Ex–mount. Previously registered as Skeleton Vellum Cat. 1200a.

Reg. no. 1932.3.17.2. From ; ; St. Vincent. Date: //. Source: E.O. Cochrane. Material: F; part/good skeleton. Sternum, two humeri and two femurs only. The skin of this bird is also in the collection, 1932.3.17.1.

Reg. no. S1952.3.228. From ; ; . Date: //. Source: Material: ; part/good skeleton. Sternum only.

Reg. no. S1985.22.1. From ; ; Captivity. Date: //1985. Source: D. Frampton. Material: ; part/good skeleton. Sternum and two femurs only.

Reg. no. E1983.9.1–4. From ; ; Captivity. Date: //1983. Source: M. Reynolds. Material: 4 ; whole/good eggs (clutch). Eggs apparently laid by the same female, between 22 June and 20 July, in Cornwall.

Reg. no. E1941.3.1.770. From ; ; Captivity. Date: //1887. Source: H. Munt coll..

Material: 1 ; whole/good egg. Laid in captivity.

Reg. no. E1984.4.35. From ; ; Captivity. Date: //1982. Source: Jersey Wildlife Preservation Trust. Material: 1 ; whole/good egg.

Reg. no. E1984.4.36. From ; ; Captivity. Date: //1982. Source: Jersey Wildlife Preservation Trust. Material: 1 ; whole/good egg. Laid in captivity.

Reg. no. E1984.4.37. From ; ; Captivity. Date: //. Source: Jersey Wildlife Preservation Trust. Material: 1 ; whole/good egg. Laid in captivity.

Reg. no. E1984.4.38. From ; ; Captivity. Date: //. Source: Jersey Wildlife Preservation Trust. Material: 1 ; whole/good egg. Laid in captivity.

Reg. no. E1984.4.39. From ; ; Captivity. Date: //1980. Source: Jersey Wildlife Preservation Trust. Material: 1 ; whole/good egg. Laid in captivity.

Reg. no. E1984.4.40. From ; ; Captivity. Date: //1982. Source: Jersey Wildlife Preservation Trust. Material: 1 ; whole/good egg. Laid in captivity.

Reg. no. 1929.6.22.1. From ; ; Captivity. Date: 20/ 6/1929. Source: Marquis of Tavistock. Material: adult ; whole/good alc.

Reg. no. 1930.9.3.1. From ; ; Captivity. Date: 11/ 9/1930. Source: Marquis of Tavistock. Material: ; part/poor alc. Dissected specimen.

Reg. no. 1934.4.4.1. From ; ; St. Vincent. Date: 18/ 2/1934. Source: J.K.B. Stephenson. Material: ; whole/good alc.

Amazona imperialis

Imperial Amazon

Confined to the mountain forests of Dominica where it is threatened with habitat destruction and capture for food and the pet trade. In 1990 only about 80 birds were thought to remain.

Reg. no. . From ; ; Dominica (South America), [Captivity]. Date: //. Source: Zoological Society. Material: adult ; skin. Probably died in captivity. Details taken from *Cat. Bds.*. Not found 1988.

Reg. no. . From ; ; St. Thomas. Date: //. Source: Material: imm ; skin. Details taken from *Cat. Bds.*. Purchased. Not found 1988. Presumably either identity or locality incorrect.

Reg. no. 1866.5.9.1. From ; ; [Dominica]. Date: //. Source: H. Whitely. Material: imm ; whole/good skin. Reg. no. was given correctly on the original label, but erroneously on the later Museum label (as 1866.9.5.1).

Reg. no. 1889.1.30.306. From ; ; Dominica. Date: 2/ 3/. Source: Zoological Society; Salvin–Godman coll.. Material: adult F; whole/good skin. The bird was received (? at the Zoo) on 6 May 1865. The reg. no. is untraced, and appears to be incorrect; this is probably 1889.1.30.307.

Reg. no. 1889.6.10.49. From ; ; Dominica. Date: 6/ 7/1888. Source: Ramage; British

Association; West India Committee, Royal Society. Material: imm ; whole/good skin.

Reg. no. 1890.6.1.184. From ; ; Dominica. Date: //1879. Source: F. Ober; P.L. Sclater. Material: adult ; part/good skin. Head missing.

Reg. no. 1937.5.8.1. From ; ; [Captivity]. Date: 13/ 5/1937. Source: Zoological Society. Material: F; whole/good skin.

Reg. no. E1962.1.636. From ; ; Captivity. Date: //. Source: Munt coll.. Material: 1 ; whole/good egg. Laid in captivity.

Strigops habroptilus

Kakapo

Survives precariously in Fjordland, South Island, New Zealand, and on Stewart and Little Barrier Islands. The total population is probably about 60. The birds are under constant threat from feral cats, and from their own erratic social and breeding system.

Reg. no. 1837.7.15.93. From [Dusky Sound]; [South Island]; [New Zealand]. Date: //. Source: Turner. Material: adult ; whole/good skin. Holotype. Ex-mount. G.R. Gray, 1845, *The Genera of Birds*, 2: (unpaginated, 426–7), pl. 105. Previously said to be from 'one of the islands in the South Pacific'. See *Proc. zool. Soc. Lond.* 1847: 61–2 and *Ibis* 1913: 426–7.

Reg. no. 1852.1.17.1. From Milford Sound; ; New Zealand. Date: / 3/1851. Source: Lyall; Stokes. Material: adult M; whole/good skin.

Reg. no. 1852.1.17.2. From Dusky Bay; South Island; New Zealand. Date: //. Source: Lyall; Stokes. Material: F; whole/good skin.

Reg. no. 1852.1.17.3. From ; ; New Zealand. Date: //. Source: Stokes. Material: imm ; whole/good skin.

Reg. no. 1852.1.17.4. From ; ; New Zealand. Date: //. Source: Lyall; Stokes. Material: chick ; whole/good skin.

Reg. no. 1852.1.17.5. From ; ; New Zealand. Date: //. Source: Lyall; Stokes. Material: chick ; whole/good skin.

Reg. no. 1852.1.20.1. From ; ; New Zealand. Date: //. Source: Sir G. Grey. Material: ; skin. Not in the collection; exchanged with the Calcutta Museum.

Reg. no. 1852.1.20.2. From ; ; New Zealand. Date: //. Source: G. Grey. Material: adult ; whole/good skin. Specimen f. in *Cat. Bds.*.

Reg. no. 1852.1.20.3. From ; ; New Zealand. Date: //. Source: G. Grey. Material: adult ; whole/good skin.

Reg. no. 1853.6.9.1. From ; ; New Zealand. Date: //. Source: Sir G. Grey. Material: adult ; whole/good skin. Holotype of *S. greyii*. G.R. Gray, *Ibis* 1862: 230. The reg. no. was erroneously given on the label as 1855.6.9.1. Miss Warren in the catalogue of type-specimens substituted 1852.1.20.2, but this refers to another specimen in the collection. Damaged: tail detached.

Reg. no. 1899.2.10.26. From ; ; New Zealand. Date: //. Source: Metcalfe Smith. Material: ; whole/good skin.

Reg. no. 1901.10.22.1. From Wellington; [North Island]; New Zealand. Date: //. Source: Miss Potter. Material: adult ; whole/good skin. Holotype of *S.h. innominatus*. Mathews & Iredale, *Ibis* 1913: 427.

Reg. no. 1902.2.21.48. From ; South Island; New Zealand. Date: / 1/1901. Source: Earl of Ranfurly. Material: ; whole/good skin.

Reg. no. 1902.2.21.49. From ; South Island; New Zealand. Date: / 1/1901. Source: Earl of Ranfurly. Material: ; whole/good skin.

Reg. no. 1903.7.17.30. From Dusky Bay; [South Island]; New Zealand. Date: //. Source: R. Henry; Earl of Ranfurly. Material: ; whole/good skin.

Reg. no. 1904.8.8.6. From ; ; [New Zealand]. Date: //. Source: F.W. Forrester. Material: ; whole/good skin.

Reg. no. 1913.11.3.2. From ; ; New Zealand. Date: //. Source: H. Pearse. Material: ; whole/good skin.

Reg. no. 1921.3.26.1. From ; ; New Zealand. Date: //. Source: G. Lawson. Material: ; whole/good skin.

Reg. no. 1926.10.10.29. From Caswell Sound, Otago; South Island; New Zealand. Date: /10/1878. Source: W.R. Wason. Material: adult M; whole/good skin.

Reg. no. 1926.10.10.30. From Caswell Sound, Otago; South Island; New Zealand. Date: /10/1878. Source: W.R. Wason. Material: ; whole/good skin.

Reg. no. 1927.12.18.1. From Dusky Sound; South Island; New Zealand. Date: //1889. Source: Lister. Material: F; whole/good skin.

Reg. no. 1927.12.18.123. From Preservation Inlet, w. coast sounds,; [South Island]; New Zealand. Date: / 1/1889. Source: J.J. Lister. Material: ; whole/good skin.

Reg. no. 1934.11.14.1. From Martins Bay; [South Island]; New Zealand. Date: 12/ 8/1870. Source: Rowley coll.; Stevens' Sale Room. Material: ; whole/good skin.

Reg. no. 1939.12.9.785. From ; ; [New Zealand]. Date: //. Source: Rothschild coll.. Material: ; whole/good skin.

Reg. no. 1939.12.9.786. From ; ; [New Zealand]. Date: //. Source: Rothschild coll.. Material: ; whole/good skin.

Reg. no. 1939.12.9.787. From ; ; [New Zealand]. Date: //. Source: Rothschild coll.. Material: ; whole/good skin.

Reg. no. 1939.12.9.788. From ; ; [New Zealand]. Date: //. Source: Rothschild coll.. Material: ; whole/good skin.

Reg. no. 1939.12.9.789. From ; ; [New Zealand]. Date: //. Source: Rothschild coll.. Material: ; whole/good skin.

Reg. no. 1939.12.9.790. From ; ; [New Zealand]. Date: //. Source: Rothschild coll.. Material: ; whole/good skin.

Reg. no. 1939.12.9.791. From ; ; [New Zealand]. Date: //. Source: Rothschild coll.. Material: ; whole/good skin.

Reg. no. 1939.12.9.792. From ; ; [New Zealand]. Date: //. Source: Rothschild coll.. Material: ; whole/good skin.

Reg. no. 1939.12.9.793. From ; ; [New Zealand]. Date: //. Source: Rothschild coll.. Material: ; whole/good skin.

Reg. no. 1939.12.9.794. From ; ; [New Zealand]. Date: //. Source: Rothschild coll.. Material: ; whole/good skin.

Reg. no. 1945.58.1. From ; ; [New Zealand]. Date: //. Source: Larkworthy. Material: ; skin. Details taken from the register. Not found 1988.

Reg. no. 1945.58.2. From ; ; [New Zealand]. Date: //. Source: Larkworthy. Material: ; whole/good skin. Mount.

Reg. no. 1945.58.3. From ; ; [New Zealand]. Date: //. Source: Larkworthy. Material: ; whole/good skin. Mount.

Reg. no. 1945.58.4. From ; ; [New Zealand]. Date: //. Source: Larkworthy. Material: ; whole/good skin. Mount.

Reg. no. 1949.64.147. From Dusky Sound; South Island; New Zealand. Date: /12/1884. Source: A. Reischek; G.M. Mathews. Material: F; whole/good skin.

Reg. no. 1958.3.6. From ; South Island; New Zealand. Date: //. Source: Dundee Museum. Material: ; whole/good skin.

Reg. no. 1988.21.4. From ; ; [New Zealand]. Date: //. Source: Material: ; whole/good skin.

Reg. no. 1988.21.5. From ; ; [New Zealand]. Date: //. Source: Material: ; whole/good skin.

Reg. no. 1988.21.66. From ; ; [New Zealand]. Date: //. Source: Rothschild coll.. Material: adult ; whole/good skin. Mount in Tring gallery.

Reg. no. 1852.8.10.1. From ; ; New Zealand. Date: //. Source: Sir George Grey. Material: adult ; whole/good skeleton. Previously registered as Skeleton Vellum Cat. 878a. Specimen i. in *Cat. Bds.*. Not listed in the Skeleton Register, although included in the General Register.

Reg. no. 1881.1.17.71. From ; ; [New Zealand]. Date: //. Source: T.C. Eyton; E. Gerrard. Material: adult ; whole/good skeleton. Specimen k. in *Cat. Bds.*.

Reg. no. S1955.5.27. From ; ; New Zealand. Date: //. Source: Rothschild coll.. Material: adult ; whole/good skeleton.

Reg. no. S1972.1.71. From Lake Te Anau; South Island; New Zealand. Date: / 1/1875. Source: Rothschild coll.. Material: adult ; whole/good skeleton.

Reg. no. S1952.2.410. From ; ; [New Zealand]. Date: //. Source: Material: adult ; whole/good skeleton. No history. Not listed in *Cat. Bds.*. Probably a former gallery specimen.

Reg. no. 1852.1.16.12. From Thomson's Sound; (Middle Island); New Zealand. Date: / 2/1851. Source: Dr Lyall. Material: 1 ; whole/good egg. Collected by Dr Lyall of H.M. Steamship Acheron.

Reg. no. 1853.3.7.14. From ; ; New Zealand. Date: //. Source: Capt. Stokes. Material: 1 ; whole/good egg.

Reg. no. 1901.12.15.689. From ; ; New Zealand. Date: //. Source: Leopold Field coll.; Crowley coll.. Material: 1 ; whole/good egg.

Reg. no. E1941.3.1.880. From ; ; New Zealand. Date: //. Source: (Lyall); H.F. Walter; Munt. Material: 1 ; whole/good egg. Labelling of this egg is identical with one collected by Dr Lyall on Middle Island (1852.1.16.12). Munt gives a date, 1855, but does not make it clear to what this refers.

Reg. no. E1962.1.585. From ; ; New Zealand. Date: //. Source: Material: 1 ; damaged egg. From the Old Collection. Damaged at the small end.

Reg. no. 1887.3.9.10. From ; ; New Zealand. Date: //. Source: Commissioner for New Zealand. Material: juv ; whole/good alc. Colonials Indian Exhibition.

Reg. no. A1971.33.5. From Dusky Bay; ; New Zealand. Date: / 6/1901. Source: Earl of Ranfurley. Material: adult ; whole/good alc.

Reg. no. A1971.33.6. From Dusky Bay; ; New Zealand. Date: / 6/1901. Source: Earl of Ranfurley. Material: adult ; part/good alc. Dissected specimen with wing detached.

Tauraco bannermani

Bannerman's Turaco

Restricted to the Bamenda–Banso Highlands of western Cameroon, where it is seriously threatened by forest clearance.

Reg. no. 1923.10.26.75. From 6,000 ft., Banso Mountains, n. of Kumbo; ; n. Cameroon. Date: 18/ 9/1922. Source: G.L. Bates. Material: adult F; whole/good skin. Holotype. Bates, *Bull. Brit. Orn. Club* 43 (1923): 140–1. Collected in a forested ravine.

Reg. no. 1926.8.8.78. From Bambolue Lake, nr. Bamendo; ; Cameroon. Date: 2/ 2/1925. Source: G.L. Bates. Material: adult F; whole/good skin. Iris hazel, feet black, bill yellow, culmen dark red–brown.

Reg. no. 1926.8.8.79. From Santa, nr. Bamendo; ; Cameroon. Date: 21/ 2/1925. Source: G.L. Bates. Material: adult F; whole/good skin. Iris light brown, feet black, bill yellow and dark red.

Reg. no. 1926.8.8.80. From Oku, w. of Kumbo; ; Cameroon. Date: 9/ 2/1925. Source: G.L. Bates. Material: adult F; whole/good skin. Iris light red–brown, feet blackish–brown, culmen dark red, rest of bill yellow.

Reg. no. 1951.34.228. From Bamenda, Bambulue Lake; ; Cameroon. Date: 24/

5/1948. Source: W. Serle. Material: adult M; whole/good skin. Iris brown, eyelids red, feet black, bill yellow, culmen red. One of two males in a single wood on the mountain top.

Reg. no. 1951.34.229. From Bamenda, Bambulue Lake; ; Cameroon. Date: 24/5/1948. Source: W. Serle. Material: adult M; whole/good skin. Iris brown, eyelids red, feet black, bill yellow, culmen red. One of two males in a single wood on the mountain top.

Neomorphus geoffroyi dulcis
Snethlage's Ground-Cuckoo
Restricted to southeastern Brazil, where it is very rare and may already be extinct. It may survive at one or two sites in Espirito Santo.

Reg. no. 1888.8.23.55. From ; Bahia; Brazil. Date: //. Source: Wucherer; Salvin-Godman coll.. Material: adult ; whole/good skin. Subspecies not critically determined, but this specimen comes from within the range claimed for *N.g. dulcis*. There are further Brazilian specimens in the collection.

Reg. no. 1888.8.23.56. From ; Bahia; Brazil. Date: //. Source: Wucherer; Salvin-Godman coll.. Material: adult ; whole/good skin. Subspecies not critically determined, but this specimen comes from within the range claimed for *N.g. dulcis*. There are further Brazilian specimens in the collection.

Neomorphus radiolosus
Banded Ground-Cuckoo
Confined to wet forest in the foothills and lower slopes of the Andes in southwestern Colombia and northwestern Ecuador. It is very rare and there are few recent records.

Reg. no. 1888.8.23.61. From Intac (Intaj); ; Ecuador. Date: /12/1877. Source: C. Buckley; Salvin-Godman coll.. Material: adult ; whole/good skin. Holotype. P.L. Sclater & Salvin, *Proc. zool. Soc. Lond.* 1878: 439, pl. 27.

Coua delalandei
Snail-eating Coua
Humid forests of northern Madagascar. Not definitely recorded since 1834, but there were unconfirmed reports of birds having been seen between 1920 and 1930.

Reg. no. 1837.6.10.466. From ; ; Madagascar. Date: //. Source: Baron Laugier de Chartrouse. Material: ; whole/good skin.

Reg. no. 1887.12.19.301. From ; ; Madagascar. Date: //. Source: Seebohm coll.. Material: ; whole/good skin.

Tyto soumagnei
Madagascar Red Owl
Rainforests of east-central Madagascar. Seen only once since 1934.

Reg. no. 1879.3.5.25. From Tananarive (Antananarivo); ; Madagascar. Date: //. Source: Lorimer; Higgins. Material: ; whole/good skin.

Reg. no. 1955.6.N.20.4928. From ; ; Madagascar. Date: / 3/1893. Source: J. Wills; J.H. Gurney coll.. Material: adult ; whole/good skin.

Reg. no. 1955.6.N.20.4929. From Merimitatra; ; Madagascar. Date: 9/ 1/1895. Source: J. Wills; J.H. Gurney coll.. Material: ; whole/good skin.

Otus ireneae

Sokoke Scops-Owl

Endemic to the Arabuko-Sokoke Forest in coastal Kenya. The population was estimated as 1,000 pairs in 1984 but decreasing due to illegal forest clearance.

Reg. no. 1971.16.18. From Sokoke Forest; ; Kenya. Date: 15/ 4/1970. Source: P.R. Colston; A. Forbes Watson. Material: adult M; whole/good skin. Weight 50 grams. The trunk is in the spirit collection, A1971.19.51.

Reg. no. A1971.19.51. From Sokoke Forest; ; Kenya. Date: 15/ 4/1970. Source: P.R. Colston; A. Forbes Watson. Material: adult M; part/good alc. Trunk only from a skin 1971.16.18, also in the collection.

Reg. no. A1971.19.52. From Sokoke Forest; ; Kenya. Date: 15/ 4/1970. Source: A. Forbes Watson; P.R. Colston. Material: adult M; part/good alc. Trunk only, from a skin in Nairobi Museum.

Reg. no. A1971.19.53. From Sokoke Forest; ; Kenya. Date: 15/ 4/1970. Source: A. Forbes Watson; P.R. Colston. Material: adult M; part/good alc. Trunk only, from a skin in Nairobi Museum.

Otus rutilus capnodes

Anjouan Scops-Owl

Known only from Anjouan in the Comoro Islands. The subspecies was not seen between the 1880s, when the original series of specimens had been collected, and 1992, when the bird was rediscovered. The population is estimated to be at least several tens of pairs and perhaps as many as 200. Searches in 1906-1907 and 1959 had failed to find any birds at all.

Reg. no. 1888.5.4.2. From ; Anjouan Island (Johanna Island); Comoro Islands. Date: //. Source: G.A. Frank. Material: ; whole/good skin.

Reg. no. 1888.5.4.3. From ; Anjouan Island (Johanna Island); Comoro Islands. Date: //. Source: G.A. Frank. Material: ; whole/good skin.

Reg. no. 1888.5.4.4. From ; Anjouan Island (Johanna Island); Comoro Islands. Date: //. Source: G.A. Frank. Material: ; whole/good skin.

Reg. no. 1890.1.30.11. From ; Anjouan Island; Comoro Islands. Date: //. Source: A. Boucard. Material: adult ; whole/good skin.

Reg. no. 1890.1.30.12. From ; Anjouan Island; Comoro Islands. Date: //. Source: A. Boucard. Material: adult ; whole/good skin.

Reg. no. 1890.1.30.13. From ; Anjouan Island; Comoro Islands. Date: //. Source: A. Boucard. Material: adult ; whole/good skin.

Reg. no. 1890.1.30.14. From ; Anjouan Island; Comoro Islands. Date: //. Source: A. Boucard. Material: adult ; whole/good skin.

Reg. no. 1894.8.15.35. From ; Anjouan Island; Comoro Islands. Date: //. Source: Shelley Museum; Humblot coll.. Material: M; whole/good skin.

Reg. no. 1894.8.15.36. From ; Anjouan Island; Comoro Islands. Date: //. Source: Shelley Museum; Humblot coll.. Material: F; whole/good skin.

Reg. no. 1955.6.N.20.3848. From forest of e. Imerina; Anjouan Island; Comoro Islands. Date: //. Source: G.A. Frank; J.H. Gurney coll.. Material: adult ; whole/good skin. Syntype. Ex-mount. Gurney, *Ibis* 1889: 104. Two other syntypes are in the collection.

Reg. no. 1955.6.N.20.3849. From ; Anjouan Island; Comoro Islands. Date: //. Source: G.A. Frank; J.H. Gurney coll.. Material: adult ; whole/good skin. Syntype. Iris pale yellow.

Reg. no. 1955.6.N.20.3850. From ; Anjouan Island; Comoro Islands. Date: //. Source: G.A. Frank; J.H. Gurney coll.. Material: adult ; whole/good skin. Syntype.

Ninox novaeseelandiae albaria

Lord Howe Island Boobook

A form of the Boobook from Lord Howe Island, where it has been extinct since 1940.

Reg. no. 1881.5.1.561. From ; ; Lord Howe Island. Date: 6/ 9/1853. Source: J. MacGillivray, voyage of H.M.S. Herald; Gould coll.. Material: adult ; whole/good skin.

Reg. no. 1919.7.15.285. From ; ; Lord Howe Island. Date: 13/12/1913. Source: G.M. Mathews. Material: adult M; whole/good skin. Bill blue–black, iris golden–yellow, feet light yellow.

Sceloglaux albifacies albifacies

South Island Laughing Owl

Laughing Owls formerly occurred on both North and South Islands of New Zealand, but have been extinct on North Island since 1889 and on South Island since 1950. Only one specimen has been preserved from North Island, the type of *S.a. rufifacies*, but it is not clear if this represents a valid race.

Reg. no. 1845.1.13.5. From Waikouaiti; South Island; New Zealand. Date: //. Source: P. Earl, voyage of H.M.S. Erebus & Terror. Material: imm ; whole/good skin. Ex–mount. Holotype. G.R. Gray, 1844, *The Zoology of the Voyage of H.M.S. Erebus and Terror*, Birds, p. 2, pl. 1.

Reg. no. 1846.7.16.15. From ; ; New Zealand. Date: //. Source: [Massena coll.]; E. Wilson. Material: ; whole/good skin. Ex–mount. Damaged: left scapulars detached.

Reg. no. 1852.1.17.13. From ; South Island; New Zealand. Date: //. Source: Stokes. Material: ; whole/good skin. Ex–mount.

Reg. no. 1939.12.9.539. From ; ; [New Zealand]. Date: //. Source: Rothschild coll.. Material: ; whole/good skin. Mount.

Reg. no. 1955.6.N.20.4771. From ; ; New Zealand. Date: //. Source: J.H. Gurney coll.. Material: adult ; whole/good skin. Mount.

Reg. no. 1955.6.N.20.4772. From ; South Island; New Zealand. Date: //. Source: Mr Coles; Sir F.G.M. Boileau; Gurney coll.. Material: adult F; whole/good skin. Ex-mount. Bought from Coles, Naturalist, Christchurch, N.Z., in March 1895. Probably a female. See Buller, *A History of the Birds of New Zealand*, 1st ed. (1873), p. 22 & plate; 2nd ed. (1888), p.198–205. There is a ms letter by Boileau concerning this specimen in the Gurney register.

Reg. no. E1955.5.148. From ; South Island; New Zealand. Date: 1/ 8/1894. Source: Coles; Sir F.G.M. Boileau; J.H. Gurney coll.. Material: 1 ; whole/good egg. There is a ms letter in the Gurney skin register which refers to this egg.

Reg. no. E1988.4.26. From Canterbury; South Island; New Zealand. Date: //. Source: Sir W. Buller; Rothschild coll.. Material: 1 ; whole/good egg.

***Sceloglaux albifacies* subspecies**

Reg. no. E1955.5.149. From ; ; New Zealand. Date: //. Source: F. Stalchsmith; J.H. Gurney coll.. Material: 2 ; whole/good eggs. Bought at a sale in July 1897. Race not determined.

Reg. no. . From ; ; [New Zealand]. Date: 13/11/1895. Source: Prof. Newton. Material: M; part/good alc.

Reg. no. . From ; ; [New Zealand]. Date: 13/11/1895. Source: Prof. Newton. Material: M; part/good alc.

Athene blewitti

Forest Owlet

Endemic to central India, but no specimens have been collected since 1914, though there have been unconfirmed sight records subsequently.

Reg. no. 1886.2.1.543. From Busnah, Phooljan State; ; central India. Date: 14/12/1872. Source: F.R. Blewitt; Hume coll.. Material: adult F; whole/good skin. Holotype. Hume, *Stray Feathers* 1 (1873): 468–9.

Reg. no. 1886.2.1.544. From Khandeish; ; India. Date: 20/ 4/1881. Source: J. Davidson; Hume coll.. Material: adult F; whole/good skin.

Reg. no. 1886.2.1.545. From Khandeish; ; India. Date: 5/12/1880. Source: J. Davidson; Hume coll.. Material: adult M; skin. Not in the collection; exchanged with the American Museum of Natural History in 1928.

Reg. no. 1886.2.1.546. From Khandeish; ; India. Date: 5/12/1880. Source: J. Davidson; Hume coll.. Material: adult M; whole/good skin.

Reg. no. 1965.M.5230. From Mandvi, Tapti River, 140 miles n. of Bombay; ; India. Date: 9/10/1914. Source: Meinertzhagen coll.. Material: M; whole/good skin.

Speotyto cunicularia amaura

Antigua Burrowing Owl

Known only from 5 specimens collected on Antigua, Nevis and St. Kitts in the West Indies. Extinction occurred at the end of the

19th century as a result of the introduction of the mongoose.

Reg. no. 1839.5.30.3. From ; ; St. Nevis. Date: //. Source: T. Cottle. Material: adult ; whole/good skin. The reg. no. was erroneously given as 1839.6.6.3 on the label.

Siphonorhis americanus

Jamaican Pauraque

Endemic to Jamaica. The last known specimen was collected in 1859. It had been rare and local for many years before that, and was not known to Gosse in the 1830s. It is believed to have been exterminated by rats.

Reg. no. 1890.2.18.193. From Savanna la Mar, Westmoreland; ; Jamaica. Date: 21/ 8/1858. Source: Osburn; Sclater coll.. Material: adult ; whole/good skin.

Reg. no. 1890.2.18.194. From Trelawny; ; Jamaica. Date: 14/ 9/1859. Source: Osburn; Sclater coll.. Material: adult ; whole/good skin.

Eleothreptus anomalus

Sickle-winged Nightjar

Occurs in Paraguay, Uruguay, southeastern Brazil and northern Argentina in gallery forest, marshland and campo. Very little is known about it and the species is probably extremely rare.

Reg. no. 1885.11.20.333. From Quilmes, nr. Buenos Aires; ; [Argentina]. Date: 31/ 3/1877. Source: H. Durnford; Miss Smith. Material: adult F; whole/good skin.

Reg. no. 1885.11.20.334. From ; Patagonia; [Argentina]. Date: //. Source: H. Durnford; [Miss Smith]. Material: ; skin. Details taken from the register. Not found 1988.

Reg. no. 1888.8.1.107. From ; ; s. Brazil. Date: //. Source: J. Natterer; Salvin–Godman coll.. Material: adult F; whole/good skin.

Reg. no. 1890.2.18.183. From 'Bogota'; ; [South America]. Date: //. Source: Sclater coll.. Material: ; skin. Details taken from the register. Not found 1988.

Reg. no. 1890.2.18.184. From [Irisanga]; ; Brazil. Date: 19/ 1/. Source: J. Natterer; Sclater coll.. Material: adult F; whole/good skin. Some data taken from *Cat. Bds.*; no date or locality on the label.

Reg. no. 1925.7.3.25. From Caravini, Villarica; ; Paraguay. Date: / 4/1924. Source: F. Schade. Material: adult M; whole/good skin.

Reg. no. VEL.31.71a. From [Sao Paulo] (Demerara); ; Brazil. Date: //. Source: J.E. Gray. Material: adult M; whole/good skin. Holotype. Ex-mount. Gould, *Proc. zool. Soc. Lond.* 1837 (publ. 1838): 105. See *Revta Mus. paul.* 32 (1938): 237 concerning the locality.

Reg. no. E1951.7.6. From Cuirango Estado do Rio Grande do Sul; ; Brazil. Date: 28/11/1925. Source: [H.A. Schwarz]; Dr K. Jordan coll.. Material: 2 ; damaged eggs. Both eggs were broken in 1972 during the move of the collections from London to Tring.

Glaucis dohrnii

Hook-billed Hermit

Restricted to parts of southeastern Brazil where its status is obscure.

Reg. no. . From ; ; Brazil. Date: //. Source: Salvin-Godman coll.. Material: adult M; skin. Details taken from *Cat. Bds.*, specimen c. Not found 1988. The skin was said to have been of the 'Rio make' (the style which came from Rio).

Reg. no. 1888.7.25.60. From ; ; Ecuador. Date: //. Source: Gould coll.. Material: adult F; whole/good skin. Holotype. Ex-mount. Bourcier & Mulsant, *Annls Sci. Phys. nat. Lyon*, series 2, 4 (1852): 139.

Reg. no. 1888.7.25.61. From ; ; [Brazil]. Date: //. Source: Parzudaki; Gould coll.. Material: adult M; whole/good skin. Holotype of *Gryphus spixi*. Gould, *Proc. zool. Soc. Lond.* 1860: 304. The locality is 'Supposed to be Brazil'.

Reg. no. 1988.21.40. From Espirito Santo; ; Brazil. Date: //. Source: de Gaud; Gould coll.. Material: adult M; whole/good skin.

Eupherusa poliocerca

White-tailed Hummingbird

Forest edge and clearings in Guerrero and western Oaxaca, Mexico. It is endangered by habitat destruction.

Reg. no. 1988.21.41. From Putla; ; Mexico. Date: //. Source: Rebouch; Gould coll.. Material: adult M; whole/good skin.

Reg. no. 1988.21.42. From Omilteme, Chilpancingo, Sierra Madre del Sur; ; Mexico. Date: / 7/. Source: D.W. Smith; Mrs H.H. Smith; Salvin-Godman coll.. Material: adult M; whole/good skin. The locality is given as Chilpancingo on the label, and Sierra Madre del Sur in *Cat. Bds.*.

Reg. no. 1988.21.43. From Omilteme, Chilpancingo, Sierra Madre del Sur; ; Mexico. Date: / 7/. Source: D.W. Smith; Mrs H.H. Smith; Salvin-Godman coll.. Material: adult M; whole/good skin. The locality is given as Chilpancingo on the label, and Sierra Madre del Sur in *Cat. Bds.*.

Hylonympha macrocerca

Scissor-tailed Hummingbird

This species was long known only from trade-skins of questionable origin. Comparatively recently it was discovered to be endemic to the Paria Peninsula of Venezuela, where it is now endangered by forest clearance.

Reg. no. 1887.3.22.951. From ; ; Brazil. Date: / 6/1879. Source: H. Whitely; Salvin-Godman coll.. Material: adult M; whole/good skin.

Reg. no. 1887.3.22.952. From ; ; Brazil. Date: /12/1877. Source: H. Whitely; Salvin-Godman coll.. Material: adult M; whole/good skin.

Reg. no. 1887.3.22.953. From ; ; Brazil. Date: //. Source: H. Whitely; Salvin-Godman coll.. Material: adult M; skin. Not in the collection. Exchanged with Acad. Nat. Sci. Philadelphia, in 1938.

Reg. no. 1887.3.22.954. From ; ; Brazil. Date: //1877. Source: H. Whitely; Salvin–Godman coll.. Material: M; whole/good skin.

Reg. no. 1887.3.22.955. From ; ; Brazil. Date: //. Source: H. Whitely; Salvin–Godman coll.. Material: adult F; whole/good skin.

Reg. no. 1888.7.25.157. From ; ; [n.e. Venezuela]. Date: //. Source: H. Whitely; Gould coll.. Material: adult M; whole/good skin. Holotype. Gould, *Ann. Mag. nat. Hist.* (4) 12 (1873): 429. The locality was unknown to Gould, but the species has since been found on the Paria Peninsula, n.e. Venezuela (see *Auk* 65 (1948): 62–6).

Reg. no. 1988.21.44. From ; ; n. Brazil. Date: //. Source: Gould coll.. Material: adult M; whole/good skin.

Reg. no. 1988.21.45. From ; ; n. Brazil. Date: //. Source: Gould coll.. Material: adult M; whole/good skin.

Reg. no. 1988.21.46. From ; ; n. Brazil. Date: //. Source: Gould coll.. Material: adult M; whole/good skin.

Reg. no. 1988.21.47. From ; ; n. Brazil. Date: //. Source: H. Whitely; Gould coll.. Material: adult F; whole/good skin.

Eriocnemis nigrivestis

Black-breasted Puffleg

Known only from Mt. Pichincha and Mt. Atacazo in Ecuador. Threatened by habitat destruction spreading from the city of Quito.

Reg. no. 1887.3.14.374. From Mt. Pichincha; ; Ecuador. Date: //. Source: Whymper; Sclater coll.. Material: adult M; skin. Details taken from the register and *Cat. Bds.*. Not found 1988.

Reg. no. 1887.3.14.375. From Mt. Pichincha; ; Ecuador. Date: //1882. Source: Whymper; Sclater coll.. Material: adult F; whole/good skin.

Reg. no. 1887.3.22.1649. From Intac (Intaj); ; Ecuador. Date: /12/1877. Source: C. Buckley; Salvin–Godman coll.. Material: adult F; whole/good skin.

Reg. no. 1887.3.22.1659. From Intac; ; Ecuador. Date: /12/1877. Source: C. Buckley; Salvin–Godman coll.. Material: adult M; whole/good skin.

Reg. no. 1887.3.22.1660. From Intac (Intaj); ; Ecuador. Date: /12/1877. Source: C. Buckley; Salvin–Godman coll.. Material: adult M; whole/good skin.

Reg. no. 1887.3.22.1661. From Sarayacu; ; Ecuador. Date: //. Source: C. Buckley; Salvin–Godman coll.. Material: adult F; whole/good skin.

Reg. no. 1887.3.22.1662. From Sarayacu; ; Ecuador. Date: / 2/1880. Source: C. Buckley; Salvin–Godman coll.. Material: adult M; whole/good skin.

Reg. no. 1887.3.22.1663. From Sarayacu; ; Ecuador. Date: / 2/1880. Source: C. Buckley; Salvin–Godman coll.. Material: adult F; whole/good skin.

Reg. no. 1887.3.22.1664. From ; ; Ecuador. Date: //. Source: Verreaux; Salvin–Godman coll.. Material: imm M; whole/good skin.

Reg. no. 1897.11.12.100. From Nono, w. side Mt. Pichincha; ; Ecuador. Date: //. Source: L. Soderstrom. Material: adult M; whole/good skin. Collected between March and May.

Reg. no. 1897.11.12.99. From Nono, w. side Mt. Pichincha; ; Ecuador. Date: //. Source: L. Soderstrom. Material: adult F; whole/good skin. Collected between March and June. Previously said to have been a male, but apparently female.

Reg. no. 1913.3.20.170. From ; ; Ecuador. Date: //. Source: R.J. Balston coll.. Material: adult M; whole/good skin.

Reg. no. 1913.3.20.171. From ; ; Ecuador. Date: //. Source: R.J. Balston coll.. Material: adult M; skin. Not in the collection; exchanged with Mr Elwes.

Reg. no. 1920.10.31.53. From 9,500 ft., w. side Mt. Pichincha; ; Ecuador. Date: 30/ 4/1903. Source: W.F.H. Rosenberg. Material: adult M; whole/good skin.

Reg. no. 1920.10.31.54. From 9,500 ft., w. side Mt. Pichincha; ; Ecuador. Date: //. Source: W.F.H. Rosenberg. Material: adult M; whole/good skin.

Reg. no. 1920.10.31.55. From 9,500 ft., w. side Mt. Pichincha; ; Ecuador. Date: //. Source: W.F.H. Rosenberg. Material: adult M; whole/good skin.

Reg. no. 1920.10.31.56. From 9,500 ft., w. side Mt. Pichincha; ; Ecuador. Date: //. Source: W.F.H. Rosenberg. Material: adult F; whole/good skin.

Reg. no. 1925.12.24.159. From 14,000 ft., Mt. Pichincha; ; Ecuador. Date: / 2/1915. Source: W. Goodfellow; E.J. Brook coll.; Lt.Col. B.E. Spragge. Material: adult M; whole/good skin.

Reg. no. 1925.12.24.160. From 14,000 ft., Mt. Pichincha; ; Ecuador. Date: / 2/1915. Source: W. Goodfellow; E.J. Brook coll.; Lt.Col. B.E. Spragge. Material: adult M; whole/good skin.

Reg. no. 1925.12.24.161. From 14,000 ft., Mt. Pichincha; ; Ecuador. Date: / 2/1915. Source: W. Goodfellow; E.J. Brook coll.; Lt.Col. B.E. Spragge. Material: adult F; whole/good skin.

Reg. no. 1925.12.24.162. From 14,000 ft., Mt. Pichincha; ; Ecuador. Date: / 2/1915. Source: W. Goodfellow; E.J. Brook coll.; Lt.Col. B.E. Spragge. Material: adult M; whole/good skin.

Reg. no. 1925.12.24.163. From 14,000 ft., Mt. Pichincha; ; Ecuador. Date: / 2/1915. Source: W. Goodfellow; E.J. Brook coll.; Lt.Col. B.E. Spragge. Material: adult M; whole/good skin.

Reg. no. 1953.68.179. From Tambo, Mt. Pichincha; ; Ecuador. Date: 6/ 2/1940. Source: L. Gomez. Material: adult F; whole/good skin.

Reg. no. 1953.68.231. From Verde Cocha, Mt. Pichincha; ; Ecuador. Date: 12/ 6/1941.

Source: L. Gomez. Material: adult M; whole/good skin.

Reg. no. 1969.37.96. From Verde Cocha, Mt. Pichincha; ; Ecuador. Date: / 6/1937. Source: L. Gomez; Hewitt coll.. Material: adult M; whole/good skin.

Reg. no. 1969.37.97. From Verde Cocha, Mt. Pichincha; ; Ecuador. Date: / 6/1937. Source: L. Gomez; Hewitt coll.. Material: adult M; whole/good skin.

Reg. no. 1969.37.98. From Verde Cocha, Mt. Pichincha; ; Ecuador. Date: / 6/1937. Source: L. Gomez; Hewitt coll.. Material: adult F; whole/good skin.

Reg. no. 1988.21.49. From ; ; Ecuador. Date: //. Source: Gould coll.. Material: adult F; whole/good skin.

Reg. no. 1988.21.50. From ; ; Ecuador. Date: //. Source: Gould coll.. Material: adult F; whole/good skin.

Reg. no. 1988.21.51. From ; ; Ecuador. Date: //. Source: Gould coll.. Material: adult M; whole/good skin.

Reg. no. 1988.21.52. From ; ; Ecuador. Date: //. Source: Gould coll.. Material: adult M; whole/good skin.

Reg. no. 1988.21.53. From ; ; Ecuador. Date: //. Source: Gould coll.. Material: adult M; whole/good skin.

Reg. no. 1988.21.54. From ; ; Ecuador. Date: //. Source: Gould coll.. Material: adult M; whole/good skin.

Reg. no. 1988.21.55. From Cochabamba, nr. Otobila, Quito; ; Ecuador. Date: //. Source: Smith; Gould coll.. Material: adult M; whole/good skin.

Eriocnemis godini

Turquoise-throated Puffleg

Possibly extinct. Known only from northwestern Ecuador and Bogota trade skins. There are no recent records, and the range and status of the species is not understood. It is distinctly different in appearance from *E. vestitus*, *contra* Collar & Andrew (1988a).

Reg. no. 1887.3.22.1644. From ; ; Ecuador. Date: //1883. Source: H. Whitely; Salvin-Godman coll.. Material: adult M; whole/good skin.

Reg. no. 1887.3.22.1645. From ; ; Ecuador. Date: //. Source: Salvin-Godman coll.. Material: adult ; skin. Details taken from the register. Not found 1988.

Reg. no. 1888.7.25.176. From ravines of the valley of Guayabamba; ; Ecuador. Date: //. Source: J. Bourcier; Gould coll.. Material: adult M; whole/good skin. Syntype. Ex-mount. Bourcier, *C. r. hebd. Seanc. Acad. Sci. Paris* 32 (1851): 186-7. There is a female syntype in the collection.

Reg. no. 1988.21.48. From ; ; Ecuador. Date: //. Source: Gould coll.. Material: adult F; whole/good skin. Ex-mount. This appears to be the supposed female syntype, but there is no evidence in support.

Acestrura bombus

Little Woodstar

Occurs in forest and scrub in Ecuador and Peru, where it has become very rare in recent years for no discernible reason.

Reg. no. . From ; ; [South America]. Date: //. Source: O.T. Baron; Rothschild coll.. Material: adult M; whole/good skin. Mount in Tring gallery.

Reg. no. . From ; ; [South America]. Date: //. Source: O.T. Baron; Rothschild coll.. Material: ; whole/good skin. Mount in Tring gallery.

Reg. no. . From ; ; [South America]. Date: //. Source: O.T. Baron; Rothschild coll.. Material: ; whole/good skin. Mount in Tring gallery.

Reg. no. 1887.3.14.203. From Tambillo; ; Peru. Date: 22/12/1877. Source: Stoltzmann; Sclater coll.. Material: adult M; whole/good skin. The Sclater label gives a date of 1879, possibly the year in which he received the specimen.

Reg. no. 1887.3.22.1172. From Balzar Mountains; ; Ecuador. Date: / 3/1880. Source: Illingworth; Salvin–Godman coll.. Material: adult F; whole/good skin.

Reg. no. 1887.3.22.1173. From Jima; ; Ecuador. Date: //. Source: C. Buckley; Salvin–Godman coll.. Material: adult M; skin. Details taken from the register and *Cat. Bds.*, where it is specimen h. Listed as an adult female in the register, but as an adult male in *Cat. Bds.*. Not in the collection. Exchanged with A.L. Butler in 1924.

Reg. no. 1887.3.22.1174. From Jima; ; Ecuador. Date: / 3/1877. Source: C. Buckley; Salvin–Godman coll.. Material: adult F; whole/good skin. The specimen has an unusual bronzy tinge to the green back.

Reg. no. 1887.3.22.1175. From Jima; ; Ecuador. Date: / 3/1877. Source: C. Buckley; Salvin–Godman coll.. Material: adult F; whole/good skin.

Reg. no. 1887.3.22.1176. From Jima; ; Ecuador. Date: / 3/1877. Source: C. Buckley; Salvin–Godman coll.. Material: imm M; whole/good skin.

Reg. no. 1887.3.22.1177. From Baisa; ; Ecuador. Date: /12/1877. Source: C. Buckley; Salvin–Godman coll.. Material: adult M; whole/good skin. The reg. no. was erroneously given on the label as 1887.3.22.1179.

Reg. no. 1887.3.22.1178. From Baisa; ; Ecuador. Date: /12/1877. Source: C. Buckley; Salvin–Godman coll.. Material: adult M; whole/good skin.

Reg. no. 1887.3.22.1223. From ; ; Peru. Date: / 7/1880. Source: H. Whitely; Salvin–Godman coll.. Material: ; whole/good skin. Previously identified as *Myrtis (Eulidia) yarrellii*, subsequently reidentified as *A. bombus*. If the latter is correct, the bird is in an unusual plumage. Specimen p. in *Cat. Bds.* under *bombus*.

Reg. no. 1888.7.25.204. From Citado; ; Ecuador. Date: //1870. Source: Buckley; Gould coll.. Material: adult M; whole/good skin. Syntype. Gould, *Proc. zool. Soc.*

Lond. 1870: 803–4. There are several other syntypes in the collection.

Reg. no. 1912.7.1.217. From Suecha, 9,000 ft., Huamachuco; ; Peru. Date: / 3/1894. Source: O.T. Baron; Salvin–Godman coll.. Material: adult F; whole/good skin.

Reg. no. 1912.7.1.218. From Guayabamba; ; Peru. Date: 19/ 9/1894. Source: O.T. Baron; Salvin–Godman coll.. Material: adult F; whole/good skin.

Reg. no. 1913.3.20.568. From ; ; Ecuador. Date: //. Source: R.J. Balston coll.. Material: adult M; skin. Details taken from the register. Not found 1988.

Reg. no. 1913.3.20.569. From ; ; Ecuador. Date: //. Source: R.J. Balston coll.. Material: adult M; whole/good skin.

Reg. no. 1920.10.31.65. From Bahahoyo; ; Ecuador. Date: //. Source: W.F.H. Rosenberg. Material: adult M; whole/good skin.

Reg. no. 1920.10.31.66. From Naranjito; ; Ecuador. Date: //. Source: W.F.H. Rosenberg. Material: adult M; whole/good skin.

Reg. no. 1925.12.24.251. From Santo Domingo; ; w. Ecuador. Date: / 8/1914. Source: W. Goodfellow. Material: adult M; whole/good skin.

Reg. no. 1953.68.150. From Tunguilla valley; ; Ecuador. Date: 11/ 7/1939. Source: L. Gomez. Material: adult F; whole/good skin.

Reg. no. 1953.68.151. From Macas; ; e. Ecuador. Date: 5/10/1940. Source: L. Gomez. Material: imm M; whole/good skin. Previously identified as a female.

Reg. no. 1969.37.132. From Tunguilla valley; ; Ecuador. Date: 28/ 2/1940. Source: L. Gomez; Hewitt coll.. Material: adult M; whole/good skin.

Reg. no. 1969.37.133. From Tunguilla valley; ; Ecuador. Date: 11/ 7/1939. Source: L. Gomez; Hewitt coll.. Material: adult F; whole/good skin.

Reg. no. 1969.37.134. From Macas; ; e. Ecuador. Date: / 9/1937. Source: L. Gomez; Hewitt coll.. Material: imm M; whole/good skin.

Reg. no. 1969.37.135. From El Bosque, El Oro; ; Ecuador. Date: 18/ 7/1939. Source: L. Gomez; Hewitt coll.. Material: adult M; whole/good skin.

Reg. no. 1969.41.1451. From Tunguilla valley; ; Ecuador. Date: 8/ 7/1939. Source: L. Gomez; Hewitt coll.. Material: adult F; whole/good skin.

Reg. no. 1988.21.56. From Citado; ; Ecuador. Date: //. Source: C. Buckley; Gould coll.. Material: adult F; whole/good skin. Syntype.

Reg. no. 1988.21.57. From Citado; ; Ecuador. Date: //. Source: Buckley; Gould coll.. Material: adult M; whole/good skin. Syntype.

Reg. no. 1988.21.58. From ; ; . Date: //. Source: Gould coll.. Material: adult M; whole/good skin.

Reg. no. 1988.21.59. From Citado; ; Ecuador. Date: //. Source: Buckley; Gould coll.. Material: adult M; whole/good skin. Syntype.

Reg. no. 1988.21.60. From Citado; ; Ecuador. Date: //. Source: Buckley; Gould coll.. Material: ; whole/good skin. Syntype. Adult female or immature.

Reg. no. 1988.21.61. From ; ; . Date: //. Source: Warszewicz; Gould coll.. Material: F; whole/good skin. The collector is queried on the label.

Reg. no. 1988.21.62. From Quito; ; Ecuador. Date: //. Source: Gould coll.. Material: adult F; whole/good skin. Previously identified as *Acestrura heliodor*.

Halcyon cinnamomina cinnamomina
Guam Kingfisher

Known only from the forests of the island of Guam in the Marianas. It was still common in 1945, but declined disastrously and by 1976 the total population was thought to be no more than 150 pairs. It is threatened by the introduced brown tree snake. In 1989 there were about 30–35 birds in captivity.

Reg. no. 1854.1.17.56. From ; ; Marianas Islands. Date: //. Source: Verreaux. Material: ; whole/good skin.

Reg. no. 1855.12.19.240. From ; ; Marianas Islands. Date: //. Source: Zoological Society. Material: ; whole/good skin.

Reg. no. 1863.7.8.1. From ; ; [Marianas Islands]. Date: //. Source: Leadbeater; Linnean Society. Material: adult ; whole/good skin. Holotype. Ex-mount. Swainson, 1821, *Zoological Illustrations*, first series, part 2: (unpaginated page), pl. 67.

Reg. no. 1896.7.16.13. From ; ; Guam. Date: 12/ 3/1895. Source: W. Rothschild. Material: M; whole/good skin.

Reg. no. 1896.7.16.14. From ; ; Guam. Date: 7/ 2/1895. Source: W. Rothschild. Material: M; whole/good skin.

Reg. no. 1896.7.16.15. From ; ; Guam. Date: 2/12/1894. Source: W. Rothschild. Material: F; whole/good skin.

Reg. no. 1896.7.16.16. From ; ; Guam. Date: 6/12/1894. Source: W. Rothschild. Material: F; whole/good skin.

Reg. no. 1969.40.91. From ; ; Guam. Date: 15/12/1894. Source: Hewitt coll.. Material: M; whole/good skin.

Electron carinatum
Keel-billed Motmot

Humid forest on the Caribbean slopes from Mexico to Costa Rica. The species is locally distributed throughout its range.

Reg. no. 1888.8.10.100. From ; ; Guatemala. Date: //. Source: Bouvier;

Salvin–Godman coll.. Material: adult ; whole/good skin.

Reg. no. 1888.8.10.98. From nr. the city of Belize; ; Belize (British Honduras). Date: / 1/1881. Source: F.J.S. Blancaneaux; Salvin–Godman coll.. Material: adult ; whole/good skin.

Reg. no. 1888.8.10.99. From Chontales; ; Nicaragua. Date: //. Source: T. Belt; Salvin–Godman coll.. Material: adult ; whole/good skin.

Reg. no. 1890.4.20.39. From ; ; Guatemala. Date: //1872. Source: Bouvier; Sclater coll.. Material: adult ; whole/good skin.

Reg. no. 1890.4.20.40. From Lake of Yojoa; ; [Honduras]. Date: //. Source: G.C. Taylor; Sclater coll.. Material: adult ; whole/good skin.

Reg. no. 1899.6.30.582. From La Libertad, Chontales; ; Nicaragua. Date: //1892. Source: W.B. Richardson; Salvin–Godman coll.. Material: adult M; whole/good skin.

Reg. no. 1900.6.20.63. From Rio Coco; ; Nicaragua. Date: 18/ 3/1898. Source: W.B. Richardson; Salvin–Godman coll.. Material: adult M; whole/good skin.

Reg. no. 1900.6.20.64. From Rio Coco; ; Nicaragua. Date: 24/ 3/1898. Source: W.B. Richardson; Salvin–Godman coll.. Material: adult M; whole/good skin.

Reg. no. 1900.6.20.65. From Rio Grande; ; Nicaragua. Date: 8/ 2/1898. Source: W.B. Richardson; Salvin–Godman coll.. Material: adult M; whole/good skin.

Reg. no. 1913.10.14.54. From Rio Makal, nr. San Filipi; Campeche; [Mexico]. Date: 22/ 5/1888. Source: F.J.S. Blancaneaux; Salvin–Godman coll.. Material: adult M; whole/good skin.

Jacamaralcyon tridactyla

Three-toed Jacamar

Occurs only in southeastern Brazil, where it is very local.

Reg. no. . From ; ; s.e. Brazil. Date: //. Source: Gould coll.. Material: adult ; skin. Mount. Specimen g. in the *Cat. Bds.*. Not found 1988.

Reg. no. 1888.8.21.92. From vicinity of Bahia; [Bahia]; Brazil. Date: //1868. Source: Wucherer; Salvin–Godman coll.. Material: adult F; whole/good skin. Erroneously stated in *Cat. Bds.* to be from the 'vicinity of Bolivia'.

Reg. no. 1888.8.21.93. From Novo Fribourgo; Rio de Janeiro; Brazil. Date: //1868. Source: J. Youds; Salvin–Godman coll.. Material: adult M; whole/good skin.

Reg. no. 1888.8.21.94. From Novo Fribourgo; Rio de Janeiro; Brazil. Date: //1868. Source: J. Youds; Salvin–Godman coll.. Material: adult M; whole/good skin.

Reg. no. 1890.8.12.64. From Santa Fe, Minas; ; Brazil. Date: //1869. Source: Rogers; Sclater coll.. Material: adult M; whole/good skin.

Reg. no. 1890.8.12.65. From ; ; s.e. Brazil. Date: //. Source: Leadbeater; Sclater coll.. Material: adult F; whole/good skin.

Reg. no. 1895.4.1.354. From Rio de Janeiro; ; [Brazil]. Date: //. Source: A. Fry. Material: adult M; whole/good skin.

Reg. no. 1895.4.1.355. From Rio de Janeiro; ; [Brazil]. Date: //. Source: A. Fry. Material: adult F; whole/good skin.

Reg. no. 1988.21.64. From ; ; Brazil. Date: //. Source: Material: adult ; whole/good skin. No history.

Reg. no. 1881.1.17.32. From ; ; s.e. Brazil. Date: //. Source: Eyton coll.; Gerrard. Material: ; whole/good skeleton.

Reg. no. 1940.12.8.197. From ; ; [South America]. Date: //. Source: Rothschild coll.. Material: adult ; part/good alc. Partly dissected.

Reg. no. A1988.30.1. From ; ; [South America]. Date: //. Source: Material: ; whole/good alc. No history.

Reg. no. A1988.30.2. From ; ; [South America]. Date: //. Source: Material: ; whole/good alc. No history.

Colaptes auratus rufipileus

Guadalupe Flicker

Forests of Guadalupe, off Baja California. Extinct since 1906, probably as a result of habitat destruction by goats and predation by feral cats.

Reg. no. 1898.3.10.56. From Guadalupe Island; ; Mexico. Date: 15/ 5/1885. Source: A. Forrer; E. Hargitt. Material: adult M; whole/good skin.

Dryocopus galeatus

Helmeted Woodpecker

Occurs in southern Brazil, Paraguay and Argentina, but now greatly reduced in range and no longer occurring in many areas.

Reg. no. 1846.7.16.68. From ; ; Brazil. Date: //. Source: [? Massena]; E. Wilson. Material: adult F; whole/good skin.

Reg. no. 1855.12.19.18. From ; ; Brazil. Date: //. Source: Zoological Society. Material: adult F; whole/good skin. Ex-mount. The *Cat. Bds.* indicates two females from the Zoological Society. The register states that 1855.12.19.18 was given as a duplicate to Nottingham Museum in 1895. Both specimens may have been registered with the same number.

Reg. no. 1898.3.10.2739. From Joinville; Santa Catarina (Santa Catharina); s. Brazil. Date: //. Source: F.W. Killmann; Berlepsch coll.; Hargitt coll.. Material: adult M; whole/good skin.

Reg. no. 1905.10.12.127. From Sapucay; ; Paraguay. Date: 22/ 8/1904. Source: W. Foster. Material: adult F; whole/good skin. Irides reddish-brown, bill creamy-grey, feet and legs slate colour.

Reg. no. 1905.10.12.128. From Sapucay; ; Paraguay. Date: 7/ 7/1904. Source: W. Foster. Material: adult M; whole/good skin. Irides reddish-brown, bill creamy-grey at point, greyish at base, feet and legs slate colour.

Dryocopus javensis richardsi
Tristram's Woodpecker

Parts of Korea and formerly the islands of Tsushima in the Korean Strait, where it has not been seen since 1920. It still persists in a few localities in central Korea.

Reg. no. 1891.10.13.20. From ; ; Korea. Date: 2/ 2/1891. Source: C.W. Campbell. Material: ; whole/good skin.

Reg. no. 1897.10.30.638. From Tsushima Island; ; Japan. Date: 16/10/1891. Source: P.A. Holst; Seebohm coll.. Material: adult M; whole/good skin.

Reg. no. 1897.10.30.639. From Tsushima Island; ; Japan. Date: 21/10/1891. Source: P.A. Holst; Seebohm coll.. Material: adult M; whole/good skin.

Reg. no. 1897.10.30.640. From Tsushima Island; ; Japan. Date: 30/11/1891. Source: P.A. Holst; Seebohm coll.. Material: adult F; whole/good skin.

Reg. no. 1897.10.30.641. From Tsushima Island; ; Japan. Date: 14/11/1891. Source: P.A. Holst; Seebohm coll.. Material: adult F; whole/good skin.

Reg. no. 1901.9.18.4. From ; ; Korea. Date: 28/10/1900. Source: E. Scarlett. Material: adult F; whole/good skin.

Campephilus principalis principalis
Ivory-billed Woodpecker

The species occurs in Cuba (*C.p. bairdii*) and in southeast USA (*C.p. principalis*). The Cuban race survives only in the eastern highlands; the mainland form has not definitely been seen for a long time, but unconfirmed sightings persist.

Reg. no. . From De Soto Co.; Florida; United States. Date: //. Source: W.E.D. Scott; Salvin-Godman coll.. Material: adult ; skin. Five specimens, adult males and females, February, May and December. Details taken from *Cat. Bds.*. Not found 1988.

Reg. no. . From Argo; Florida; United States. Date: //. Source: W.E.D. Scott; Salvin-Godman coll.. Material: adult F; skin. Details taken from *Cat. Bds.* Not found 1988.

Reg. no. 1888.10.10.7527. From ; Florida; United States. Date: //1874. Source: E.R. Same; C. Hart Merriam; Henshaw coll.; Salvin-Godman coll.. Material: adult M; whole/good skin.

Reg. no. 1898.3.10.2482. From ; Florida; United States. Date: //. Source: G.B. Sennett; E. Hargitt coll.. Material: adult F; whole/good skin.

Reg. no. 1898.3.10.2483. From ; Florida; United States. Date: //. Source: G.A Boardman; Sennett coll.; E. Hargitt coll.. Material: adult M; whole/good skin.

Reg. no. 1898.3.10.2484. From ; Florida; United States. Date: / 1/. Source: G.B. Sennett; E. Hargitt coll.. Material: adult F; whole/good skin. Ex-mount.

Reg. no. 1898.3.10.2485. From Caloosahatchee River; Florida; United States. Date: //1881. Source: G.B. Sennett;

E. Hargitt coll.. Material: adult M; whole/good skin.

Reg. no. 1898.3.14.874. From De Soto Co.; Florida; United States. Date: 3/ 2/1887. Source: W.R. Dean; Salvin–Godman coll.. Material: adult F; whole/good skin.

Reg. no. 1898.3.14.875. From De Soto Co.; Florida; United States. Date: 1/ 2/1887. Source: W.R. Dean; Salvin–Godman coll.. Material: adult F; whole/good skin.

Reg. no. 1898.3.14.876. From De Soto Co.; Florida; United States. Date: 17/ 5/1887. Source: W.R. Dean; Salvin–Godman coll.. Material: adult M; whole/good skin.

Reg. no. 1898.3.14.877. From De Soto Co.; Florida; United States. Date: 12/12/1886. Source: W.R. Dean; Salvin–Godman coll.. Material: adult M; whole/good skin.

Reg. no. 1898.3.14.878. From De Soto Co.; Florida; United States. Date: 12/12/1886. Source: W.R. Dean; Salvin–Godman coll.. Material: adult F; whole/good skin.

Reg. no. 1898.3.14.879. From Argo; Florida; United States. Date: 30/ 3/1889. Source: W.E.D. Scott; Salvin–Godman coll.. Material: adult F; whole/good skin.

Reg. no. 1898.3.14.880. From Tarpon Springs; Florida; United States. Date: 15/ 3/1887. Source: W.E.D. Scott; Salvin–Godman coll.. Material: chick F; whole/good skin.

Reg. no. 1906.12.7.920. From ; Florida; United States. Date: //. Source: G.S. Miller. Material: ; whole/good skin.

Reg. no. 1914.12.1.490. From Charlotte Harbour; Florida; United States. Date: 29/ 3/1904. Source: P.R. Lowe. Material: M; whole/good skin.

Reg. no. 1914.12.1.491. From Charlotte Harbour; Florida; United States. Date: 29/ 3/1904. Source: P.R. Lowe. Material: F; whole/good skin.

Reg. no. 1939.12.9.549. From ; ; North America. Date: //. Source: Rothschild coll.. Material: M; whole/good skin. Mount.

Reg. no. 1939.12.9.550. From ; ; North America. Date: //. Source: Rothschild coll.. Material: F; whole/good skin. Mount.

Reg. no. VEL.23.170a. From ; ; North America. Date: //. Source: Hardwicke coll.. Material: adult M; whole/good skin.

Campephilus imperialis

Imperial Woodpecker

Formerly occurred in pine forests in parts of Mexico, but may now be extinct as a result of forest clearance and killing for food.

Reg. no. . From ; ; Mexico. Date: //. Source: Zoological Society. Material: adult F; skin. Mount. Details taken from *Cat. Bds.*. Not found 1988.

Reg. no. . From ; ; Mexico. Date: //. Source: Material: adult M; skin. Mount. Details taken from *Cat. Bds.*. Not found 1988.

Reg. no. 1888.8.5.60. From ; ; Mexico. Date: //. Source: J. Gould; Salvin–Godman coll.. Material: adult F; whole/good skin. Syntype. The sex is given as male in the register.

Reg. no. 1888.8.5.61. From ; ; Mexico. Date: //1876. Source: J. Gould; Salvin-Godman coll.. Material: adult M; whole/good skin. Syntype. Gould, *Proc. zool. Soc. Lond.* 1832: 140. The other syntype is also in the collection. See Salvin & Godman, 1895, *Biologia Centrali-Americana*, Aves, 2: 444-6, concerning the probable collector and locality.

Reg. no. 1888.8.5.62. From Cuidad; Durango; Mexico. Date: / 1/1882. Source: A. Forrer; Salvin-Godman coll.. Material: adult F; whole/good skin. The reg. no. is erroneously given as 1888.8.1.63 on the label.

Reg. no. 1888.8.5.63. From Cuidad; Durango; Mexico. Date: / 1/1882. Source: A. Forrer; Salvin-Godman coll.. Material: adult M; whole/good skin. The reg. no. is erroneously given as 1888.8.1.62 on the label.

Reg. no. 1898.3.10.2486. From Ciudad; Durango; Mexico. Date: 15/ 2/. Source: A. Forrer; E. Hargitt coll.. Material: adult M; whole/good skin.

Reg. no. 1898.3.10.2487. From Ciudad; Durango; Mexico. Date: 15/ 2/. Source: A. Forrer; E. Hargitt coll.. Material: adult F; whole/good skin.

Reg. no. 1988.21.6. From ; ; Mexico. Date: //. Source: J. Gould. Material: adult M; whole/good skin.

Sapheopipo noguchii

Okinawa Woodpecker

Endemic to Okinawa where it is found in hill forest in the north of the island. It is now rare as a result of continuing habitat loss. Its numbers are estimated to be between 40 and 200.

Reg. no. 1889.5.1.1. From Okinawa; Ryukyu Islands; Japan. Date: / 8/1886. Source: H. Pryer; Seebohm coll.. Material: juv M; whole/good skin. Holotype. Seebohm, *Ibis* 1887: 178, pl. 7.

Reg. no. 1897.10.30.647. From Okinawa; ; Japan. Date: 10/ 3/1892. Source: P.A. Holst; Seebohm coll.. Material: F; whole/good skin. The locality is incorrectly given as Tsushima in the register.

Reg. no. 1897.10.30.648. From Okinawa; ; Japan. Date: 14/ 3/1892. Source: P.A. Holst; Seebohm coll.. Material: adult M; whole/good skin. The locality is incorrectly given as Tsushima in the register.

Reg. no. 1897.10.30.649. From Okinawa; ; Japan. Date: 28/ 2/1892. Source: P.A. Holst; Seebohm coll.. Material: adult F; whole/good skin. The locality is incorrectly given as Tsushima in the register.

Geositta excelsior aricomae

Stout-billed Cinclodes

Known from a few mountain woodlands in Peru and Bolivia, and now exceptionally rare.

Reg. no. 1889.5.14.56. From Tilotilo, Prov. Yungas; ; Bolivia. Date: //1876. Source: C. Buckley; Salvin-Godman coll.. Material:

adult ; whole/good skin. Previously ascribed to *Cinclodes nigrofumosus*; of uncertain identity. This specimen is from outside the range given in Peters' *Check-list* for *G.e. aricomae*.

Myrmotherula erythronotos

Black-hooded Antwren

Formerly known only from Nova Friburgo in Rio de Janeiro, Brazil, where it had been thought to be extinct as a result of forest destruction. It was rediscovered at another locality in 1987, but is still exceedingly rare.

Reg. no. . From ; ; s.e. Brazil. Date: //. Source: Gould coll.. Material: adult M; skin. Details from *Cat. Bds.*. Not found 1988.

Reg. no. 1855.12.17.66. From ; Rio de Janeiro; Brazil. Date: //. Source: Rio de Janeiro Museum. Material: ; skin. Not in the collection. Specimen a. in *Cat. Bds.*. Exchanged with Museo National Rio de Janeiro in 1984.

Reg. no. 1889.9.20.277. From ; ; Brazil. Date: //. Source: Sclater coll.. Material: M; skin. Not in the collection; exchanged with the American Museum of Natural History in 1921.

Reg. no. 1889.9.20.278. From ; ; Brazil. Date: //. Source: Sclater coll.. Material: F; whole/good skin.

Reg. no. 1895.4.1.656. From ; Rio de Janeiro; Brazil. Date: //. Source: A. Fry. Material: M; whole/good skin.

Pyriglena atra

Fringe-backed Fire-eye

Known only from around Salvador in Bahia, Ecuador. It is threatened by habitat destruction. Its present status seems to be unknown.

Reg. no. 1857.11.4.8. From ; ; Brazil. Date: //. Source: Zoological Society. Material: M; whole/good skin.

Reg. no. 1889.9.20.442. From Para; lower Amazon; Brazil. Date: //. Source: Layard; Sclater coll.. Material: M; skin. Details taken from the register. Not found 1988.

Reg. no. 1889.9.20.443. From Para; lower Amazon; Brazil. Date: //. Source: Layard; Sclater coll.. Material: F; skin. Details taken from the register. Not found 1988.

Reg. no. 1889.9.20.444. From ; Matto Grosso; Brazil. Date: //. Source: Natterer; Sclater coll.. Material: M; skin. Details taken from the register. Not found 1988.

Reg. no. 1889.9.20.445. From ; ; Brazil. Date: //. Source: Sclater coll.. Material: M; whole/good skin.

Reg. no. A1969.13.36. From ; ; [Brazil]. Date: //. Source: Material: ; whole/good alc.

Grallaria gigantea hylodroma

Giant Antpitta

The species occurs in three races in the Andes of Ecuador and the eastern slope of the central Andes of Colombia. All are very rare, and threatened by habitat loss. The Colombian form (no specimens in the collection) may already be extinct.

Reg. no. 1916.8.24.50. From Mt. Pichincha; ; Ecuador. Date: /12/1914. Source: Walter Goodfellow; E.J. Brook. Material: adult F; whole/good skin. Iris brown, feet brown, upper mandible dark brown, lower mandible paler and slightly yellow at tip. In the register, the month of collection is given as November, and the age as juvenile.

Reg. no. 1938.12.20.1. From Lloa, nr. Mt. Pichincha; ; Ecuador. Date: 2/ 8/1937. Source: C.S. Webb. Material: adult M; whole/good skin. From 0.03S 78.07W.

Reg. no. 1940.12.5.314. From Mindo, Mt. Pichincha; ; Ecuador. Date: 28/ 2/1939. Source: L. Gomez. Material: adult M; whole/good skin.

Reg. no. 1940.12.5.315. From Mindo, Mt. Pichincha; ; Ecuador. Date: 27/ 2/1939. Source: L. Gomez. Material: adult F; skin. Details taken from the register. Not found 1988.

Reg. no. 1953.68.–. From San Jose; ; Ecuador. Date: //. Source: L. Gomez. Material: adult F; whole/good skin. Reg. no. incomplete and untraced.

Grallaria rufocinerea

Bicolored Antpitta

Very little is known about this species which is recorded from a few widely separated localities in the Central Andes of Colombia. It inhabits forest undergrowth.

Reg. no. 1889.9.20.618. From Santa Elena, Antioquia; ; Colombia. Date: //. Source: T.K. Salmon; Sclater coll.. Material: adult ; whole/good skin. Holotype. P.L. Sclater & Salvin, *Proc. zool. Soc. Lond.* 1879: 526–7.

Grallaria milleri

Brown–banded Antpitta

Probably extinct. It has not been seen since about 1942. Formerly occurred in two cloud forest areas in the Andes of Colombia. Most of its habitat has been cleared.

Reg. no. 1921.7.3.61. From La Guneta, w. Quindio Andes; Cauca; Colombia. Date: 31/ 8/1911. Source: A.A. Allen & L.E. Miller; American Museum of Natural History. Material: adult F; whole/good skin. See also *Bull. Am. Mus. nat. Hist.* 31 (1912): 147.

Iodopleura pipra leucopygia

Buff–throated Purpletuft

Known only from these two specimens, said to have come from Guyana, but this has been questioned.

Reg. no. 1888.1.20.969. From ; ; Guyana (British Guiana). Date: //1877. Source: H. Whitely, sen.; Salvin–Godman coll.. Material: adult ; whole/good skin. Syntype. Salvin, *Ibis* 1885: 305. The other syntype is in the collection.

Reg. no. 1888.1.20.970. From ; ; Guyana (British Guiana). Date: //1877. Source: H. Whitely, sen.; Salvin–Godman coll.. Material: adult ; whole/good skin. Syntype.

Iodopleura pipra pipra

Buff-throated Purpletuft

Endemic to eastern Brazil where it has been found in old second growth and forest treetops.

Reg. no. 1844.9.24.4. From ; ; Brazil. Date: //. Source: Leadbeater. Material: adult ; whole/good skin. Ex-mount.

Reg. no. 1881.5.1.3744. From ; ; Brazil. Date: //. Source: Gould coll.. Material: adult F; whole/good skin.

Reg. no. 1888.1.13.1667. From ; ; Brazil. Date: //. Source: Argent; Sclater coll.. Material: adult M; whole/good skin.

Reg. no. 1888.1.13.1668. From ; ; Brazil. Date: //. Source: Argent; Sclater coll.. Material: adult F; whole/good skin.

Reg. no. 1888.1.13.1669. From ; ; Brazil. Date: //. Source: Verreaux; Sclater coll.. Material: imm ; whole/good skin.

Reg. no. 1888.1.20.967. From ; ; Brazil. Date: //. Source: Salvin–Godman coll.. Material: adult M; whole/good skin.

Reg. no. 1888.1.20.968. From Novo Fribourgo; Rio de Janeiro; s.e. Brazil. Date: //. Source: J. Youds; Salvin–Godman coll.. Material: adult M; whole/good skin.

Reg. no. 1895.4.1.735. From Rio de Janeiro; ; [Brazil]. Date: //. Source: A. Fry. Material: adult ; whole/good skin.

Reg. no. 1895.4.1.736. From Rio de Janeiro; ; [Brazil]. Date: //. Source: A. Fry. Material: adult M; whole/good skin.

Reg. no. 1895.4.1.737. From Rio de Janeiro; ; [Brazil]. Date: //. Source: A. Fry. Material: adult F; whole/good skin.

Reg. no. VEL.12.177a. From ; ; Brazil. Date: //. Source: Lord Stuart. Material: ; skin. Details taken from the register. Not found 1988, and not in the *Cat. Bds.*, so presumably missing by then (1888).

Reg. no. VEL.12.177b. From ; ; Brazil. Date: //. Source: Gould. Material: adult ; whole/good skin. Ex-mount.

Xenicus longipes stokesii

North Island Bush Wren

Formerly occurred in North Island, New Zealand, where it was mainly ground haunting and prey to rats and other imported predators. It was recorded only from the Rimataka Hills and is almost certainly now extinct. It was last seen in 1949.

Reg. no. 1852.1.17.38. From Rimutaku Hills, Wellington; North Island; New Zealand. Date: //. Source: Lyall, Stokes, voyage of H.M.S. Acheron. Material: imm ; whole/good skin. Holotype. G.R. Gray, *Ibis* 1862: 219. See also *Bull. Brit. Orn. Club* 15 (1904): 15. The register gives the locality as Queen Charlotte Sound.

Reg. no. 1852.1.17.39. From Rimutaku Hills, Wellington; North Island; New Zealand. Date: //. Source: Stokes. Material: ; whole/good skin. The register gives the locality as Rima Taken Hills.

Xenicus longipes longipes
South Island Bush Wren

Mountain forests of South Island, New Zealand, where it is now probably extinct, having been seen last in 1972.

Reg. no. 1875.7.2.21. From Milford Sound; South Island; New Zealand. Date: 7/ 1/1893. Source: Colonial Museum. Material: adult M; whole/good skin.

Reg. no. 1881.5.1.255. From ; ; New Zealand. Date: //. Source: Gould coll.. Material: adult ; whole/good skin.

Reg. no. 1902.2.21.37. From Dusky Sound; South Island; New Zealand. Date: / 3/1901. Source: Ranfurly. Material: M; whole/good skin.

Reg. no. 1902.2.21.38. From Dusky Sound; South Island; New Zealand. Date: / 3/1901. Source: Ranfurly. Material: M; whole/good skin.

Reg. no. 1903.12.10.20. From Nelson; South Island; New Zealand. Date: 20/10/1896. Source: H.H. Travers; Rosenberg. Material: F; whole/good skin.

Reg. no. 1903.12.10.21. From Nelson; South Island; New Zealand. Date: 20/10/1896. Source: H.H. Travers; Rosenberg. Material: adult M; whole/good skin.

Reg. no. 1903.12.10.22. From Karamea Range; South Island; New Zealand. Date: //1895. Source: H.H. Travers; Rosenberg. Material: M; whole/good skin.

Reg. no. 1904.8.2.8. From Nelson; South Island; New Zealand. Date: 1/ 5/1897. Source: Ranfurly. Material: M; whole/good skin.

Reg. no. 1904.8.2.9. From Nelson; South Island; New Zealand. Date: / 1/1901. Source: Ranfurly. Material: F; whole/good skin.

Reg. no. 1927.12.18.305. From West Coast Sound, George Sound; ; New Zealand. Date: / 1/1889. Source: J.J. Lister coll.. Material: M; whole/good skin.

Reg. no. 1945.54.3. From ; ; New Zealand. Date: //. Source: W.J.P. Wardlaw. Material: F; whole/good skin.

Reg. no. 1969.41.1452. From Karamea Ranges; South Island; New Zealand. Date: //. Source: Hewitt coll.. Material: adult M; whole/good skin.

Reg. no. . From Dusky Bay; ; New Zealand. Date: //. Source: Material: ; painting. Holotype. Gmelin, *Systema Naturae*, 1, pt. 2, (1789): 979. Based on a drawing in the NHM library, by Forster, (old) no. 165. See also *Bull. Br. Mus. nat. Hist. (Hist.)* 1 (1959): 309–10.

Reg. no. 1940.12.8.148. From Nelson; South Island; New Zealand. Date: 25/ 5/1896. Source: H.H. Travers; Rothschild coll.. Material: adult ; whole/good alc.

Xenicus longipes variabilis
Stewart Island Bush Wren

Formerly occurred on Stewart Island (where extinct since 1967) and on small islets off its southwestern coast where the race may still survive.

Reg. no. 1932.10.25.2. From Solomon Island; Stewart Island; New Zealand. Date: /12/1932. Source: E.F. Stead. Material: ; whole/good skin.

***Xenicus longipes* subspecies**

Reg. no. S1972.1.107. From ; ; [New Zealand]. Date: //. Source: Rothschild coll.. Material: ; whole/good skeleton. See also S1972.1.108: only one skeleton is listed in *Bull. Br. Mus. nat. Hist. (Zool.)* 34 (1979): 158.

Reg. no. S1972.1.108. From ; ; [New Zealand]. Date: //. Source: Rothschild coll.. Material: ; whole/good skeleton. See also S1972.1.107: only one skeleton is listed in *Bull. Br. Mus. nat. Hist. (Zool.)* 34 (1979): 158.

Reg. no. 1940.12.8.89-93. From ; South Island; New Zealand. Date: //. Source: Rothschild coll.. Material: adult ; whole/good alc. Six specimens. There is a note in the register opposite no. 89 which reads "Part also made into flat skin (spirit and skin collection)" [writing identified as that of G.S. Cowles].

Xenicus lyalli
Stephens Island Wren

Occurred only on Stephens Island in Cook Strait, New Zealand. All the known specimens were collected in 1894 by the lighthouseman's cat, which also exterminated the species.

Reg. no. 1895.10.17.13. From ; Stephens Island; New Zealand. Date: //. Source: H.H. Travers; Rothschild. Material: adult F; whole/good skin. Details taken from the register.

Reg. no. 1939.12.9.76. From ; Stephens Island; New Zealand. Date: //1894. Source: Rothschild coll.. Material: adult M; whole/good skin.

Reg. no. 1939.12.9.77. From ; Stephens Island; New Zealand. Date: //1894. Source: Rothschild coll.. Material: adult F; whole/good skin.

Atrichornis clamosus
Noisy Scrub-Bird

Formerly occurred through southwestern Australia, but was not recorded from 1889 to 1961 and was thought to be extinct. In 1961 a small population was rediscovered. Numbers in some areas have been increasing in recent years through careful conservation.

Reg. no. 1880.12.31.68. From Albany, King George's Sound; Western Australia; Australia. Date: 4/ 6/1843. Source: J. Gould; T.C. Eyton; E. Gerrard. Material: M; whole/good skin.

Reg. no. 1886.1.5.3. From Perth; Western Australia; Australia. Date: //. Source: Colonial Exhibition. Material: ; whole/good skin.

Calandrella razae

Razo Lark

Occurs only on the small island of Razo in the Cape Verde Islands, where the size of the population apparently varies with the climate, having been only about 20 pairs in the early 1980s, but by 1985 had risen to about 150 birds. The taxonomic position of this species has been the subject of much controversy, it has been placed in *Alauda* and in the monotypic genus *Razocorys*.

Reg. no. 1897.7.30.5. From ; Razo Island (Raza Island); Cape Verde Islands. Date: 28/ 4/1897. Source: B. Alexander. Material: adult F; whole/good skin.

Reg. no. 1897.7.30.6. From ; Razo Island (Raza Island); Cape Verde Islands. Date: 28/ 4/1897. Source: B. Alexander. Material: adult M; whole/good skin.

Reg. no. 1900.5.26.92. From ; Razo Island (Raza Island); Cape Verde Islands. Date: 11/10/1897. Source: B. Alexander. Material: adult F; whole/good skin.

Reg. no. 1911.12.23.2929. From ; Razo Island (Raza Island); Cape Verde Islands. Date: / 5/1897. Source: B. Alexander. Material: adult M; whole/good skin.

Reg. no. 1911.12.23.2930. From ; Razo Island (Raza Island); Cape Verde Islands. Date: / 5/1897. Source: B. Alexander. Material: adult F; whole/good skin.

Reg. no. 1911.12.23.2931. From ; Razo Island (Raza Island); Cape Verde Islands. Date: / 5/1897. Source: B. Alexander. Material: adult M; whole/good skin.

Reg. no. 1911.12.23.2932. From ; Razo Island (Raza Island); Cape Verde Islands. Date: / 5/1897. Source: B. Alexander. Material: adult M; whole/good skin. Syntype. Alexander, *Ibis* 1898: 107–8, pl. 3. This is the author's selected male. His selected female and several other syntypes are also in the collection.

Reg. no. 1911.12.23.2933. From ; Razo Island (Raza Island); Cape Verde Islands. Date: 8/10/1897. Source: B. Alexander. Material: adult F; whole/good skin.

Reg. no. 1911.12.23.2934. From ; Razo Island (Raza Island); Cape Verde Islands. Date: 11/10/1897. Source: B. Alexander. Material: adult M; whole/good skin.

Reg. no. 1911.12.23.2935. From ; Razo Island (Raza Island); Cape Verde Islands. Date: 11/10/1897. Source: B. Alexander. Material: adult M; whole/good skin.

Reg. no. 1911.12.23.2936. From ; [Razo Island]; Cape Verde Islands. Date: 11/10/1897. Source: B. Alexander. Material: adult M; whole/good skin.

Reg. no. 1911.12.23.2937. From ; Razo Island (Raza Island); Cape Verde Islands. Date: 11/10/1897. Source: B. Alexander. Material: adult M; whole/good skin.

Reg. no. 1911.12.23.2938. From ; Razo Island (Raza Island); Cape Verde Islands. Date: 11/10/1897. Source: B. Alexander. Material: adult M; whole/good skin.

Reg. no. 1911.12.23.2939. From ; Razo Island (Raza Island); Cape Verde Islands.

Date: 11/10/1897. Source: B. Alexander. Material: adult M; whole/good skin.

Reg. no. 1911.12.23.2940. From ; Razo Island (Raza Island); Cape Verde Islands. Date: 11/10/1897. Source: B. Alexander. Material: adult F; whole/good skin.

Reg. no. 1911.12.23.2941. From ; Razo Island (Raza Island); Cape Verde Islands. Date: 11/10/1897. Source: B. Alexander. Material: adult M; whole/good skin.

Reg. no. 1911.12.23.2942. From ; Razo Island (Raza Island); Cape Verde Islands. Date: 11/10/1897. Source: B. Alexander. Material: adult F; whole/good skin.

Reg. no. 1911.12.23.2943. From ; Razo Island (Raza Island); Cape Verde Islands. Date: 11/10/1897. Source: B. Alexander. Material: adult F; whole/good skin.

Reg. no. 1911.12.23.2944. From ; Razo Island (Raza Island); Cape Verde Islands. Date: 11/10/1897. Source: B. Alexander. Material: adult F; whole/good skin.

Reg. no. 1911.12.23.2945. From ; Razo Island (Raza Island); Cape Verde Islands. Date: 11/10/1897. Source: B. Alexander. Material: adult F; whole/good skin.

Reg. no. 1936.2.21.740. From ; Razo Island (Raza Island); Cape Verde Islands. Date: 11/10/1897. Source: B. Alexander. Material: adult F; whole/good skin.

Reg. no. 1936.2.21.741. From ; Razo Island (Raza Island); Cape Verde Islands. Date: 11/10/1897. Source: B. Alexander. Material: adult M; whole/good skin.

Reg. no. 1936.2.21.742. From ; Razo Island (Raza Island); Cape Verde Islands. Date: 11/10/1897. Source: B. Alexander. Material: adult F; whole/good skin.

Reg. no. 1936.2.21.744. From ; Razo Island (Raza Island); Cape Verde Islands. Date: 11/10/1897. Source: B. Alexander. Material: adult M; skin. Details taken from the register. Not in the collection; exchanged with Meinertzhagen in 1949, but apparently not returned with his collection.

Reg. no. 1936.2.21.745. From ; Razo Island (Raza Island); Cape Verde Islands. Date: / 5/1897. Source: B. Alexander. Material: adult M; whole/good skin.

Reg. no. 1965.M.7325. From ; Razo Island (Raza Island); Cape Verde Islands. Date: 11/10/1897. Source: B. Alexander; Meinertzhagen coll.. Material: adult M; whole/good skin. Previously registered as 1936.2.21.743. Exchanged with Meinertzhagen in 1949, and subsequently returned with that collection.

Coracinia striata cebuensis

Bar-bellied Cuckoo-Shrike

Endemic to Cebu where it became extinct in 1906. Nothing appears to be known about it.

Reg. no. 1887.12.30.40. From ; Cebu; Philippines. Date: / 4/1877. Source: A.H. Everett; Tweeddale coll. Material: adult M; whole/good skin. Syntype. Ogilvie Grant, *Ibis* 1896: 535. The female syntype is also in the collection.

Reg. no. 1887.12.30.41. From ; Cebu; Philippines. Date: / 4/1877. Source: A.H. Everett; Tweeddale coll. Material: imm F; whole/good skin.

Reg. no. 1887.12.30.42. From ; Cebu; Philippines. Date: / 4/1877. Source: A.H. Everett; Tweeddale coll. Material: imm M; whole/good skin.

Reg. no. 1887.12.30.43. From ; Cebu; Philippines. Date: / 4/1877. Source: A.H. Everett; Tweeddale coll. Material: adult F; whole/good skin. Syntype. Damaged: head loose.

Coracina coerulescens altera

Blackish Cuckoo-Shrike

Endemic to Cebu where it is believed to have been extinct since 1906, though Hachisuka, writing in the 1930s, recorded that it was still common.

Reg. no. 1887.12.30.130. From ; Cebu; Philippines. Date: / 4/1877. Source: A.H. Everett; Tweeddale coll. Material: adult M; whole/good skin. Syntype. Ramsay, *Ibis* 1881: 34. There are three other syntypes in the collection.

Reg. no. 1887.12.30.131. From ; Cebu; Philippines. Date: / 4/1877. Source: A.H. Everett; Tweeddale coll. Material: adult F; whole/good skin. Syntype.

Reg. no. 1887.12.30.132. From ; Cebu; Philippines. Date: / 4/1877. Source: A.H. Everett; Tweeddale coll. Material: adult M; whole/good skin. Syntype.

Reg. no. 1887.12.30.133. From ; Cebu; Philippines. Date: / 4/1877. Source: A.H. Everett; Tweeddale coll. Material: adult F; whole/good skin. Syntype.

Reg. no. 1887.12.30.134. From ; Cebu; Philippines. Date: / 4/1877. Source: A.H. Everett; Tweeddale coll. Material: adult F; skin. Syntype. Not in the collection. Exchanged with Delaware Museum in 1972.

Lalage leucopyga leucopyga

Norfolk Island Triller

Known only from Norfolk Island where it became extinct in 1962.

Reg. no. 1856.8.30.24. From ; ; Norfolk Island. Date: //. Source: Dr Rayner, voyage of H.M.S. Herald. Material: adult M; whole/good skin.

Reg. no. 1856.8.30.25. From ; ; Norfolk Island. Date: //. Source: Dr Rayner, voyage of H.M.S. Herald. Material: adult F; whole/good skin.

Reg. no. 1870.3.31.89. From ; ; Norfolk Island. Date: / 6/1865. Source: J. Brenchley. Material: adult M; whole/good skin. Eye black.

Reg. no. 1881.5.1.3726. From ; ; [Norfolk Island]. Date: //. Source: Gould. Material: adult ; whole/good skin.

Reg. no. 1897.7.2.161. From ; ; Norfolk Island. Date: //. Source: Seebohm coll.. Material: adult M; whole/good skin.

Reg. no. 1897.7.2.162. From ; ; Norfolk Island. Date: /10/1879. Source: E.L. Layard; Seebohm coll.. Material: adult ; whole/good skin. Bill black, legs black. Formerly in alcohol.

Reg. no. 1919.7.15.23. From Ball Bay Reserve; ; Norfolk Island. Date: 19/ 7/1912. Source: G.M. Mathews. Material: adult M; whole/good skin. Iris brown, feet black tinged greenish, bill black.

Reg. no. 1919.7.15.24. From Steel's Point; ; Norfolk Island. Date: 11/11/1912. Source: G.M. Mathews. Material: adult M; whole/good skin. Iris dark brown, feet black, bill black.

Reg. no. 1919.7.15.25. From Steel's Point; ; Norfolk Island. Date: 23/ 1/1913. Source: G.M. Mathews. Material: adult M; whole/good skin. Iris dark grey, feet black, bill black.

Reg. no. 1919.7.15.26. From nr. Cascades; ; Norfolk Island. Date: 8/ 3/1913. Source: G.M. Mathews. Material: adult M; whole/good skin. Iris dark brown, feet blue–black, bill bluish–black.

Reg. no. 1919.7.15.27. From Dixey; ; Norfolk Island. Date: 29/ 3/1913. Source: G.M. Mathews. Material: adult M; whole/good skin. Iris dark brown, feet blue–black, bill black.

Reg. no. 1919.7.15.28. From Dixey; ; Norfolk Island. Date: 29/ 3/1913. Source: G.M. Mathews. Material: adult F; whole/good skin. Iris dark brown, feet blue–black, bill black.

Reg. no. 1919.7.15.29. From Dixey; ; Norfolk Island. Date: 23/ 4/1913. Source: G.M. Mathews. Material: adult F; whole/good skin. Iris dark brown, feet blue–black, bill bluish–black.

Reg. no. 1919.7.15.30. From nr. Cascades; ; Norfolk Island. Date: 30/ 4/1913. Source: G.M. Mathews. Material: adult F; whole/good skin. Iris dark brown, feet blue–black, bill black and yellow.

Reg. no. 1919.7.15.31. From Dixey; ; Norfolk Island. Date: 1/ 5/1913. Source: G.M. Mathews. Material: adult F; whole/good skin. Iris dark brown, feet bluish–black, bill black and greyish.

Reg. no. 1919.7.15.32. From Dixey; ; Norfolk Island. Date: 1/ 5/1913. Source: G.M. Mathews. Material: adult M; whole/good skin. Iris dark brown, feet blue–black, bill black and yellow.

Reg. no. 1919.7.15.33. From nr. Kingston; ; Norfolk Island. Date: 9/ 5/1913. Source: G.M. Mathews. Material: adult ; whole/good skin. Iris dark brown, feet bluish–black, bill black tinged greyish.

Reg. no. 1919.7.15.34. From nr. Middlegate; ; Norfolk Island. Date: 22/ 5/1913. Source: G.M. Mathews. Material: M; whole/good skin. Iris dark brown, feet blue–black, bill black and yellow.

Reg. no. 1939.12.9.2295. From Dixey; ; Norfolk Island. Date: 29/ 4/1913. Source: G.M. Mathews; Rothschild coll.. Material: adult F; whole/good skin. Iris dark brown, feet bluish–black, bill black and yellow.

Reg. no. 1939.12.9.2296. From nr. Middlegate; ; Norfolk Island. Date: 28/ 4/1913. Source: G.M. Mathews; Rothschild coll.. Material: adult F; whole/good skin. Iris dark brown or grey, feet greyish-black, bill black and yellow.

Reg. no. 1939.12.9.2297. From Dixey; ; Norfolk Island. Date: 28/ 4/1913. Source: G.M. Mathews; Rothschild coll.. Material: adult F; whole/good skin. Iris dark grey, feet greyish-black, bill black and yellow.

Reg. no. 1939.12.9.2298. From ; ; Norfolk Island. Date: //. Source: Rothschild coll.. Material: adult ; whole/good skin.

Reg. no. E1961.1.436. From ; ; Norfolk Island. Date: //. Source: J.B. Gould coll.. Material: 2 ; whole/good eggs (clutch).

Lanius newtoni

Newton's Fiscal

Endemic to Sao Tome where it is found in primary forest. The species was rediscovered in 1990, when one bird was observed. Prior to that, it had not been seen since 1928.

Reg. no. 1891.8.20.8. From ; ; Sao Tome (Sao Thome). Date: /10/1890. Source: Prof. B.du Bocage. Material: adult ; whole/good skin. There are two labels, one stating the bird to be a male, the other, female.

Cinclus cinclus olympicus

Cyprus Dipper

A large dipper known only from the mountains of Cyprus. It has not been seen since 1939.

Reg. no. 1903.8.2.5. From Troodos; ; Cyprus. Date: 5/11/1901. Source: D.M.A. Bate. Material: F; whole/good skin.

Reg. no. 1909.8.7.41. From Troodos; ; Cyprus. Date: / 5/1909. Source: C.B. Horsbrugh. Material: adult M; whole/good skin.

Reg. no. 1909.8.7.42. From Troodos; ; Cyprus. Date: / 6/1909. Source: C.B. Horsbrugh. Material: adult F; whole/good skin.

Reg. no. 1909.8.7.43. From Troodos; ; Cyprus. Date: / 5/1909. Source: C.B. Horsbrugh. Material: juv M; whole/good skin.

Reg. no. 1909.8.7.44. From Troodos; ; Cyprus. Date: 30/ 5/1909. Source: C.B. Horsbrugh. Material: juv M; whole/good skin.

Reg. no. 1909.8.7.45. From Troodos; ; Cyprus. Date: / 5/1909. Source: C.B. Horsbrugh. Material: juv M; whole/good skin. The month of collection is given as June in the register.

Reg. no. 1909.8.7.46. From Troodos; ; Cyprus. Date: / 6/1909. Source: C.B. Horsbrugh. Material: juv M; whole/good skin. The age is erroneously given as adult in the register.

Reg. no. 1951.13.1734. From Troodos; ; Cyprus. Date: 12/ 5/1909. Source: W.A. Payn. Material: adult M; whole/good skin.

Hylorchilus sumichrasti

Sumichrast's Wren

Occurs only in a few areas of dense lowland forest in Mexico and is threatened with habitat destruction although it can survive in shaded coffee plantations.

Reg. no. 1925.12.27.2. From Precidio; ; Mexico. Date: 26/ 3/1925. Source: W.W. Brown; J.E. Thayer; Mus. Comp. Zool.. Material: adult M; whole/good skin.

Reg. no. 1925.12.27.3. From Precidio; ; Mexico. Date: 1/ 4/1925. Source: W.W. Brown; J.E. Thayer; Mus. Comp. Zool.. Material: adult M; whole/good skin.

Reg. no. 1925.12.27.4. From Precidio; ; Mexico. Date: 29/ 4/1925. Source: W.W. Brown; Mus. Comp. Zool.. Material: adult M; whole/good skin.

Thryomanes bewickii leucophrys

San Clemente Wren

Known only from the island of San Clemente off California, where it became extinct in 1968 as a result of habitat destruction by goats.

Reg. no. 1898.7.12.247. From San Clemente Island; California; [United States]. Date: 31/ 3/1897. Source: C.K. Worthen; Salvin-Godman coll.. Material: adult M; whole/good skin.

Reg. no. 1898.7.12.248. From San Clemente Island; California; [United States]. Date: 28/ 3/1897. Source: C.K. Worthen; Salvin-Godman coll.. Material: adult F; whole/good skin.

Thryomanes bewickii brevicauda

Guadalupe Wren

Endemic to the island of Guadalupe off Baja California where it has not been seen since 1897, and is probably extinct. Its scrub habitat was devastated by goats, and it was probably hunted by feral cats.

Reg. no. 1888.10.10.1455. From Guadalupe Island; Baja California; [Mexico]. Date: //. Source: A.J. Grayson; H.W. Henshaw coll.. Material: adult M; whole/good skin.

Ferminia cerverai

Zapata Wren

Known only from the Zapata swamp in Cuba. It almost became extinct in the 1960s and 1970s, probably as a result of habitat destruction. There are hopes that it may be recovering.

Reg. no. 1927.7.14.3. From Santo Tomas, Peninsula de Zapata; ; Cuba. Date: 3/ 3/1927. Source: Museum of Comparative Zoology. Material: adult M; whole/good skin.

Troglodytes aedon guadeloupensis

Guadeloupe House-Wren

Mountain forests of Guadeloupe in the West Indies. It was rediscovered in 1969 after being believed extinct since 1914. It is still rare, and its original decline was probably due to predation by rats, cats and mongooses.

Reg. no. 1915.7.12.34. From ; ; Guadeloupe. Date: 17/ 2/1902. Source: S. Branch; Rothschild coll.. Material: adult M; whole/good skin. The date given above is from the collector's label; it is noted as 14 November and 17 November 1902 on later labels.

Troglodytes aedon martinicensis
Martinique House-Wren
Martinique. The last specimen was taken in 1886, and it is believed that cats, rats and mongooses were responsible for its extinction.

Reg. no. 1886.9.15.155. From ; ; Martinique. Date: //. Source: Verreaux; P.L. Sclater. Material: ; whole/good skin.

Reg. no. 1886.9.15.156. From ; ; Martinique. Date: //. Source: Parzudaki; P.L. Sclater. Material: adult ; whole/good skin. Syntype. P.L. Sclater, *Proc. zool. Soc. Lond.* 1866: 321.

Troglodytes aedon mesoleucus
St. Lucia House-Wren
St. Lucia. Formerly widespread but now confined to a single small valley in the north eastern coastal area.

Reg. no. 1886.9.15.160. From ; ; St. Lucia. Date: //1875. Source: J. Semper; Sclater coll.. Material: adult M; whole/good skin. Holotype. P.L. Sclater, *Proc. zool. Soc. Lond.* 1876: 14.

Reg. no. 1889.8.5.17. From ; ; St. Lucia. Date: 24/ 3/1889. Source: West Indian Committee. Material: adult M; whole/good skin.

Reg. no. 1889.8.5.18. From ; ; St. Lucia. Date: 9/ 4/1889. Source: West Indian Committee. Material: adult F; whole/good skin.

Mimodes graysoni
Socorro Mockingbird
Endemic to Socorro Island in the Revillagigedos Islands, Mexico, where it was apparently still common up to the late 1950s but, by the early 1980s was severely at risk, due to cats and habitat degradation by introduced sheep. Only one bird could be found in 1984, but several were discovered in August 1987 and more subsequently.

Reg. no. 1897.11.10.10. From ; Socorro Island; [Mexico]. Date: 13/ 5/1897. Source: A.W. Anthony. Material: adult F; whole/good skin.

Reg. no. 1897.11.10.9. From ; Socorro Island; [Mexico]. Date: 13/ 5/1897. Source: A.W. Anthony. Material: adult F; whole/good skin.

Reg. no. 1898.7.12.238. From ; Socorro Island; [Mexico]. Date: 20/ 5/1897. Source: C.K. Worthen; Salvin-Godman coll.. Material: adult M; whole/good skin.

Reg. no. 1898.7.12.239. From ; Socorro Island; [Mexico]. Date: 10/ 5/1897. Source: C.K. Worthen; Salvin-Godman coll.. Material: adult F; whole/good skin.

Cinclocerthia ruficauda gutturalis
Grey Trembler
Martinique. Now restricted to woodlands in the south of the island and very rare indeed. The decline is thought to be due to

a mixture of habitat destruction and predation by introduced mammals.

Reg. no. 1856.3.12.10. From ; ; Martinique. Date: //. Source: Verreaux. Material: ; whole/good skin.

Reg. no. 1886.8.2.190. From ; ; Martinique. Date: //. Source: Verreaux ; P.L. Sclater. Material: ; whole/good skin.

Ramphocinclus brachyurus brachyurus
White-breasted Thrasher
Endemic to, and formerly common on, Martinique, where only 40 pairs still exist and are threatened by increasing habitat destruction and fragmentation.

Reg. no. 1881.12.28.1. From ; ; Martinique. Date: //. Source: Ober; Smithsonian Institution. Material: M; whole/good skin.

Reg. no. 1885.3.2.366. From ; ; Martinique. Date: / 2/1879. Source: Ober; Smithsonian Inst.; Salvin-Godman coll.. Material: M; whole/good skin.

Reg. no. 1886.8.2.187. From ; ; Martinique. Date: / 8/1879. Source: Ober; Smithsonian Inst.; P.L. Sclater coll.. Material: M; whole/good skin.

Reg. no. 1886.8.2.188. From ; ; Martinique. Date: //. Source: E. Parzudaki; P.L. Sclater. Material: ; whole/good skin.

Reg. no. 1893.12.16.10. From ; ; Martinique. Date: 31/10/1887. Source: W.B. Richardson; P.L. Sclater coll.. Material: M; whole/good skin.

Reg. no. 1905.6.28.131. From ; ; Martinique. Date: / 4/1863. Source: Cavendish Taylor coll.. Material: ; whole/good skin.

Ramphocinclus brachyurus sanctaeluciae
White-breasted Thrasher
Endemic to, and formerly common on, St. Lucia, where it is confined to narrow bands of riverine forest. A maximum of 58 pairs were estimated in 1987 and 46 in 1992. As with the last race, this one is threatened by increasing habitat destruction and fragmentation.

Reg. no. 1885.3.2.367. From ; ; St. Lucia. Date: //. Source: Semper; Salvin-Godman coll.. Material: ; whole/good skin.

Reg. no. 1885.3.2.368. From ; ; St. Lucia. Date: //. Source: Semper; Salvin-Godman coll.. Material: ; whole/good skin.

Reg. no. 1886.8.2.189. From ; ; St. Lucia. Date: //1871. Source: Semper; P.L. Sclater. Material: ; whole/good skin.

Reg. no. 1889.8.5.13. From ; ; St. Lucia. Date: 26/ 3/1889. Source: West Indian Committee. Material: adult M; whole/good skin.

Reg. no. 1889.8.5.14. From ; ; St. Lucia. Date: 26/ 3/1889. Source: West Indian Committee. Material: adult F; whole/good skin.

Alethe choloensis choloensis

Cholo Alethe

This race is known only from scattered mountain sites in southern Malawi. It is threatened by habitat clearance.

Reg. no. 1927.2.2.7. From Thyolo Mountain (Cholo Mountain); ; Malawi (Nyasaland). Date: 3/10/1926. Source: C.F. Belcher. Material: adult F; whole/good skin. Holotype. W.L. Sclater, *Bull. Brit. Orn. Club* 47 (1927): 86–7. Not breeding. The day of collection is erroneously given as 30th in the description. Stomach contained driver ants, a larva, bits of beetle. Iris rich deep brown, bill black, legs and feet very pale flesh.

Reg. no. 1933.3.1.955. From Thyolo Mountain (Cholo Mountain); ; Malawi (Nyasaland). Date: 3/ 9/1932. Source: J. Vincent. Material: adult M; whole/good skin. Sex organs: adult male right down to winter condition. Bill black, feet very pallid flesh with some faint brownish discolouration, iris dark burnt umber.

Reg. no. 1933.3.1.956. From Thyolo Mountain (Cholo Mountain); ; Malawi (Nyasaland). Date: 3/ 9/1932. Source: J. Vincent. Material: adult M; whole/good skin. Sex organs: adult male, testes half size and going down. Bill black, feet pallid flesh, iris burnt umber.

Reg. no. 1933.3.1.957. From Thyolo Mountain (Cholo Mountain); ; Malawi (Nyasaland). Date: 4/ 9/1932. Source: J. Vincent. Material: adult F; whole/good skin. Sex organs: adult female organs somewhat enlarged and well started up. Bill black, feet pallid flesh, iris burnt umber.

Reg. no. 1934.3.4.43. From Thyolo Mountain (Cholo Mountain); ; Malawi (Nyasaland). Date: 16/ 7/1933. Source: D.W.K. Macpherson. Material: adult M; whole/good skin. Sex organs: not breeding condition. Bill black, feet pale flesh, iris dark brown.

Reg. no. 1934.3.4.44. From Thyolo Mountain (Cholo Mountain); ; Malawi (Nyasaland). Date: 6/ 8/1933. Source: D.W.K. Macpherson. Material: adult F; whole/good skin. Sex organs: not breeding. Bill black, feet pale flesh, iris brown.

Reg. no. 1934.3.4.45. From Thyolo Mountain (Cholo Mountain); ; Malawi (Nyasaland). Date: 16/ 7/1933. Source: D.W.K. Macpherson. Material: adult M; whole/good skin. Sex organs: not breeding. Bill black, feet pale flesh, iris brown. Stomach contained ants and a large beetle.

Reg. no. 1937.12.19.173. From Mlanje Forest; ; Malawi (Nyasaland). Date: 12/ 7/1936. Source: C.W. Benson. Material: adult M; whole/good skin. Bill black, feet pale lilac, iris rich brown.

Reg. no. 1937.12.19.174. From Mlanje Forest; ; Malawi (Nyasaland). Date: 12/ 7/1936. Source: C.W. Benson. Material: adult M; skin. Details taken from the register. Not in the collection; exchanged with Institut Royal des Sciences Naturelles de Belgique in 1985.

Reg. no. 1937.12.19.175. From Mangoche Mt.; ; Malawi (Nyasaland). Date: 11/ 6/1936. Source: C.W. Benson. Material: adult M; whole/good skin. Sexual organs:

undeveloped. Bill black, feet pale flesh, tinged violet, iris dark brown.

Reg. no. 1946.5.469. From Soche Mt., Blantyre; ; Malawi (Nyasaland). Date: 6/ 1/1945. Source: C.W. Benson. Material: juv M; whole/good skin. Sexual organs: small. Bill black, keel of lower mandible yellow, gape yellow, feet pale brown. Skull ossification not started. Stomach: ant and beetle remains.

Reg. no. 1946.5.496. From Mlanje; ; Malawi (Nyasaland). Date: 18/ 8/1944. Source: C.W. Benson. Material: adult ; whole/good skin. Sexual organs destroyed by shot. Bill black.

Reg. no. 1946.5.497. From Soche Mt., Blantyre; ; Malawi (Nyasaland). Date: 22/ 3/1944. Source: C.W. Benson. Material: adult M; whole/good skin. Sexual organs: small. Bill black, feet pale lilac, soles buffy-white, iris brown.

Reg. no. 1950.61.12. From Chiperoni District; ; Mozambique (Portuguese East Africa). Date: 27/ 7/1950. Source: C.W. Benson. Material: adult F; whole/good skin. Sex organs small.

Reg. no. 1950.61.13. From Chiperoni District; ; Mozambique (Portuguese East Africa). Date: 27/ 7/1950. Source: C.W. Benson. Material: adult M; whole/good skin. Sex organs small.

Alethe choloensis namuli

Cholo Alethe

This race is known only from scattered mountain sites in Mozambique. It is threatened by habitat clearance.

Reg. no. 1933.3.1.3. From 5,600 ft., Namuli Mountain, Quelimane Province; ; Mozambique (Portuguese East Africa). Date: 25/ 7/1932. Source: J. Vincent. Material: adult F; whole/good skin. Holotype. Vincent, *Bull. Brit. Orn. Club* 53 (1933): 138. Collected at 15.21S 37.04E. See also *Ibis* 1935: 498–9.

Reg. no. 1933.3.1.958. From Namuli Mountain, Quelimane Province; ; Mozambique (Portuguese East Africa). Date: 25/ 7/1932. Source: J. Vincent. Material: adult M; whole/good skin. Sex organs: adult male in winter condition. Bill black, feet ashen pallid flesh, iris sepia.

Copsychus sechellarum

Seychelles Magpie–Robin

Survives only on Fregate in the Seychelles, having been exterminated by feral cats on other islands in the group. Only 23 birds remained in 1991, but an active conservation programme is now underway.

Reg. no. 1868.6.16.8. From ; Marianne Island; Seychelles. Date: 12/ 2/1867. Source: E. Newton. Material: adult M; whole/good skin.

Reg. no. 1888.4.20.96. From ; Marianne Island; Seychelles. Date: //1879. Source: H.L. Harry; Tristram coll.; Tweeddale coll.. Material: ; whole/good skin. The reg. no.

was erroneously given on the label as 1888.4.1.96.

Reg. no. 1895.5.1.909. From ; Aride Island; Seychelles. Date: //. Source: Shelley coll.. Material: M; whole/good skin.

Reg. no. 1895.5.1.910. From ; ; Seychelles. Date: //. Source: Shelley coll.. Material: F; whole/good skin.

Reg. no. 1895.5.1.911. From ; ; Seychelles. Date: //. Source: Shelley coll.. Material: ; whole/good skin.

Reg. no. 1909.4.6.30. From ; ; Seychelles. Date: //. Source: R. Ward. Material: adult ; whole/good skin.

Reg. no. 1909.4.6.31. From ; ; Seychelles. Date: //. Source: R. Ward. Material: adult ; whole/good skin.

Reg. no. 1927.12.18.383. From ; Marianne Island; Seychelles. Date: 14/ 2/1888. Source: J.J. Lister coll.. Material: adult M; whole/good skin.

Reg. no. 1927.12.18.384. From ; Marianne Island; Seychelles. Date: 14/ 2/1888. Source: J.J. Lister. Material: adult F; whole/good skin.

Reg. no. 1927.12.18.385. From ; Marianne Island; Seychelles. Date: 14/ 2/1888. Source: J.J. Lister. Material: M; whole/good skin.

Reg. no. 1946.75.16a. From ; Frigate Island; Seychelles. Date: / 4/1940. Source: Goodfellow; A.D. Sapsworth. Material: adult F; whole/good skin.

Reg. no. 1946.75.17a. From ; Frigate Island; Seychelles. Date: / 4/1940. Source: Goodfellow; A.D. Sapsworth. Material: adult F; whole/good skin.

Reg. no. 1946.75.18a. From ; Frigate Island; Seychelles. Date: / 4/1940. Source: Goodfellow; A.D. Sapsworth. Material: adult F; whole/good skin.

Reg. no. 1955.63.4. From ; Frigate Island; Seychelles. Date: 23/ 7/1955. Source: R. Percy & M. Ridley. Material: F; whole/good skin.

Copsychus niger cebuensis

Black Shama

Endemic to Cebu, where it was thought to be extinct as a result of the total destruction of forest, but was recently found in bamboo thickets.

Reg. no. 1891.4.25.48. From ; Cebu; Philippines. Date: / 3/1888. Source: Worcester, Steere Expedition; E.L. Moseley. Material: adult F; whole/good skin. Eye brown, bill black, legs and feet pale brown. Pulling leaves for nest when shot.

Reg. no. 1896.6.6.272. From the road between Barili and Carcao; Cebu; Philippines. Date: //. Source: J.B. Steere; Steere Expedition. Material: adult M; whole/good skin. Holotype. Steere, 1890, *A List of the Birds and Mammals collected by the Steere Expedition to the Philippines* ..., p. 20; see also *Ibis* 1891: 314.

Saxicola dacotiae murielae

Muriel's Chat

Allegranza and Montana Clara in the Canary Islands. Not seen since the original type series was collected in 1913. The reasons for its extinction are totally unknown. Collar and Stuart (1985) consider the validity of the race to be questionable.

Reg. no. 1913.10.22.104. From ; Allegranza; Canary Islands. Date: 10/ 6/1913. Source: D.A. Bannerman. Material: adult F; whole/good skin. Syntype. Iris dark hazel, bill and feet black.

Reg. no. 1913.10.22.105. From ; Allegranza; Canary Islands. Date: 12/ 6/1913. Source: D.A. Bannerman. Material: adult M; whole/good skin. Syntype. Bannerman, *Bull. Brit. Orn. Club* 33 (1913): 37–8. See also *Ibis* 1914: 74–7, 86, pl. 5. The female syntype is also in the collection.

Reg. no. 1913.10.22.106. From ; Allegranza; Canary Islands. Date: 10/ 6/1913. Source: D.A. Bannerman. Material: imm M; whole/good skin. Iris dark hazel, bill and feet black.

Reg. no. 1913.10.22.107. From ; Allegranza; Canary Islands. Date: 10/ 6/1913. Source: D.A. Bannerman. Material: imm F; whole/good skin. Iris dark hazel, bill and feet black.

Reg. no. 1913.10.22.108. From ; Allegranza; Canary Islands. Date: 12/ 6/1913. Source: D.A. Bannerman. Material: imm F; whole/good skin. Iris dark hazel, bill and feet black.

Reg. no. 1913.10.22.109. From ; Montana Clara; Canary Islands. Date: 9/ 6/1913. Source: D.A. Bannerman. Material: adult M; whole/good skin. Iris dark hazel, bill and feet black.

Reg. no. 1913.10.22.110. From ; Montana Clara; Canary Islands. Date: 9/ 6/1913. Source: D.A. Bannerman. Material: adult M; whole/good skin. Iris dark hazel, bill and feet black. The reg. no. is apparently incorrect and untraced; the number on the label was assigned to a different species.

Reg. no. 1919.8.15.72. From ; Allegranza; Canary Islands. Date: 12/ 6/1913. Source: D.A. Bannerman. Material: imm F; whole/good skin. Iris dark hazel, bill and feet black.

Reg. no. 1919.8.15.73. From ; Allegranza; Canary Islands. Date: 12/ 6/1913. Source: D.A. Bannerman. Material: imm ; whole/good skin. Iris dark hazel, bill and feet black.

Cichlherminia lherminieri sanctaeluciae

St. Lucia Forest–Thrush

St. Lucia. Now very rare and confined to a few wooded ravines which are gradually being cleared.

Reg. no. 1886.8.2.175. From ; ; St. Lucia. Date: //1871. Source: J.E. Semper; P.L. Sclater coll.. Material: adult ; whole/good skin. Syntype. P.L. Sclater, *Ibis* 1880: 73. The other syntype is also in the collection.

Reg. no. 1886.8.2.176. From ; ; St. Lucia. Date: //1871. Source: J.E. Semper; P.L.

Sclater coll.. Material: ; whole/good skin. Syntype.

Reg. no. 1889.8.5.5. From ; ; St. Lucia. Date: 30/ 3/1889. Source: West Indian Committee. Material: adult F; whole/good skin. The day of collection is almost illegible on the collector's label; it has been interpreted as 20th on the Museum label and 30th in the register. The latter is most likely correct. This specimen and 1889.8.5.6 were registered the wrong way round.

Reg. no. 1889.8.5.6. From ; ; St. Lucia. Date: 23/ 3/1889. Source: West Indian Committee. Material: adult M; whole/good skin. This specimen and 1889.8.5.5 were registered the wrong way round.

Phaeornis obscurus myadestinus

Kamao

Endemic to Kauai where it occurs in the Alakai Swamp forest, and has become exceedingly rare in recent years. It may now be extinct.

Reg. no. 1890.10.3.18. From ; Kauai; Hawaiian Islands. Date: //. Source: V. Knudsen; Christiania Museum; Collett. Material: ; whole/good skin.

Reg. no. 1895.10.16.10. From ; Kauai; Hawaiian Islands. Date: //. Source: V. Knudsen; Christiania Museum. Material: ; whole/good skin.

Reg. no. 1895.10.16.9. From ; Kauai; Hawaiian Islands. Date: //. Source: V. Knudsen; Christiania Museum. Material: ; whole/good skin.

Reg. no. 1895.7.20.57. From ; Kauai; Hawaiian Islands. Date: / 5/1894. Source: R.C.L. Perkins, Royal Society & British Association. Material: adult M; whole/good skin.

Reg. no. 1895.7.20.58. From ; Kauai; Hawaiian Islands. Date: / 5/1894. Source: R.C.L. Perkins, Royal Society & British Association. Material: adult F; whole/good skin.

Reg. no. 1897.10.28.64. From Kaholuamano; Kauai; Hawaiian Islands. Date: /10/1895. Source: R.C.L. Perkins, Royal Society & British Association. Material: imm M; whole/good skin.

Reg. no. 1897.10.28.65. From Kaholuamano; Kauai; Hawaiian Islands. Date: /10/1895. Source: R.C.L. Perkins, Royal Society & British Association. Material: adult F; whole/good skin.

Reg. no. 1897.10.28.66. From Kaholuamano; Kauai; Hawaiian Islands. Date: / 5/1895. Source: R.C.L. Perkins, Royal Society & British Association. Material: adult M; whole/good skin.

Reg. no. 1897.7.2.1052. From ; Kauai; Hawaiian Islands. Date: / 9/1888. Source: S.B. Wilson; Seebohm coll.. Material: adult M; whole/good skin.

Reg. no. S1961.11.15. From ; [Kauai]; [Hawaiian Islands]. Date: //. Source: Rothschild coll.. Material: ; whole/good skeleton. 'Probably a female somewhat decayed and not able to make sure' (Gronwald).

Reg. no. 1940.12.8.22. From ; ; Hawaiian Islands (Sandwich Islands). Date: //. Source: Palmer coll; Rothschild coll.. Material: adult ; whole/good alc. Collected between 1890–1893. The identity is not certain.

Phaeornis obscurus oahensis

Amaui

Oahu, Hawaiian Islands. Known only from a single skin collected in 1823, although fossil remains have recently been found.

Reg. no. 1988.21.2. From ; [Oahu]; [Hawaiian Islands]. Date: //1823. Source: [Bloxam]. Material: juv ; whole/good skin. This is possibly the lost type and only known specimen of this race collected by Bloxam in 1825 (S.L. Olson). It bears the number 150a on a late 19th century label, but this is untraced.

Phaeornis obscurus rutha

Molokai Olomao

Confined to the island of Molokai where it was last seen in 1988 and may be extinct.

Reg. no. 1895.7.20.105. From ; Molokai; Hawaiian Islands. Date: / 6/1893. Source: R.C.L. Perkins, Royal Society & British Association. Material: adult F; whole/good skin.

Reg. no. 1895.7.20.106. From ; Molokai; Hawaiian Islands. Date: / 5/1893. Source: R.C.L. Perkins, Royal Society & British Association. Material: adult M; whole/good skin.

Reg. no. 1895.7.20.107. From ; Molokai; Hawaiian Islands. Date: 10/ 9/1893. Source: R.C.L. Perkins, Royal Society & British Association. Material: juv M; whole/good skin.

Phaeornis obscurus lanaiensis

Lanai Olomao

Confined to the island of Lanai where it became extinct between 1931 and 1933.

Reg. no. 1895.7.20.73. From Lanaihale; Lanai; Hawaiian Islands. Date: / 7/1894. Source: R.C.L. Perkins; Royal Society & British Association. Material: adult F; whole/good skin.

Reg. no. 1895.7.20.74. From ; Lanai; Hawaiian Islands. Date: / 2/1894. Source: R.C.L. Perkins; Royal Society & British Association. Material: adult M; whole/good skin.

Reg. no. 1895.7.20.75. From ; Lanai; Hawaiian Islands. Date: /12/1893. Source: R.C.L. Perkins; Royal Society & British Association. Material: adult F; whole/good skin.

Phaeornis palmeri

Puaiohi

Confined to a small part of the Alakai Swamp on Kauai. It is very secretive, inhabiting dense undergrowth, and has always been rare. As few as 20 remain.

Reg. no. 1897.10.28.68. From Halemanu; Kauai; Hawaiian Islands. Date: / 5/1895. Source: R.C.L. Perkins, Royal Society &

British Association. Material: adult M; whole/good skin.

Reg. no. 1897.10.28.69. From Halemanu; Kauai; Hawaiian Islands. Date: / 5/1895. Source: R.C.L. Perkins, Royal Society & British Association. Material: adult F; whole/good skin.

Reg. no. 1897.10.28.70. From Halemanu; Kauai; Hawaiian Islands. Date: / 5/1895. Source: R.C.L. Perkins, Royal Society & British Association. Material: adult M; whole/good skin.

Turdus helleri

Taita Olive Thrush

Confined to the forest on the Taita Hills and on Mount Kasigau in southeastern Kenya. This forest is now under serious threat.

Reg. no. 1938.2.11.29. From Taita Hills; ; Kenya. Date: 11/11/1937. Source: R.E. Moreau. Material: adult M; whole/good skin. Bill orange-red, gape and rim round eye orange-red, feet orange-brown.

Reg. no. 1938.2.11.30. From Taita Hills; ; Kenya. Date: 8/11/1937. Source: R.E. Moreau. Material: adult M; whole/good skin. Bill orange-red, gape and rim round eye orange-red, feet orange-brown.

Reg. no. 1938.2.11.31. From Taita Hills; ; Kenya. Date: 4/11/1937. Source: R.E. Moreau. Material: adult M; whole/good skin. Bill orange-red, gape and rim round eye orange-red, feet orange-brown.

Turdus poliocephalus pritzbueri

Lifu Island Thrush

Endemic to the island of Lifu in the Loyalty Islands. Now probably extinct.

Reg. no. 1879.3.5.7. From ; Lifu; Loyalty Islands. Date: 28/ 8/1878. Source: E.L. Layard; Tristram coll.. Material: M; whole/good skin.

Reg. no. 1879.3.5.8. From ; Lifu; Loyalty Islands. Date: 28/ 8/1878. Source: E.L. Layard; Tristram coll.. Material: F; whole/good skin.

Reg. no. 1880.12.31.32. From Port Resolution; Tana [Tanna]; New Hebrides. Date: //1860. Source: J. MacGillivray; Eyton coll.; Gerrard. Material: adult F; whole/good skin. The Museum label states 'H.M.S. Herald' and 'Norfolk Island', but this information is not on the older Eyton label. The *Cat. Bds.* gives the date of collection of what appears to be this bird as 22 August 1859, also apparently in error.

Reg. no. 1888.2.20.568. From ; Lifu; Loyalty Islands. Date: 19/ 8/1878. Source: E.L. Layard; Tweeddale coll.. Material: adult M; whole/good skin.

Reg. no. 1888.2.20.569. From ; Lifu; Loyalty Islands. Date: 9/ 9/1878. Source: E.L. Layard; Tweeddale coll.. Material: adult F; whole/good skin.

Reg. no. 1896.6.1.628. From ; Lifu; Loyalty Islands. Date: 23/ 8/1878. Source: E.L. Layard; Seebohm coll.. Material: adult M; whole/good skin.

Reg. no. 1896.6.1.629. From ; Lifu; Loyalty Islands. Date: 20/ 8/1878. Source: E.L. Layard; Seebohm coll.. Material: adult M; whole/good skin. This specimen is noted as 'Type of species' in the register, and three other Seebohm skins in the collection are similarly annotated. The description of this form was published by Layard in April 1878, and these birds cannot have been before him when he wrote it.

Reg. no. 1896.6.1.630. From ; Lifu; Loyalty Islands. Date: 20/ 8/1878. Source: E.L. Layard; Seebohm coll.. Material: adult F; whole/good skin.

Reg. no. 1896.6.1.631. From ; Lifu; Loyalty Islands. Date: 19/ 8/1878. Source: E.L. Layard; Seebohm coll.. Material: adult F; whole/good skin.

Reg. no. 1896.6.1.632. From ; Lifu; Loyalty Islands. Date: //. Source: E.L. Layard; Seebohm coll.. Material: M; skin. Apparently seen in 1986, but not found 1988.

Reg. no. 1896.6.1.633. From ; Lifu; Loyalty Islands. Date: 28/ 8/1878. Source: E.L. Layard; Seebohm coll.. Material: juv M; whole/good skin.

Reg. no. 1896.6.1.634. From ; Lifu; Loyalty Islands. Date: 23/ 8/1878. Source: E.L. Layard; Seebohm coll.. Material: adult F; whole/good skin.

Reg. no. 1896.6.1.635. From ; Lifu; Loyalty Islands. Date: 20/ 8/1878. Source: E.L. Layard; Seebohm coll.. Material: juv M; whole/good skin.

Reg. no. 1896.6.1.636. From ; Tana; New Hebrides. Date: 25/ 7/1878. Source: E.L. Layard; Seebohm coll.. Material: M; whole/good skin.

Turdus poliocephalus mareensis

Mare Island Thrush

Endemic to Mare in the Loyalty Islands. Now probably extinct.

Reg. no. 1880.8.10.11. From ; Mare; Loyalty Islands. Date: //. Source: E.L. Layard; Tristram coll.. Material: M; skin. Not in the collection. Exchanged with the American Museum of Natural History about 1939.

Reg. no. 1880.8.10.12. From ; Mare; Loyalty Islands. Date: /10/1879. Source: E.L. Layard; Tristram coll.. Material: F; whole/good skin.

Reg. no. 1896.6.1.572. From ; Mare; Loyalty Islands. Date: /10/1879. Source: E.L. Layard; Seebohm coll.. Material: adult M; whole/good skin. In the register this specimen was marked 'Type of species' although it was subsequently crossed out.

Reg. no. 1896.6.1.573. From ; Mare; Loyalty Islands. Date: /10/1879. Source: E.L. Layard; Seebohm coll.. Material: adult F; whole/good skin.

Reg. no. 1896.6.1.574. From ; Mare; Loyalty Islands. Date: /11/1881. Source: E.L. Layard; Seebohm coll.. Material: adult M; skin. Apparently seen in 1986, but not found 1988.

Reg. no. 1896.6.1.575. From ; Mare; Loyalty Islands. Date: 7/ 6/1882. Source: E.L. Layard; Seebohm coll.. Material: adult M; whole/good skin.

Reg. no. 1896.6.1.576. From ; Mare; Loyalty Islands. Date: /10/1879. Source: E.L. Layard; Seebohm coll.. Material: adult M; whole/good skin.

Reg. no. 1896.6.1.577. From ; Mare; Loyalty Islands. Date: 7/ 6/1882. Source: E.L. Layard; Seebohm coll.. Material: adult F; whole/good skin. The day of collection is given as 4th in the register, apparently in error.

Reg. no. 1896.6.1.578. From ; Mare; Loyalty Islands. Date: /11/1881. Source: E.L. Layard; Seebohm coll.. Material: adult F; whole/good skin.

Reg. no. 1896.6.1.579. From ; Mare; Loyalty Islands. Date: 7/ 6/1882. Source: E.L. Layard; Seebohm coll.. Material: adult F; whole/good skin.

Reg. no. 1896.6.1.580. From ; Mare; Loyalty Islands. Date: /11/1881. Source: E.L. Layard; Seebohm coll.. Material: adult F; whole/good skin.

Reg. no. 1896.6.1.581. From ; Mare; Loyalty Islands. Date: /11/1881. Source: E.L. Layard; Seebohm coll.. Material: adult M; whole/good skin.

Reg. no. 1896.6.1.582. From ; Mare; Loyalty Islands. Date: 7/ 6/1882. Source: E.L. Layard; Seebohm coll.. Material: adult M; whole/good skin.

Reg. no. 1896.6.1.583. From ; Mare; Loyalty Islands. Date: /10/1879. Source: E.L. Layard; Seebohm coll.. Material: adult F; whole/good skin. The date of collection is given as 7 June 1882 in the register.

Turdus poliocephalus poliocephalus **Norfolk Island Thrush**

Endemic to Norfolk Island where it is threatened as a result of habitat destruction and competition with the introduced Blackbird *Turdus merula*.

Reg. no. 1838.1.10.269. From ; ; Norfolk Island. Date: //. Source: Material: ; whole/good skin. The reg. no. is untraced. The correct number may (or may not) be 1838.1.19.209.

Reg. no. 1856.8.30.27. From ; ; Norfolk Island. Date: //. Source: Dr F.M. Rayner, voyage of H.M.S. Herald. Material: adult M; whole/good skin.

Reg. no. 1856.8.30.28. From ; ; Norfolk Island. Date: //. Source: Dr F.M. Rayner, voyage of H.M.S. Herald. Material: M; whole/good skin. The register gives the sex as female.

Reg. no. 1870.3.31.131. From ; ; Norfolk Island. Date: / 6/1865. Source: J. Brenchley. Material: adult M; whole/good skin. Eye dark brown. Food, guava seeds and small brown seeds.

Reg. no. 1870.3.31.132. From ; ; Norfolk Island. Date: / 6/1865. Source: J. Brenchley. Material: adult F; whole/good skin. Eye dark brown. Food, guava seeds and small brown seeds.

Reg. no. 1880.2.20.571. From ; ; Norfolk Island. Date: / 6/1865. Source: H.E. Dresser; Seebohm coll.; Tweeddale coll.. Material: adult M; whole/good skin.

Reg. no. 1888.2.20.572. From ; ; Norfolk Island. Date: / 6/1865. Source: Tweeddale coll.. Material: adult F; whole/good skin.

Reg. no. 1888.2.20.573. From ; ; Norfolk Island. Date: / 6/1865. Source: Tweeddale coll.. Material: adult M; whole/good skin.

Reg. no. 1888.2.20.574. From ; ; Norfolk Island. Date: //. Source: Tweeddale coll.. Material: adult ; whole/good skin.

Reg. no. 1896.6.1.648. From ; ; Norfolk Island. Date: / 6/1865. Source: J. Brenchley; H.E. Dresser; Seebohm coll.. Material: adult M; whole/good skin.

Reg. no. 1896.6.1.649. From ; ; Norfolk Island. Date: 17/ 8/1880. Source: E.L. Layard; Seebohm coll.. Material: adult M; whole/good skin.

Reg. no. 1896.6.1.650. From ; ; Norfolk Island. Date: /10/1879. Source: E.L. Layard; Seebohm coll.. Material: adult F; whole/good skin.

Reg. no. 1896.6.1.651. From ; ; Norfolk Island. Date: / 4/1880. Source: E.L. Layard; Seebohm coll.. Material: adult M; whole/good skin.

Reg. no. 1896.6.1.652. From ; ; Norfolk Island. Date: //1880. Source: E.L. Layard; Seebohm coll.. Material: juv M; whole/good skin.

Reg. no. 1896.6.1.653. From ; ; Norfolk Island. Date: /10/1879. Source: E.L. Layard; Seebohm coll.. Material: adult M; whole/good skin.

Reg. no. 1896.6.1.654. From ; ; Norfolk Island. Date: / 6/1865. Source: J. Brenchley; Tweeddale coll.; Seebohm coll.. Material: adult F; whole/good skin.

Reg. no. 1896.6.1.655. From ; ; Norfolk Island. Date: //. Source: T.C. Eyton; Seebohm coll.. Material: adult ; whole/good skin.

Reg. no. 1919.7.15.100. From ; ; Norfolk Island. Date: 9/ 5/1913. Source: G.M. Mathews. Material: imm F; whole/good skin.

Reg. no. 1919.7.15.101. From Dixey; ; Norfolk Island. Date: 9/ 5/1913. Source: G.M. Mathews. Material: adult M; whole/good skin.

Reg. no. 1919.7.15.102. From ; ; Norfolk Island. Date: 23/ 5/1913. Source: G.M. Mathews. Material: adult F; whole/good skin.

Reg. no. 1919.7.15.103. From ; ; Norfolk Island. Date: 29/ 5/1913. Source: G.M. Mathews. Material: adult M; whole/good skin.

Reg. no. 1919.7.15.104. From Cascades; ; Norfolk Island. Date: 31/ 5/1913. Source: G.M. Mathews. Material: adult F; whole/good skin.

Reg. no. 1919.7.15.105. From Cascades; ; Norfolk Island. Date: 1/ 6/1913. Source:

G.M. Mathews. Material: adult F; whole/good skin.

Reg. no. 1919.7.15.106. From Cascades; ; Norfolk Island. Date: 2/ 6/1913. Source: G.M. Mathews. Material: adult F; whole/good skin.

Reg. no. 1919.7.15.95. From nr. Middlegate; ; Norfolk Island. Date: 4/ 3/1913. Source: G.M. Mathews. Material: adult M; whole/good skin.

Reg. no. 1919.7.15.96. From nr. Middlegate; ; Norfolk Island. Date: 20/ 3/1913. Source: G.M. Mathews. Material: adult M; whole/good skin.

Reg. no. 1919.7.15.97. From nr. Cascades; ; Norfolk Island. Date: 19/ 4/1913. Source: G.M. Mathews. Material: adult F; whole/good skin. Iris dark brown, feet bright brownish-yellow, bill bright dark yellow.

Reg. no. 1919.7.15.98. From Cascades; ; Norfolk Island. Date: 19/ 4/1913. Source: G.M. Mathews. Material: adult F; whole/good skin.

Reg. no. 1919.7.15.99. From ; ; Norfolk Island. Date: 5/ 5/1913. Source: G.M. Mathews. Material: adult M; whole/good skin.

Reg. no. 1939.12.9.2283. From nr. Middlegate; ; Norfolk Island. Date: 4/ 3/1913. Source: G.M. Mathews; Rothschild coll.. Material: M; whole/good skin.

Reg. no. 1939.12.9.2284. From nr. Middlegate; ; Norfolk Island. Date: 27/ 3/1913. Source: G.M. Mathews; Rothschild coll.. Material: adult F; whole/good skin.

Reg. no. 1939.12.9.2285. From nr. Middlegate; ; Norfolk Island. Date: 11/ 4/1913. Source: G.M. Mathews; Rothschild coll.. Material: adult M; whole/good skin.

Reg. no. 1939.12.9.2286. From nr. Middlegate; ; Norfolk Island. Date: 27/ 3/1913. Source: G.M. Mathews; Rothschild coll.. Material: adult M; whole/good skin.

Reg. no. 1939.12.9.2287. From Ball Bay Reserve; ; Norfolk Island. Date: 22/ 7/1912. Source: G.M. Mathews; Rothschild coll.. Material: adult F; whole/good skin. The day is given as 27th in the register.

Reg. no. 1939.12.9.2288. From nr. Kingston; ; Norfolk Island. Date: 8/ 5/1913. Source: G.M. Mathews; Rothschild coll.. Material: adult M; whole/good skin.

Reg. no. 1939.12.9.2289. From nr. Kingston; ; Norfolk Island. Date: 8/ 5/1913. Source: G.M. Mathews; Rothschild coll.. Material: adult F; whole/good skin.

Reg. no. 1939.12.9.2290. From Ball Bay Reserve; ; Norfolk Island. Date: 19/ 7/1912. Source: G.M. Mathews; Rothschild coll.. Material: adult F; whole/good skin.

Reg. no. 1939.12.9.2291. From Ball Bay Reserve; ; Norfolk Island. Date: 12/ 7/1912. Source: G.M. Mathews; Rothschild coll.. Material: adult M; whole/good skin.

Reg. no. 1939.12.9.2292. From Steels Point; ; Norfolk Island. Date: 30/10/1912.

Source: G.M. Mathews; Rothschild coll.. Material: adult F; whole/good skin.

Reg. no. 1939.12.9.2293. From nr. Cascades; ; Norfolk Island. Date: 28/ 5/1913. Source: Rothschild coll.. Material: adult M; whole/good skin.

Reg. no. 1939.12.9.2294. From Burnt Pine; ; Norfolk Island. Date: 28/ 5/1913. Source: Rothschild coll.. Material: adult F; whole/good skin.

Reg. no. 1988.21.7. From ; ; Norfolk Island. Date: //. Source: Material: ; whole/good skin. No history. Damaged: legs loose.

Reg. no. 1887.7.9.18–21. From ; ; Norfolk Island. Date: /10/1885. Source: Dr Crowfoot. Material: 4 ; whole/good eggs.

Reg. no. 1902.4.5.975–6. From ; ; Norfolk Island. Date: /11/1882. Source: Dr Crowfoot; Crowley. Material: 2 ; whole/good eggs.

Reg. no. E1941.9.4.1606. From ; ; Norfolk Island. Date: //. Source: Rothschild. Material: 2 ; whole/good eggs.

Reg. no. E1941.9.4.1607. From ; ; Norfolk Island. Date: //. Source: Rothschild. Material: 2 ; whole/good eggs.

Reg. no. E1941.9.4.1608. From ; ; Norfolk Island. Date: //. Source: Rothschild. Material: 2 ; whole/good eggs.

Reg. no. E1941.9.4.1609. From ; ; Norfolk Island. Date: //. Source: Rothschild. Material: 2 ; whole/good eggs.

Reg. no. E1941.2.1.116. From ; ; Norfolk Island. Date: 6/11/1912. Source: Mathews coll.; Rothschild. Material: 2 ; eggs (clutch). Not found <1988, believed stolen during 1975–1979.

Reg. no. E1941.2.1.117. From ; ; Norfolk Island. Date: 5/11/1912. Source: Mathews coll.; Rothschild. Material: 2 ; whole/good eggs (clutch).

Reg. no. E1941.2.1.118. From ; ; Norfolk Island. Date: 5/11/1912. Source: Mathews coll.; Rothschild. Material: 2 ; whole/good eggs (clutch).

Reg. no. E1941.2.1.119. From ; ; Norfolk Island. Date: 2/11/1912. Source: Mathews coll.; Rothschild. Material: 2 ; whole/good eggs (clutch).

Reg. no. E1941.2.1.120. From ; ; Norfolk Island. Date: 11/11/1912. Source: Mathews coll.; Rothschild. Material: 2 ; whole/good eggs (clutch).

Reg. no. E1941.2.1.121. From ; ; Norfolk Island. Date: 12/11/1912. Source: Mathews coll.; Rothschild. Material: 2 ; whole/good eggs (clutch).

Reg. no. E1941.2.1.122. From ; ; Norfolk Island. Date: 8/11/1912. Source: Mathews coll.; Rothschild. Material: 2 ; whole/good eggs (clutch).

Reg. no. E1941.2.1.123. From ; ; Norfolk Island. Date: 19/11/1912. Source: Mathews coll.; Rothschild. Material: 2 ; whole/good eggs (clutch).

Reg. no. E1941.2.1.124. From ; ; Norfolk Island. Date: 4/11/1912. Source: Mathews coll.; Rothschild. Material: 2 ; whole/good eggs (clutch).

Reg. no. E1941.2.1.125. From ; ; Norfolk Island. Date: 4/11/1912. Source: Mathews coll.; Rothschild. Material: 2 ; whole/good eggs (clutch).

Reg. no. E1941.2.1.126. From ; ; Norfolk Island. Date: 21/ 9/1912. Source: Mathews coll.; Rothschild. Material: 2 ; whole/good eggs (clutch).

Reg. no. E1941.2.1.128. From ; ; Norfolk Island. Date: 1/10/1912. Source: Mathews coll.; Rothschild. Material: 2 ; whole/good eggs (clutch).

Reg. no. E1941.2.1.129. From ; ; Norfolk Island. Date: 29/ 4/1912. Source: Mathews coll.; Rothschild. Material: 2 ; whole/good eggs (clutch).

Reg. no. E1941.2.1.130. From ; ; Norfolk Island. Date: 23/11/1912. Source: Mathews coll.; Rothschild. Material: 2 ; whole/good eggs (clutch).

Reg. no. E1941.2.1.131. From ; ; Norfolk Island. Date: 24/11/1912. Source: Mathews coll.; Rothschild. Material: 2 ; whole/good eggs (clutch).

Reg. no. E1941.2.1.132. From ; ; Norfolk Island. Date: 16/11/1912. Source: Mathews coll.; Rothschild. Material: 2 ; whole/good eggs (clutch).

Reg. no. E1941.2.1.133. From ; ; Norfolk Island. Date: 23/12/1912. Source: Mathews coll.; Rothschild. Material: 2 ; whole/good eggs (clutch).

Reg. no. E1941.2.1.134. From ; ; Norfolk Island. Date: 22/11/1912. Source: Mathews coll.; Rothschild. Material: 3 ; whole/good eggs (clutch).

Reg. no. E1941.2.1.135. From ; ; Norfolk Island. Date: 27/11/1912. Source: Mathews coll.; Rothschild. Material: 3 ; whole/good eggs (clutch).

Reg. no. E1941.2.1.136. From ; ; Norfolk Island. Date: 19/11/1912. Source: Mathews coll.; Rothschild. Material: 4 ; whole/good eggs (clutch). These eggs differ from the others in the collection in having a bluish–green rather than white ground.

Reg. no. E1941.2.1.137. From ; ; Norfolk Island. Date: 19/11/1912. Source: Mathews coll.; Rothschild. Material: 2 ; whole/good eggs (clutch).

Reg. no. E1941.2.1.138. From ; ; Norfolk Island. Date: 5/12/1912. Source: Mathews coll.; Rothschild. Material: 2 ; whole/good eggs (clutch).

Reg. no. E1941.2.1.139. From ; ; Norfolk Island. Date: 17/12/1912. Source: Mathews coll.; Rothschild. Material: 2 ; whole/good eggs (clutch).

Reg. no. E1941.2.1.140. From ; ; Norfolk Island. Date: 19/12/1912. Source: Mathews coll.; Rothschild. Material: 2 ; whole/good eggs (clutch).

Reg. no. E1941.2.1.141. From ; ; Norfolk Island. Date: 25/ 4/1913. Source: Mathews coll.; Rothschild. Material: 2 ; whole/good eggs (clutch).

Reg. no. E1941.2.1.142. From ; ; Norfolk Island. Date: 8/ 5/1913. Source: Mathews coll.; Rothschild. Material: 2 ; whole/good eggs (clutch).

Reg. no. . From ; ; Norfolk Island. Date: //. Source: Gould coll.. Material: 1 ; whole/good egg. This specimen bears the untraced number G.552.

Reg. no. 1902.4.5.973. From ; ; Norfolk Island. Date: //1881. Source: E.L. Layard; Tristram coll.; Crowley. Material: 1 ; whole/good egg. One of two eggs received with nest and parent bird from E.L. Layard. 'The nest which is exactly like that of the English Blackbird but smaller, was in a cave.' The second egg, the nest and the bird were in the collection of H. Seebohm.

Reg. no. 1902.4.5.974. From ; ; Norfolk Island. Date: 9/11/1882. Source: ?Tristram coll.; Crowley. Material: 1 ; whole/good egg.

Reg. no. E1901.1.2.8388. From ; ; Norfolk Island. Date: 9/10/1879. Source: Seebohm coll.. Material: 1 ; whole/good egg. See 1902.4.5.973, but the dates do not match.

Reg. no. E1941.2.1.127. From ; ; Norfolk Island. Date: 9/11/1912. Source: Mathews coll.; Rothschild. Material: 1 ; whole/good egg.

Turdus poliocephalus vinitinctus

Doctorbird or Lord Howe Island Thrush

Endemic to Lord Howe Island where it was formerly common, but from 1909 onwards it became rarer as a result of predation by cats and dogs. It became extinct in 1918 after rats came ashore from a wreck.

Reg. no. 1855.10.23.4. From ; ; Lord Howe Island. Date: //. Source: J. MacGillivray, voyage of H.M.S. Herald; Mus. of Economic Geology. Material: F; whole/good skin. Syntype.

Reg. no. 1855.10.23.5. From ; ; Lord Howe Island. Date: //. Source: J. MacGillivray, voyage of H.M.S. Herald; Mus. of Economic Geology. Material: adult M; whole/good skin. Syntype. Gould, *Proc. zool. Soc. Lond.* 1855: 165. The female syntype is also in the collection.

Reg. no. 1888.10.5.5. From ; ; Lord Howe Island. Date: //. Source: E. Saunders. Material: adult ; whole/good skin. The location is given as Norfolk Island in the register.

Reg. no. 1896.6.1.1000. From ; ; Lord Howe Island. Date: //. Source: Seebohm coll.. Material: adult M; whole/good skin.

Reg. no. 1896.6.1.1001. From ; ; Lord Howe Island. Date: //. Source: Seebohm coll.. Material: adult ; whole/good skin.

Reg. no. 1919.7.15.120. From ; ; Lord Howe Island. Date: 27/11/1913. Source: G.M. Mathews. Material: adult M; whole/good skin.

Reg. no. 1919.7.15.121. From ; ; Lord Howe Island. Date: 25/11/1913. Source: G.M. Mathews. Material: adult M; whole/good skin.

Reg. no. 1919.7.15.122. From ; ; Lord Howe Island. Date: 14/ 1/1914. Source: G.M. Mathews. Material: adult F; whole/good skin.

Reg. no. 1919.7.15.123. From ; ; Lord Howe Island. Date: 24/ 7/1913. Source: G.M. Mathews. Material: adult ; whole/good skin.

Reg. no. 1919.7.15.124. From ; ; Lord Howe Island. Date: 5/ 5/1914. Source: G.M. Mathews. Material: adult M; whole/good skin.

Reg. no. 1919.7.15.125. From ; ; Lord Howe Island. Date: 27/11/1913. Source: G.M. Mathews. Material: adult F; whole/good skin.

Reg. no. 1919.7.15.126. From ; ; Lord Howe Island. Date: 25/11/1913. Source: G.M. Mathews. Material: adult F; whole/good skin.

Reg. no. 1919.7.15.127. From ; ; Lord Howe Island. Date: 4/ 8/1913. Source: G.M. Mathews. Material: adult F; whole/good skin.

Reg. no. 1919.7.15.128. From ; ; Lord Howe Island. Date: 22/11/1913. Source: G.M. Mathews. Material: adult F; whole/good skin.

Reg. no. 1919.7.15.129. From ; ; Lord Howe Island. Date: 30/ 5/1914. Source: G.M. Mathews. Material: adult M; whole/good skin.

Reg. no. 1919.7.15.130. From ; ; Lord Howe Island. Date: 6/10/1913. Source: G.M. Mathews. Material: adult M; whole/good skin.

Reg. no. 1919.7.15.131. From ; ; Lord Howe Island. Date: 11/ 5/1914. Source: G.M. Mathews. Material: adult M; whole/good skin.

Reg. no. E1941.9.4.1600. From ; ; Lord Howe Island. Date: //. Source: Rothschild. Material: 2 ; whole/good eggs.

Reg. no. E1941.2.1.143. From ; ; Lord Howe Island. Date: 5/11/1913. Source: Mathews coll.; Rothschild. Material: 2 ; whole/good eggs (clutch).

Reg. no. E1941.2.1.144. From ; ; Lord Howe Island. Date: 12/11/1913. Source: Mathews coll.; Rothschild. Material: 2 ; whole/good eggs (clutch).

Reg. no. E1941.2.1.146. From ; ; Lord Howe Island. Date: 15/11/1913. Source: Mathews coll.; Rothschild. Material: 2 ; whole/good eggs (clutch).

Reg. no. E1941.2.1.148. From ; ; Lord Howe Island. Date: 28/12/1913. Source: Mathews coll.; Rothschild. Material: 2 ; whole/good eggs (clutch).

Reg. no. E1941.2.1.149. From ; ; Lord Howe Island. Date: 29/12/1913. Source: Mathews coll.; Rothschild. Material: 2 ; whole/good eggs (clutch).

Reg. no. E1941.2.1.150. From ; ; Lord Howe Island. Date: 28/ 9/1914. Source: Mathews coll.; Rothschild. Material: 3 ; whole/good eggs (clutch).

Reg. no. E1941.2.1.145. From ; ; Lord Howe Island. Date: 7/11/1913. Source: Mathews coll.; Rothschild. Material: 1 ; whole/good egg.

Reg. no. E1941.2.1.147. From ; ; Lord Howe Island. Date: 17/12/1913. Source: Mathews coll.; Rothschild. Material: 1 ; whole/good egg.

Turdus ravidus

Grand Cayman Thrush

Formerly found on the Island of Grand Cayman, where it was restricted to the eastern and northeastern parts of the island. It was last seen in 1938 and is now presumed extinct.

Reg. no. 1925.12.27.5. From ; ; Cayman Islands. Date: 18/ 5/1911. Source: W.W. Brown; Museum of Comparative Zoology. Material: M; whole/good skin.

'Turdus' ulietensis

Bay Thrush

Known only from Georg Forster's painting, based on a specimen (no longer extant) taken on Raiatea in the Society Islands on 1 June 1774. The species has not been seen since. Its taxonomic position is obscure, but it is almost certainly not a *Turdus*.

Reg. no. . From ; Raiatea Island; Society Islands. Date: 1/ 6/1774. Source: J.R. & J.G.A. Forster. Material: F; painting. Holotype. Gmelin, *Systema Naturae*, 1, pt. 2, (1789): 815. Based on a drawing in the NHM library, by Forster, (old) no. 146, marked 'Raiatea, female, June 1, 1774'. Published in Fuller, 1987, *Extinct Birds*, p. 198, pl. 39. See *Bull. Br. Mus. nat. Hist. (Hist.)* 1 (1959): 306.

Moupinia altirostris altirostris

Jerdon's Babbler

Formerly occurred on grasslands that covered the Irrawaddy–Sittang plains of south-central Burma. Not seen since 1941 and presumed extinct. Other races survive in India and Pakistan.

Reg. no. 1862.1.17.18. From islands in the Burrampootra River, Thayetmyo; Magwe; Burma. Date: //. Source: T.C. Jerdon. Material: adult ; whole/good skin. Holotype. Jerdon, *Ibis* 1862: 22. The bird had been frequenting long grass. Food, ants and small coleoptera.

Reg. no. 1882.1.20.243. From Myitko or Myitkyo; Pegu; Burma. Date: 21/ 2/1878. Source: E.W. Oates. Material: adult M; whole/good skin. Testes very large.

Reg. no. 1882.1.20.244. From Pagina or Paghein; Pegu; Burma. Date: 15/ 8/1880. Source: E.W. Oates. Material: adult F; whole/good skin.

Reg. no. 1882.1.20.245. From Pagina or Paghein; Pegu; Burma. Date: 15/ 8/1880. Source: E.W. Oates. Material: adult F; whole/good skin.

Reg. no. 1886.10.1.4727. From Pagina; Pegu; Burma. Date: 16/ 8/1880. Source:

E.W. Oates; Hume coll.. Material: adult M; whole/good skin.

Reg. no. 1898.10.3.709. From ; lower Pegu; Burma. Date: 12/ 9/. Source: E.W. Oates; Seebohm coll.. Material: adult M; whole/good skin.

Reg. no. E1988.4.27. From Maymyo, the plains at the foot of the Kachin Hills; ; Burma. Date: 9/ 4/1914. Source: Stuart Baker coll.. Material: 5 ; whole/good eggs (clutch). Baker suggests the authentication is not beyond question.

Acrocephalus familiaris

Millerbird

Endemic to Laysan Island, where it lived in the scrub. The natural vegetation was destroyed between 1904 and 1923 after the introduction of rabbits. The bird became extinct some time during this period.

Reg. no. 1939.12.9.582. From ; Laysan Island; Hawaiian Islands. Date: 22/ 8/1896. Source: Schauinsland; Rothschild coll.. Material: adult M; whole/good skin.

Reg. no. 1939.12.9.583. From ; Laysan Island; Hawaiian Islands. Date: 22/ 8/1896. Source: Schauinsland; Rothschild coll.. Material: adult F; whole/good skin.

Reg. no. N1973.2.3. From ; Laysan Island; Hawaiian Islands. Date: //1891. Source: H.C. Palmer; Rothschild. Material: 4 ; whole/good nests. Four out of five nests collected about 1891, along with 5 eggs, which are also in the collection, E1973.20.8. See Rothschild, 1893, *The Avifauna of Laysan*, p. 1–2, pl. 41.

Reg. no. E1941.9.4.1379. From ; [Laysan Island]; [Hawaiian Islands]. Date: //. Source: Rothschild. Material: 2 ; whole/good eggs. No authentication of identity.

Reg. no. E1973.20.8. From ; Laysan Island; Hawaiian Islands. Date: //1891. Source: H.C. Palmer; Rothschild. Material: 5 ; whole/good eggs. Collected about 1891, along with nests, which are also in the collection, N1973.2.3. See Rothschild, 1893, *The Avifauna of Laysan*, p. 1–2, pl. 41; Munro, 1960, *Birds of Hawaii*, p. 79–80.

Reg. no. 1940.12.8.18. From ; Laysan Island; Hawaiian Islands. Date: //. Source: Palmer coll.; Rothschild coll.. Material: adult ; whole/good alc. Collected between 1890–1893. The label on the jar erroneously gives the reg. no. as 1940.12.8.98.

Acrocephalus caffer garretti

Huahine Reed–Warbler

Huahine, Society Islands. Known only from three specimens collected by Andrew Garrett, two of which are extant. The third, formerly in the Tristram collection, may have been lost as a result of the bombing of Liverpool Museum in World War Two.

Reg. no. 1898.9.1.2535. From ; Huahine; Society Islands. Date: //. Source: A. Garrett; Seebohm coll.. Material: adult ; whole/good skin. Holotype. Holyoak & Thibault, *Bull. Brit. Orn. Club* 98 (1978): 122–7.

Acrocephalus caffer musae

Raiatea Reed–Warbler

Known only from Georg Forster's plate and the description by J.R. Forster. No such bird has subsequently been reported on Raiatea.

Reg. no. . From ; Raiatea; Society Islands. Date: //1773. Source: Material: ; painting. Holotype. Forster, in Lichtenstein, ed., 1844, *Descriptiones Animalium* ..., p. 163–4. Based on a drawing in the NHM library, by Georg Forster, (old) no. 55. See *Bull. Br. Mus. nat. Hist. (Hist.)* 1 (1959): 286. No specimens exist. See *Bull. Brit. Orn. Club* 98 (1978): 124.

Bebrornis rodericanus

Rodriguez Brush–Warbler

Confined to a small portion of Rodriguez Island. It was common in the 19th century but, in the early 1980s, the population was thought to be less than 50 birds.

Reg. no. 1876.3.14.1. From ; ; Rodriguez. Date: //. Source: Slater; Royal Society. Material: adult ; whole/good skin.

Reg. no. 1876.3.14.2. From ; ; Rodriguez. Date: //. Source: Slater; Royal Society. Material: adult M; whole/good skin.

Reg. no. 1876.3.14.3. From ; ; Rodriguez. Date: //. Source: Slater; Royal Society. Material: adult F; whole/good skin.

Reg. no. 1876.3.14.4. From ; ; Rodriguez. Date: //. Source: Slater; Royal Society. Material: adult M; whole/good skin.

Reg. no. 1876.3.14.64. From ; ; Rodriguez. Date: //. Source: Slater; Royal Society. Material: chick M; whole/good skin. The nest is also in the collection, 1876.3.14.65.

Reg. no. 1876.3.15.7. From ; ; Rodriguez. Date: /11/1874. Source: G. Gulliver; Royal Society. Material: adult F; whole/good skin.

Reg. no. 1876.3.15.8. From ; ; Rodriguez. Date: /11/1874. Source: G. Gulliver; Royal Society. Material: adult M; whole/good skin.

Reg. no. 1876.3.14.65. From ; ; Rodriguez. Date: //. Source: Slater; Royal Society. Material: 1 ; whole/good nest. A chick from this nest is also in the collection, 1876.3.14.64.

Reg. no. N1975.4.1. From St. Gabriel; ; Rodriguez. Date: //. Source: A.S. Cheke. Material: 1 ; nest. Collected between 1972–1975. Not found 1988; thought to have been temporarily misplaced.

Reg. no. A1979.11.1. From St. Gabriel; ; Rodriguez. Date: //1974. Source: A.S. Cheke. Material: chick ; whole/good alc.

Reg. no. 1876.3.15.3. From ; ; Rodriguez. Date: //. Source: Gulliver; Royal Society. Material: 1 ; whole/good nest. This nest was later re–registered as N145.1. An egg from this nest is also in the collection, 1876.3.15.4.

Reg. no. 1876.3.15.4. From ; ; Rodriguez. Date: //. Source: Gulliver; Royal Society. Material: 1 ; whole/good egg. The nest is also in the collection, 1876.3.15.3. See

Sharpe, *Phil. Trans. R. Soc.* 168, extra volume, (1879): 461.

Nesillas aldabranus

Aldabra Brush-Warbler

Confined to a tiny coastal strip of vegetation on Aldabra. The known population was never more than five. In 1977 there were only two. Searches in 1986 failed to find it and the species may now be extinct.

Reg. no. N257.1. From Gionnet, Middle Island; ; Aldabra Islands. Date: 11/12/1967. Source: M. Penney; Royal Society Expedition to Aldabra. Material: 1 ; whole/poor nest. The clutch is in the collection, E1968.5.1. The nest was placed 2 ft. up in young *Pandanus*. The female was also collected. The nest is somewhat scruffy.

Reg. no. N257.2. From Gionnet, Middle Island; ; Aldabra Islands. Date: 29/ 1/1968. Source: C.W. Benson. Material: 1 ; whole/poor nest. An old nest collected on the date given, believed to be of this species, but not so proven.

Reg. no. N257.3. From Gionnet, Middle Island; ; Aldabra Islands. Date: 29/ 1/1968. Source: C.W. Benson. Material: 1 ; whole/poor nest. An old nest collected on the date given, believed to be of this species, but not so proven.

Reg. no. E1968.5.1. From Gionnet, Middle Island; ; Aldabra Islands. Date: 11/12/1967. Source: M. Penney; Royal Society Expedition to Aldabra. Material: 3 ; whole/good eggs (clutch). One egg is broken. The nest is in the collection, N257.1. It was placed 2 ft. up in young *Pandanus*. The female was also taken.

Sylvia melanocephala norissae

Faiyum Sardinian Warbler

This race was found only in the Faiyum area of north-eastern Egypt. It appears to be extinct, the last good record being 7 February 1939.

Reg. no. 1917.1.12.1. From north side of Lake Birket Karun (Qarun); Fayoum (Faiyum); Egypt. Date: 23/ 3/1912. Source: M.J. Nicoll; Egyptian Government Zoology Service. Material: adult M; whole/good skin. Holotype. Nicoll, *Bull. Brit. Orn. Club* 37 (1917): 28-29. Collected on a small islet in the lake.

Reg. no. 1935.1.15.18. From Gigireh el Kebar; ; Egypt. Date: 13/ 3/1917. Source: W. Raw; Egyptian Government Zoology Service. Material: adult M; whole/good skin.

Reg. no. 1941.5.30.4246. From Jel el Rusas; Fayoum (Faiyum); Egypt. Date: 26/ 1/1919. Source: W. Raw; J.L. Bonhote coll.; Ticehurst coll.. Material: adult F; whole/good skin.

Reg. no. 1941.5.30.4247. From Jel el Rusas; Fayoum (Faiyum); Egypt. Date: 26/ 1/1919. Source: W. Raw; J.L. Bonhote coll.; Ticehurst coll.. Material: adult M; whole/good skin.

Reg. no. 1941.5.30.4248. From Jel el Rusas; Fayoum (Faiyum); Egypt. Date: 26/

1/1919. Source: W. Raw; J.L. Bonhote coll.; Ticehurst coll.. Material: adult M; whole/good skin.

Reg. no. 1965.M.13981. From ; Fayoum Lake (Faiyum); Egypt. Date: 27/ 3/1923. Source: Meinertzhagen coll.. Material: adult M; whole/good skin.

Reg. no. 1965.M.13982. From ; Fayoum Lake (Faiyum); Egypt. Date: 27/ 3/1923. Source: Meinertzhagen coll.. Material: adult M; whole/good skin.

Reg. no. 1965.M.13983. From ; Fayoum Lake (Faiyum); Egypt. Date: 27/ 3/1923. Source: Meinertzhagen coll.. Material: adult M; whole/good skin.

Reg. no. 1965.M.13984. From ; Fayoum Lake (Faiyum); Egypt. Date: 27/ 3/1923. Source: Meinertzhagen coll.. Material: adult M; whole/good skin.

Reg. no. 1965.M.13985. From ; Fayoum Lake (Faiyum); Egypt. Date: 11/11/1909. Source: M.J. Nicoll; Meinertzhagen coll.. Material: adult M; whole/good skin.

Amaurocichla bocagei

Bocage's Longbill

Confined to the undisturbed primary forests of southern Sao Tome. It has not been collected since 1928. Two populations were found in 1990, the species having been considered extinct (although one may have been seen in 1987).

Reg. no. 1892.5.31.3. From San Miguel; ; w. coast of Sao Tome (St. Thomas). Date: //. Source: F. Newton; Prof. B. du Bocage. Material: adult ; whole/good skin. Holotype. Sharpe, *Proc. zool. Soc. Lond.* 1892: 227–8, pl. 20.

Bowdleria punctata rufescens

Chatham Islands Fernbird

Probably extinct. Occurred on Pitt and Mangere Islands, mainly in scrubby bush, much of which has been destroyed by burning and by rabbits and goats. It is possible that some birds survive in scrub on steep cliffs on Mangere, but this is considered unlikely.

Reg. no. 1874.7.7.2. From ; ; Chatham Islands. Date: //. Source: W. Travers; Boulton. Material: adult M; whole/good skin.

Reg. no. 1874.7.7.3. From ; ; Chatham Islands. Date: //. Source: W. Travers; Boulton. Material: adult F; whole/good skin.

Reg. no. 1939.12.9.66. From ; ; Chatham Islands. Date: //1890. Source: H.C. Palmer; Rothschild coll.. Material: ; whole/good skin.

Reg. no. 1939.12.9.67. From ; ; Chatham Islands. Date: //1890. Source: H.C. Palmer; Rothschild coll.. Material: ; whole/good skin.

Dasyornis broadbenti litoralis

Rufous Bristlebird

Known only from 2 coastal localities in extreme southwestern Australia where it has not been seen since 1940 and is presumed extinct.

Reg. no. 1905.1.24a.37. From Cape Naturaliste; Western Australia; Australia. Date: 9/ 7/1904. Source: CPC; B.H. Woodward. Material: ; whole/good skin.

Gerygone insularis

Lord Howe Island Gerygone

Lord Howe Island. Believed to have become extinct shortly after the escape of rats from a wreck in 1918. Previously it had been common.

Reg. no. 1888.10.5.12. From ; ; Lord Howe Island. Date: //. Source: E. Saunders. Material: adult ; whole/good skin.

Reg. no. 1897.7.2.1058. From ; ; Lord Howe Island. Date: / 9/1887. Source: Sydney Museum; Seebohm coll.. Material: adult M; whole/good skin.

Reg. no. 1919.7.15.138. From Mrs Nichol's; ; Lord Howe Island. Date: 25/ 7/1913. Source: G.M. Mathews. Material: adult M; whole/good skin. Iris bright pink, feet dark blue–grey, bill black.

Reg. no. 1919.7.15.139. From ; ; Lord Howe Island. Date: 6/10/1913. Source: G.M. Mathews. Material: adult M; whole/good skin. Eye bright pink, feet dark brownish–grey, bill blue–black.

Reg. no. 1919.7.15.140. From ; ; Lord Howe Island. Date: 6/11/1913. Source: G.M. Mathews. Material: adult M; whole/good skin. Eye bright pink, feet very dark blue–grey, bill blue–grey.

Reg. no. 1919.7.15.141. From the Pines; ; Lord Howe Island. Date: 12/ 5/1914. Source: G.M. Mathews. Material: adult F; whole/good skin. Eye pink, feet darkish blue–grey, bill dark grey.

Reg. no. 1919.7.15.142. From ; ; Lord Howe Island. Date: 21/11/1913. Source: G.M. Mathews. Material: adult ; whole/good skin. Eye pink, feet dark bluish–grey, bill black. Age not certain.

Reg. no. 1919.7.15.143. From Mrs Nichol's; ; Lord Howe Island. Date: 4/ 8/1913. Source: G.M. Mathews. Material: adult ; whole/good skin. Iris pink, feet dark grey, bill black. Age not certain.

Petroica traversi

Black or Chatham Islands Robin

Chatham Islands, formerly reduced to a population of 7 birds on Little Mangere but now successfully transplanted to Mangere and Rangatira. The population now numbers about 40.

Reg. no. 1874.7.7.1. From [Mangere]; ; Chatham Islands. Date: //. Source: W. Travers; Dr Bulton. Material: adult ; whole/good skin.

Reg. no. 1893.6.24.8. From Little Mangere; ; Chatham Islands. Date: //. Source: W. Hawkins; H.O. Forbes. Material: adult F; whole/good skin.

Reg. no. . From ; ; . Date: //. Source: Material: ; alc. Not in the collection. Two spirit specimens are listed in *Bull. Br. Mus. nat. Hist. (Zool.)* 34 (1979): 167, but these were recently reidentified as *P. macrocephala*.

Platysteira laticincta

Banded Wattle-eye

Restricted to the Bamenda–Banso Highlands of western Cameroon, where it is seriously threatened by forest clearance.

Reg. no. 1926.8.8.6. From 6,000 ft., Oku, w. of Kumbo; ; Cameroon. Date: 8/ 2/1925. Source: G.L. Bates. Material: adult M; whole/good skin. Holotype. Bates, *Bull. Brit. Orn. Club* 46 (1926): 91.

Reg. no. 1952.20.266. From Bamenda; ; Cameroon. Date: 28/ 5/1948. Source: W. Serle. Material: adult M; whole/good skin. Testes not enlarged. Iris greyish, sclera violet, feet blackish, bill black. One of three birds in a patch of forest on the mountain side.

Reg. no. 1952.20.267. From Bamenda; ; Cameroon. Date: 31/ 5/1948. Source: W. Serle. Material: adult M; whole/good skin. Testes small. Iris grey, wattle red, feet blackish, bill black. One of a pair in the trees on the edge of a mountain forest.

Reg. no. 1952.20.268. From Bamenda; ; Cameroon. Date: 28/ 5/1948. Source: W. Serle. Material: adult F; whole/good skin. Ovary not enlarged. Iris greyish, sclera violet, feet blackish, bill black. One of three birds in a patch of forest on the mountain side.

Reg. no. 1952.20.269. From nr. Bambulue Lake, Bamenda; ; Cameroon. Date: 18/ 9/1948. Source: W. Serle. Material: adult M; whole/good skin. Testes considerably enlarged. Wattle red, bill and feet black. One of a pair in a forested mountain ravine.

Reg. no. 1952.20.270. From nr. Bambulue Lake, Bamenda; ; Cameroon. Date: 4/ 9/1948. Source: W. Serle. Material: adult M; whole/good skin. Testes small. Iris greyish, wattle red, bill black, feet purplish-black. A single bird in a forested mountain valley.

Reg. no. 1952.20.271. From nr. Bambulue Lake, Bamenda; ; Cameroon. Date: 4/ 9/1948. Source: W. Serle. Material: M; skin. Details taken from the register. Not in the collection; exchanged with the American Museum of Natural History in 1960.

Reg. no. 1952.20.272. From nr. Bambulue Lake, Bamenda; ; Cameroon. Date: 5/ 7/1948. Source: W. Serle. Material: adult M; whole/good skin. Testes not enlarged. Iris violet-grey, wattle red, bill and feet black. One of two birds in the undershrub in a mountain forest in a ravine.

Reg. no. 1952.20.273. From nr. Bambulue Lake, Bamenda; ; Cameroon. Date: 29/ 9/1948. Source: W. Serle. Material: adult M; whole/good skin. Testes considerably enlarged. One of a pair in a forested ravine, in the undershrub.

Reg. no. 1952.20.274. From nr. Bambulue Lake, Bamenda; ; Cameroon. Date: 14/ 7/1948. Source: W. Serle. Material: adult F; whole/good skin. Ovary not enlarged. Wattle red, feet blackish, bill black. One of a pair in the trees in a patch of mountain forest.

Reg. no. 1966.16.4623. From Umuagwu; ; Nigeria. Date: 14/11/1959. Source: W. Serle. Material: adult M; whole/good skin.

Insects in stomach. A member of a mixed bird party in primary forest. The reg. no. is untraced; 1966.16.4623 was assigned to a bird of a different species.

Reg. no. 1966.16.4674. From nr. Bambulue Lake, Bamenda; ; Cameroon. Date: 18/ 9/1948. Source: W. Serle. Material: adult F; whole/good skin. Ovary greatly enlarged. Wattle red, bill and feet black. One of a pair in a forested ravine.

Reg. no. 1966.16.4675. From nr. Bambulue Lake, Bamenda; ; Cameroon. Date: 14/ 7/1948. Source: W. Serle. Material: adult M; whole/good skin. Testes not enlarged. Wattle red, feet blackish, bill black. One of a pair in the trees in a patch of mountain forest.

Reg. no. 1966.16.4676. From nr. Bambulue Lake, Bamenda; ; Cameroon. Date: 29/ 9/1948. Source: W. Serle. Material: adult F; whole/good skin. Ovary slightly enlarged. One of a pair in a forested ravine, in the undershrub.

Reg. no. 1966.16.4677. From nr. Bambulue Lake, Bamenda; ; Cameroon. Date: 9/ 8/1948. Source: W. Serle. Material: adult M; whole/good skin. Testes not enlarged. Iris brownish–grey, wattle red, feet and bill black. Flitting about the foliage in the forest by the crater lake.

Reg. no. 1966.16.4678. From Bambulue Lake, [Bamenda]; ; Cameroon. Date: 26/ 4/1955. Source: W. Serle. Material: adult F; whole/good skin. Ovary not enlarged. One of two in the undershrub in a forest ravine. Insects in stomach.

Reg. no. 1966.16.4679. From Lake Oku, Bamenda; ; Cameroon. Date: 16/10/1948. Source: W. Serle. Material: M; whole/good skin. Testes very small. Iris reddish–brown, wattle red, feet
purplish–black, bill black. One of a party in primary forest.

Reg. no. 1966.16.4680. From Bamenda; ; Cameroon. Date: 31/ 5/1948. Source: W. Serle. Material: adult F; whole/good skin. Ovary small. Iris grey, wattle red, feet blackish, bill black. One of a pair in the trees on the edge of a mountain forest.

Reg. no. 1966.16.4681. From Ndu; ; Cameroon. Date: 17/ 7/1948. Source: W. Serle. Material: adult F; whole/good skin. Ovaries not enlarged. Iris grey–brown, wattle red, feet blackish, bill black. In the undershrub in a mountain forest of mixed primary and secondary growth.

Reg. no. 1966.16.4682. From Ndu; ; Cameroon. Date: 25/ 6/1948. Source: W. Serle. Material: adult M; whole/good skin. Testes slightly enlarged. Iris greyish, wattle red, feet black, bill black. One of a pair in the foliage in a mountainside wood.

Reg. no. 1966.16.4683. From Ndu; ; Cameroon. Date: 17/ 7/1948. Source: W. Serle. Material: adult M; whole/good skin. Testes small. Iris grey–brown, wattle red, feet blackish, bill black. In the foliage of a tree in a mountain wood of mixed primary and secondary growth.

Reg. no. 1984.9.12. From Lake Oku, [Bamenda]; ; Cameroon. Date: 27/ 3/1984. Source: International Council for Bird

Preservation. Material: juv M; whole/good skin. Age not certain.

Pomarea (nigra) nigra

Tahiti Monarch

Endemic to Tahiti, but very close to extinction, if not already extinct. No more than 6 birds were seen in 1984, 2 in 1985 and none at all in 1986. The reason for the decline is unknown. *P. n. pomarea*, an extinct form from Maupiti, known only from a single specimen collected in 1823, is now regarded as a full species.

Reg. no. 1858.4.3.63. From ; Tahiti; Society Islands. Date: //. Source: Zoological Society coll.. Material: ; whole/good skin. The identity of this specimen has been questioned; it may be *P. mendozae*, from the Marquesas Islands.

Reg. no. 1869.5.10.34. From ; Tahiti; Society Islands. Date: //. Source: Verreaux. Material: adult M; whole/good skin.

Reg. no. 1928.10.27.76. From ; Tahiti; Society Islands. Date: 11/ 5/1921. Source: E.H. Quayle; R.H. Beck; American Museum of Natural History. Material: F; whole/good skin. Iris brown, bill and feet blue. Sexual organs small.

Reg. no. 1928.10.27.77. From ; Tahiti; Society Islands. Date: 10/ 5/1921. Source: E.H. Quayle; R.H. Beck; American Museum of Natural History. Material: ; whole/good skin. Iris brown, bill and feet blue.

Reg. no. 1988.25.1. From ; Tahiti; Society Islands. Date: //. Source: Rothschild coll.. Material: adult M; whole/good skin. Mount in Tring gallery.

Pomarea mendozae mendozae

Hivaoa Monarch

Confined to Tahuata (where extinct) and Hivaoa in the Marquesas Islands. Now very rare on the latter, the reasons for the decline being unknown. Only one bird was seen in 1975.

Reg. no. E1977.3.2. From Mohotani; ; Marquesas Islands. Date: 16/ 4/1975. Source: J.C. Thibault; Thibault Holyoak coll.. Material: 1 ; damaged egg. Broken shell found on ground under tree of *Eugenia variflora* containing *Pomarea* nest with nestling.

Reg. no. 1925.12.22.202. From ; Hiva Oa; Marquesas Islands. Date: 27/12/1924. Source: H.J. Kelsall, St. George Expedition. Material: adult M; whole/good skin.

Reg. no. 1925.12.22.203. From ; Hiva Oa; Marquesas Islands. Date: 27/12/1924. Source: H.J. Kelsall, St. George Expedition. Material: adult F; whole/good skin.

Pomarea mendozae nukuhivae

Nukuhiva Monarch

Found only on Nukuhiva in the Marquesas. Presumably once common on the island but, by 1922, restricted to valleys at the western end. In 1972 it was thought probably to be extinct, though there were reports of its having been seen in 1987.

Reg. no. 1846.7.29.7. From ; Nukuhiva; Marquesas Islands. Date: //. Source: Turner. Material: ; whole/good skin.

Reg. no. 1846.7.29.8. From ; Nukuhiva; Marquesas Islands. Date: //. Source: Turner. Material: ; whole/good skin.

Reg. no. 1846.7.29.9. From ; Nukuhiva; Marquesas Islands. Date: //. Source: Turner. Material: juv ; whole/good skin.

Reg. no. 1856.3.12.36. From ; Nukuhiva; Marquesas Islands. Date: //. Source: Parzudaki. Material: F; whole/good skin.

Rhipidura fuliginosa cervina

Lord Howe Island Fantail

Formerly occurred in the forests of Lord Howe Island. Disappeared between 1924 and 1928, the reason being unknown, but possibly as a result of rats which escaped from a wreck in 1918.

Reg. no. 1881.5.1.559. From ; ; Lord Howe Island. Date: 13/ 9/1853. Source: J. MacGillivray, voyage of H.M.S. Herald; Gould coll.. Material: adult M; whole/good skin. Syntype of *R. macgillivrayi*. Sharpe, *Proc. zool. Soc. Lond.* 1881: 789, pl. 67. The female syntype is also in the collection.

Reg. no. 1881.5.1.567. From ; ; Lord Howe Island. Date: 13/ 9/1853. Source: J. MacGillivray, voyage of H.M.S. Herald; Gould coll.. Material: F; whole/good skin. Syntype of *R. macgillivrayi*.

Reg. no. 1897.7.2.786. From ; ; Lord Howe Island. Date: / 9/1887. Source: J. MacGillivray; Sydney Museum; Seebohm coll.. Material: adult M; whole/good skin.

Reg. no. 1919.7.15.132. From ; ; Lord Howe Island. Date: 11/ 5/1914. Source: G.M. Mathews. Material: adult M; whole/good skin.

Reg. no. 1919.7.15.133. From ; ; Lord Howe Island. Date: 30/ 5/1914. Source: G.M. Mathews. Material: adult F; skin. Specimen not found Jan 1975.

Reg. no. 1919.7.15.134. From ; ; Lord Howe Island. Date: 11/ 5/1914. Source: G.M. Mathews. Material: adult M; whole/good skin.

Reg. no. 1919.7.15.135. From ; ; Lord Howe Island. Date: 25/ 7/1913. Source: G.M. Mathews. Material: adult ; whole/good skin.

Reg. no. 1919.7.15.136. From ; ; Lord Howe Island. Date: 24/ 1/1914. Source: G.M. Mathews. Material: adult F; whole/good skin.

Reg. no. 1919.7.15.137. From ; ; Lord Howe Island. Date: 5/ 5/1914. Source: G.M. Mathews. Material: adult F; whole/good skin.

Reg. no. 1841.2.1.105. From ; ; Lord Howe Island. Date: 6/10/1913. Source: Mathews coll.; Rothschild. Material: 2 ; whole/good eggs (clutch).

Reg. no. E1941.2.1.106. From ; ; Lord Howe Island. Date: 8/10/1913. Source: Mathews coll.; Rothschild. Material: 3 ; whole/good eggs (clutch).

Reg. no. E1941.2.1.107. From ; ; Lord Howe Island. Date: 18/11/1913. Source: Mathews coll.; Rothschild. Material: 3 ; whole/good eggs (clutch).

Reg. no. E1941.2.1.108. From ; ; Lord Howe Island. Date: 6/10/1914. Source: Mathews coll.; Rothschild. Material: 3 ; whole/good eggs (clutch).

Reg. no. E1941.2.1.109. From ; ; Lord Howe Island. Date: 10/10/1914. Source: Mathews coll.; Rothschild. Material: 3 ; whole/good eggs (clutch).

Reg. no. E1941.2.1.110. From ; ; Lord Howe Island. Date: 12/10/1914. Source: Mathews coll.; Rothschild. Material: 3 ; whole/good eggs (clutch).

Turnagra capensis tanagra

North Island Piopio

North Island, New Zealand. Formerly common, but began to disappear quickly on contact with European settlers. Extinct since 1955.

Reg. no. 1886.6.24.23. From ; North Island; New Zealand. Date: / 3/1850. Source: Strange; Earle; Jardine coll.. Material: adult ; whole/good skin.

Turnagra capensis capensis

South Island Piopio

South Island, New Zealand. It declined rapidly in the 19th century, and was extinct by 1963.

Reg. no. 1849.12.12.23. From Pigeon Bay; [South Island]; New Zealand. Date: //. Source: Capt. Stokes. Material: adult M; whole/good skin. 'Pigeon Bay' was entered in the register in pencil sometime after the original entry was made. The information is not on the label.

Reg. no. 1849.12.12.24. From ; [South Island]; New Zealand. Date: //. Source: Capt. Stokes. Material: adult F; whole/good skin.

Reg. no. 1886.6.24.22. From ; [South Island]; New Zealand. Date: / 3/1850. Source: Strange; Earle; Jardine coll.; E. Gerrard. Material: adult ; whole/good skin.

Reg. no. 1903.12.10.2. From ; Stephens Island; New Zealand. Date: //. Source: J.H. Flemming; Rosenberg. Material: adult F; whole/good skin.

Reg. no. 1926.10.10.45. From Teramakau; [South Island]; New Zealand. Date: //1878. Source: Lt.Gen. S.R. Wason. Material: adult M; whole/good skin.

Reg. no. 1932.4.9.1. From ; [South Island]; New Zealand. Date: //1880. Source: Mrs S. Wilson. Material: adult ; whole/good skin. Collected about 1880.

Reg. no. 1949.64.218. From ; South Island; New Zealand. Date: /10/1887. Source: G.M. Mathews. Material: adult M; whole/good skin.

Reg. no. . From ; [South Island]; New Zealand. Date: //. Source: New Zealand Company. Material: ; skin. Mount. Details taken from *Cat. Bds.* Not found 1988.

Reg. no. 1902.2.21.53. From ; South Island; New Zealand. Date: / 1/1901. Source: Earl of Ranfurly. Material: adult ; whole/good skin.

Reg. no. 1902.2.21.54. From ; South Island; New Zealand. Date: / 1/1901. Source: Earl of Ranfurly. Material: adult ; whole/good skin.

Reg. no. 1903.7.17.18. From Dusky Bay; [South Island]; New Zealand. Date: / 6/1901. Source: Earl of Ranfurly. Material: adult M; whole/good skin. Male by dissection. Iris, bill and legs black.

Reg. no. E1941.9.4.1591. From ; Stephens Island; New Zealand. Date: //. Source: H. Travers; W. Rothschild. Material: 2 ; whole/good eggs.

Reg. no. E1941.3.1.20. From ; ; Captivity. Date: 13/ 8/1898. Source: H. Munt coll.. Material: 1 ; whole/good egg. 'South Island Thrush', '*Turnagra crassirostris*' on eggs.

Reg. no. E1941.3.1.21. From ; ; Captivity. Date: 15/ 7/1899. Source: H. Munt coll.. Material: 1 ; whole/good egg. 'South Island Thrush', '*Turnagra crassirostris*' on eggs.

Reg. no. E1941.3.1.22. From ; ; Captivity. Date: 30/ 6/1900. Source: H. Munt coll.. Material: 1 ; whole/good egg. 'South Island Thrush', '*Turnagra crassirostris*' on eggs.

***Turnagra capensis* subspecies**

Reg. no. S1961.11.22. From ; ; [New Zealand]. Date: //. Source: Rothschild coll.. Material: adult ; whole/good skeleton. Previously identified as *Geocichla litsitsirupa*.

Dicaeum quadricolor

Cebu Flowerpecker

Endemic to Cebu in the Philippines where it lived in primary forest which was completely removed by 1906. The species was believed to have become extinct shortly thereafter but was seen again in 1992.

Reg. no. 1888.7.12.133. From ; Cebu; Philippines. Date: / 4/1877. Source: A.H. Everett; Tweeddale coll.. Material: adult ; whole/good skin. Holotype. Tweeddale, *Proc. zool. Soc. Lond.* 1877 (publ. 1878): 762, pl. 77.

Dicaeum trigonostigma pallidus

Paler Flowerpecker

Endemic to Cebu in the Philippines where it was still common in 1906. The last specimen was collected in 1910, which is now considered to be the date of extinction.

Reg. no. 1888.7.12.70. From ; Cebu; Philippines. Date: / 4/1877. Source: A.H. Everett; Tweeddale coll.. Material: M; whole/good skin.

Zosterops everetti everetti

Everett's White-eye

Endemic to Cebu in the Philippines, where it became extinct in 1906.

Reg. no. 1888.7.1.700. From ; Cebu; Philippines. Date: / 4/1877. Source: A.H. Everett; Tweeddale coll.. Material: adult M; whole/good skin. Syntype. Tweeddale, *Proc. zool. Soc. Lond.* 1877 (publ. 1878): 762. The female syntype is also in the collection. The register gives the collector as Meyer, presumably in error.

Reg. no. 1888.7.1.701. From ; Cebu; Philippines. Date: / 4/1877. Source: A.H. Everett; Tweeddale coll.. Material: F; whole/good skin. Syntype.

Reg. no. 1896.6.6.419. From ; Cebu; Philippines. Date: 13/ 3/. Source: Steere coll.. Material: M; whole/good skin.

Zosterops luteirostris luteirostris

Splendid White-eye

Confined to Gizo in the Solomon Islands, where it is now very rare as a result of habitat destruction. The race *Z.l. splendida* of Ranongga (or Ganonga) Island is thought not to be at risk.

Reg. no. 1904.7.13.36. From ; Gizo Island; Solomon Islands. Date: 11/11/1903. Source: A.S. Meek; Gerrard. Material: adult M; whole/good skin. Damaged: both legs almost detached.

Reg. no. 1904.7.13.37. From ; Gizo Island; Solomon Islands. Date: 10/11/1903. Source: A.S. Meek; Gerrard. Material: adult F; whole/good skin. Damaged: one wing very loose.

Zosterops strenua

Robust White-eye

Lord Howe Island. Common prior to 1918 after which it became extinct as a result of the introduction of rats following a wreck.

Reg. no. 1855.10.23.2. From ; ; Lord Howe Island. Date: //. Source: J. MacGillivray, voyage of H.M.S. Herald; Mus. Economic Geography. Material: adult F; whole/good skin. Holotype. Gould, *Proc. zool. Soc. Lond.* 1855: 166.

Reg. no. 1888.10.5.6. From ; ; [Lord Howe Island]. Date: //. Source: E. Saunders. Material: ; whole/good skin. The register and the label both give the locality of this specimen as Norfolk Island, but this species is endemic to Lord Howe Island.

Reg. no. 1898.9.30.213. From ; ; Lord Howe Island. Date: //. Source: Seebohm coll.. Material: adult ; whole/good skin.

Reg. no. 1898.9.30.214. From ; ; Lord Howe Island. Date: / 9/1887. Source: Seebohm coll.. Material: adult M; skin. Details taken from the register. Not found 1988.

Reg. no. 1919.7.15.159. From ; ; Lord Howe Island. Date: 10/ 1/1914. Source: G.M. Mathews. Material: adult M; whole/good skin.

Reg. no. 1919.7.15.160. From ; ; Lord Howe Island. Date: 11/ 1/1914. Source: G.M. Mathews. Material: adult M; whole/good skin.

Reg. no. 1919.7.15.161. From ; ; Lord Howe Island. Date: 6/11/1913. Source: G.M. Mathews. Material: adult F; whole/good skin.

Reg. no. 1919.7.15.162. From ; ; Lord Howe Island. Date: 12/ 5/1914. Source: G.M. Mathews. Material: adult M; whole/good skin.

Reg. no. 1919.7.15.163. From ; ; Lord Howe Island. Date: 24/11/1913. Source: G.M. Mathews. Material: adult F; whole/good skin.

Reg. no. 1919.7.15.164. From ; ; Lord Howe Island. Date: 30/ 5/1914. Source: G.M. Mathews. Material: adult F; whole/good skin.

Reg. no. 1919.7.15.165. From ; ; Lord Howe Island. Date: 11/ 5/1914. Source: G.M. Mathews. Material: adult M; whole/good skin.

Reg. no. 1919.7.15.166. From ; ; Lord Howe Island. Date: 24/ 1/1914. Source: G.M. Mathews. Material: adult F; whole/good skin.

Zosterops albogularis

White-chested White-eye

Norfolk Island. Common until the end of the 1880s, but now confined to the remote forests of the Mount Pitt Reserve. The population was 50 in 1962, and was last seen reliably in 1980. Despite several searches in the 1980s the species could not be found, although a few individuals may remain.

Reg. no. . From ; ; [Norfolk Island]. Date: //. Source: Sir T. Mitchell. Material: ; skin. Details from the *Cat. Bds.*. Not found 1988.

Reg. no. 1839.8.2.109. From ; ; [Norfolk Island] (Australia). Date: //. Source: Sir T. Mitchell. Material: ; whole/good skin.

Reg. no. 1870.3.31.93. From ; ; Norfolk Island. Date: / 6/1865. Source: J. Brenchley. Material: M; whole/good skin. Eye reddish-brown. Food, guava seeds and snails.

Reg. no. 1870.3.31.94. From ; ; Norfolk Island. Date: / 6/1865. Source: J. Brenchley. Material: adult F; whole/good skin. Eye reddish-brown. Food, guava seeds and snails.

Reg. no. 1898.9.30.197. From ; ; Norfolk Island. Date: /10/1879. Source: E.L. Layard; Seebohm coll.. Material: adult M; whole/good skin.

Reg. no. 1919.7.15.57. From Mount Pitt; ; Norfolk Island. Date: 26/ 8/1912. Source: G.M. Mathews. Material: adult F; whole/good skin. The day of collection is given as 24th in the register.

Reg. no. 1919.7.15.58. From Stockyard Creek; ; Norfolk Island. Date: 21/11/1912. Source: G.M. Mathews. Material: adult M; whole/good skin.

Reg. no. 1919.7.15.59. From Steels Point; ; Norfolk Island. Date: 28/ 2/1913. Source: G.M. Mathews. Material: adult F; whole/good skin.

Reg. no. 1919.7.15.60. From nr. Cascades; ; Norfolk Island. Date: 4/ 3/1913. Source: G.M. Mathews. Material: adult M; whole/good skin.

Reg. no. 1919.7.15.61. From nr. Middlegate; ; Norfolk Island. Date: 6/ 3/1913. Source: G.M. Mathews. Material: adult M; whole/good skin.

Reg. no. 1919.7.15.62. From nr. Middlegate; ; Norfolk Island. Date: 8/ 3/1913. Source: G.M. Mathews. Material: adult F; whole/good skin.

Reg. no. 1919.7.15.63. From nr. Cascades; ; Norfolk Island. Date: 3/ 4/1913. Source: G.M. Mathews. Material: adult M; whole/good skin.

Reg. no. 1919.7.15.64. From Dixey; ; Norfolk Island. Date: 25/ 4/1913. Source: G.M. Mathews. Material: adult F; whole/good skin.

Reg. no. 1919.7.15.65. From nr. Cascades; ; Norfolk Island. Date: 15/ 4/1913. Source: G.M. Mathews. Material: adult F; whole/good skin.

Reg. no. 1919.7.15.66. From nr. Kingston; ; Norfolk Island. Date: 29/ 4/1913. Source: G.M. Mathews. Material: adult M; whole/good skin.

Reg. no. 1919.7.15.67. From nr. Cascades; ; Norfolk Island. Date: 28/ 5/1913. Source: G.M. Mathews. Material: adult M; whole/good skin.

Reg. no. 1919.7.15.68. From Cascades; ; Norfolk Island. Date: 12/ 6/1913. Source: G.M. Mathews. Material: adult M; whole/good skin.

Reg. no. 1939.12.9.2305. From nr. Dixey; ; Norfolk Island. Date: 29/ 3/1913. Source: G.M. Mathews; Rothschild coll.. Material: adult M; whole/good skin.

Reg. no. 1939.12.9.2306. From Steels Point; ; Norfolk Island. Date: 3/10/1912. Source: G.M. Mathews; Rothschild coll.. Material: adult F; whole/good skin.

Reg. no. 1939.12.9.2307. From Middlegate; ; Norfolk Island. Date: 6/ 2/1913. Source: G.M. Mathews; Rothschild coll.. Material: adult F; whole/good skin.

Reg. no. 1939.12.9.2308. From Middlegate; ; Norfolk Island. Date: 27/ 3/1913. Source: G.M. Mathews; Rothschild coll.. Material: adult ; whole/good skin.

Reg. no. 1939.12.9.2309. From Middlegate; ; Norfolk Island. Date: 6/ 2/1913. Source: G.M. Mathews; Rothschild coll.. Material: adult M; whole/good skin.

Reg. no. 1939.12.9.2310. From nr. Stockyard Creek; ; Norfolk Island. Date: 17/12/1912. Source: Rothschild coll.. Material: adult M; whole/good skin.

Reg. no. 1939.12.9.2311. From Middlegate; ; Norfolk Island. Date: 6/ 3/1913. Source: G.M. Mathews; Rothschild coll.. Material: adult M; whole/good skin.

Reg. no. 1988.21.8. From ; ; Norfolk Island. Date: //. Source: Material: ; whole/good skin.

Zosterops mayottensis semiflava

Seychelles Yellow White-eye

Seychelles. It has not been seen since 1888 and is presumed extinct.

Reg. no. 1880.6.3.4. From ; Marianne Island; Seychelles. Date: //1879. Source: H.L. Warry; Tristram coll.. Material: ; whole/good skin.

Reg. no. 1895.5.1.2212. From ; Marianne Island; Seychelles. Date: //. Source: Shelley coll.. Material: F; whole/good skin.

Reg. no. 1927.12.18.397. From ; Marianne Island (Marie Anne Island); Seychelles. Date: 14/ 2/1888. Source: J.J. Lister coll.. Material: M; whole/good skin.

Reg. no. 1927.12.18.398. From ; Marianne Island (Marie Anne Island); Seychelles. Date: 14/ 2/1888. Source: J.J. Lister coll.. Material: M; whole/good skin.

Reg. no. 1927.12.18.399. From ; Marianne Island (Marie Anne Island); Seychelles. Date: 14/ 2/1888. Source: J.J. Lister coll.. Material: ; whole/good skin.

Reg. no. 1927.12.18.400. From ; Marianne Island (Marie Anne Island); Seychelles. Date: 14/ 2/1888. Source: J.J. Lister coll.. Material: ; whole/good skin.

Zosterops modesta

Grey White-eye

Occurs only in 3 tiny areas of mixed secondary forest in central Mahe, Seychelles. The population was around 100 in the mid-1970s since when it has been steadily declining.

Reg. no. 1895.5.1.2237. From ; ; Seychelles. Date: //. Source: Shelley coll.. Material: adult ; whole/good skin.

Reg. no. 1906.12.21.406. From ; Mahe; Seychelles. Date: 25/ 3/1906. Source: M.J. Nicoll; Earl of Crawford, voyage of R.Y.S. Valhalla. Material: adult M; whole/good skin.

Reg. no. 1906.12.21.407. From ; Mahe; Seychelles. Date: 25/ 3/1906. Source: M.J. Nicoll; Earl of Crawford, voyage of R.Y.S. Valhalla. Material: adult M; whole/good skin. Iris hazel, bill lead colour with tip blackish, feet and toes dark brown.

Reg. no. 1906.12.21.408. From ; Mahe; Seychelles. Date: 25/ 3/1906. Source: M.J. Nicoll; Earl of Crawford, voyage of R.Y.S. Valhalla. Material: adult M; skin. Details taken from the register. Not found 1988.

Reg. no. 1927.12.18.396. From Cascade; Mahe; Seychelles. Date: 3/ 1/1888. Source: J.J. Lister. Material: adult F; whole/good skin.

Reg. no. E1977.5.1. From Mission; Mahe; Seychelles. Date: 9/10/1976. Source: P.W. Greig Smith. Material: 1 ; damaged egg.

Reg. no. E1977.5.2. From Mission; Mahe; Seychelles. Date: 10/10/1976. Source: P.W. Greig Smith. Material: 1 ; damaged egg.

Rukia ruki

Truk White-eye

Known only from several islands in the lagoon of Truk, Caroline Islands. The species is now very rare except on Tol, where it is still common in a tiny patch of forest near the summit of the central mountain.

Reg. no. 1898.4.29.15. From Ruk Island; Truk Islands; Caroline Islands. Date: 16/12/1895. Source: W. Rothschild. Material: adult M; whole/good skin.

Meliphaga cassidix

Helmeted Honeyeater

Locally distributed in Western Australia where it is rapidly declining as a result of hybridization with *M. melanops*.

Reg. no. 1867.3.16.87. From ; Victoria; Australia. Date: //1866. Source: J. Gould. Material: adult M; whole/good skin.

Reg. no. 1877.12.6.6. From ; South Australia; Australia. Date: //. Source: F. Nicholson. Material: adult M; whole/good skin.

Reg. no. 1881.11.7.804. From ; South Australia; Australia. Date: //. Source: Cockerell coll.; F.D. Godman. Material: F; whole/good skin.

Reg. no. 1881.11.7.805. From ; South Australia; Australia. Date: //. Source: Cockerell coll.; F.D. Godman. Material: ; whole/good skin.

Reg. no. 1881.11.7.807. From ; South Australia; Australia. Date: //. Source: Cockerell coll.; F.D. Godman. Material: ; whole/good skin.

Reg. no. 1881.11.7.808. From ; South Australia; Australia. Date: //. Source: Cockerell coll.; F.D. Godman. Material: ; whole/good skin.

Reg. no. 1881.5.1.5576. From ; Victoria; Australia. Date: //. Source: Gould coll.. Material: adult M; whole/good skin.

Reg. no. 1881.5.1.5582. From ; Victoria; Australia. Date: //. Source: Gould coll.. Material: adult M; whole/good skin.

Reg. no. 1881.5.1.6158. From ; Victoria; Australia. Date: //. Source: J. Gould coll.. Material: adult M; whole/good skin.

Reg. no. 1881.5.1.6159. From ; Victoria; Australia. Date: //. Source: J. Gould coll.. Material: adult F; whole/good skin.

Reg. no. 1881.5.1.6160. From Port Phillip Bay; Victoria; Australia. Date: //. Source: J. Gould coll.. Material: adult M; whole/good skin. The location is given as West Port in the register.

Reg. no. 1881.5.1.6161. From Port Phillip Bay; Victoria; Australia. Date: //. Source: J. Gould coll.. Material: adult M; whole/good skin. The location is given as West Port in the register.

Reg. no. 1886.6.24.338. From ; ; Australia. Date: //1866. Source: Sanderson; W. Jardine; Gerrard. Material: adult M; whole/good skin. Ex-mount. Holotype. Gould, 1867, *Birds of Australia*, Supplement, p. 77, pl. 39.

Reg. no. 1886.6.24.339. From ; ; Australia. Date: //. Source: W. Jardine; Gerrard. Material: ; skin. Details taken from the register. Not found 1988.

Reg. no. 1888.7.1.838. From Sydney; New South Wales; Australia. Date: //. Source: Tweeddale coll.. Material: ; whole/good skin.

Reg. no. E1949.10.483. From Beaconsfield; s. Victoria; Australia. Date: 13/11/1909. Source: F.E. Wilson; G. Mathews. Material: 1 ; whole/poor egg. Egg broken.

Apalopteron familiare familiare

Mukojima Honeyeater

Known from the northern Bonin Islands where it is now probably extinct on Mukojima and certainly so on Chichijima. The present population on the latter island is the race *A.f. hahashima* of the southern Bonins which has been introduced to Chichijima by man.

Reg. no. 1898.10.3.870. From Parry Island; Bonin Islands; Japan. Date: 13/ 6/1889. Source: P.A. Holst; Seebohm coll.. Material: juv F; whole/good skin.

Reg. no. 1898.10.3.871. From Parry Island; Bonin Islands; Japan. Date: 13/ 6/1889. Source: P.A. Holst; Seebohm coll.. Material: adult M; whole/good skin.

Reg. no. 1898.10.3.872. From Parry Island; Bonin Islands; Japan. Date: 13/ 6/1889. Source: P.A. Holst; Seebohm coll.. Material: F; whole/good skin. Sex given as male in the register.

Reg. no. 1898.10.3.873. From Parry Island; Bonin Islands; Japan. Date: 13/ 6/1889. Source: P.A. Holst; Seebohm coll.. Material: F; whole/good skin. Sex given as male in the register.

Reg. no. 1898.10.3.874. From Parry Island; Bonin Islands; Japan. Date: 15/ 6/1889. Source: P.A. Holst; Seebohm coll.. Material: adult F; whole/good skin.

Reg. no. 1898.10.3.875. From Parry Island, Mukosima; Bonin Islands; Japan. Date: 10/10/1890. Source: P.A. Holst; Seebohm coll.. Material: adult M; whole/good skin.

Moho braccatus

Kauai Oo

Confined to the Alakai Swamp on Kauai in the Hawaiian Islands. Formerly common, it has now been reduced to a few pairs. Avian malaria is believed to have been the cause.

Reg. no. 1890.10.3.17. From ; Kauai; Hawaiian Islands. Date: //. Source: V. Knudsen; Collett. Material: ; whole/good skin.

Reg. no. 1895.10.16.7. From ; Kauai; Hawaiian Islands. Date: //. Source: Knudsen coll.; Collett. Material: ; whole/good skin.

Reg. no. 1895.10.16.8. From ; Kauai; Hawaiian Islands. Date: //. Source: Knudsen coll.; Collett. Material: ; whole/good skin.

Reg. no. 1895.7.20.55. From ; Kauai; Hawaiian Islands. Date: / 5/1894. Source: R.C.L. Perkins. Material: adult M; whole/good skin.

Reg. no. 1895.7.20.56. From ; Kauai; Hawaiian Islands. Date: / 5/1894. Source: R.C.L. Perkins. Material: adult F; whole/good skin.

Reg. no. 1897.10.28.4. From Kaholuamano; Kauai; Hawaiian Islands. Date: /10/1895. Source: R.C.L. Perkins; Royal Society & British Association. Material: adult F; whole/good skin.

Reg. no. 1897.10.28.5. From Halemanu; Kauai; Hawaiian Islands. Date: / 5/1895. Source: [R.C.L. Perkins]; Royal Society &

British Association. Material: juv ; whole/good skin.

Reg. no. 1898.9.30.91. From ; Kauai; Hawaiian Islands. Date: / 9/1888. Source: S.B. Wilson; Seebohm coll.. Material: adult F; whole/good skin.

Reg. no. 1898.9.30.92. From ; Kauai; Hawaiian Islands. Date: / 9/1888. Source: S.B. Wilson; Seebohm coll.. Material: adult ; whole/good skin.

Reg. no. 1925.2.16.103. From ; Kauai; Hawaiian Islands. Date: //. Source: S.B. Wilson. Material: M; whole/good skin.

Reg. no. 1925.2.16.104. From ; Kauai; Hawaiian Islands. Date: / 4/1891. Source: S.B. Wilson. Material: M; whole/good skin.

Reg. no. 1939.12.9.323. From ; Kauai; Hawaiian Islands. Date: 1/ 1/1891. Source: H.C. Palmer coll.; Rothschild coll.. Material: adult M; whole/good skin.

Reg. no. 1939.12.9.324. From ; Kauai; Hawaiian Islands. Date: 26/ 3/1891. Source: H.C. Palmer coll.; Rothschild coll.. Material: adult M; whole/good skin.

Reg. no. 1939.12.9.325. From ; Kauai; Hawaiian Islands. Date: 12/ 4/1891. Source: H.C. Palmer coll.; Rothschild coll.. Material: adult F; whole/good skin.

Reg. no. 1939.12.9.326. From ; Kauai; Hawaiian Islands. Date: 31/ 1/1891. Source: H.C. Palmer coll.; Rothschild coll.. Material: adult F; whole/good skin.

Reg. no. 1939.12.9.327. From ; Kauai; Hawaiian Islands. Date: 2/ 2/1891. Source: H.C. Palmer coll.; Rothschild coll.. Material: adult F; whole/good skin.

Reg. no. 1939.12.9.328. From ; Kauai; Hawaiian Islands. Date: 26/ 3/1891. Source: H.C. Palmer coll.; Rothschild coll.. Material: adult F; whole/good skin.

Reg. no. 1939.12.9.329. From ; Kauai; Hawaiian Islands. Date: 26/ 3/1891. Source: H.C. Palmer coll.; Rothschild coll.. Material: adult F; whole/good skin.

Reg. no. 1939.12.9.330. From ; Kauai; Hawaiian Islands. Date: 24/ 3/1891. Source: H.C. Palmer coll.; Rothschild coll.. Material: adult ; whole/good skin.

Reg. no. 1939.12.9.331. From ; Kauai; Hawaiian Islands. Date: 14/ 4/1891. Source: H.C. Palmer coll.; Rothschild coll.. Material: adult F; whole/good skin.

Moho bishopi

Bishop's Oo

Extinct on Molokai where it was last seen in 1904. A bird either of this species, or of another very closely related and as yet unnamed, was discovered in 1981 on Maui.

Reg. no. 1895.7.20.90. From ; Molokai; Hawaiian Islands. Date: 7/ 9/1893. Source: R.C.L. Perkins. Material: adult M; whole/good skin.

Reg. no. 1895.7.20.91. From ; Molokai; Hawaiian Islands. Date: 30/ 8/1893. Source: R.C.L. Perkins. Material: adult M; whole/good skin.

Reg. no. 1895.7.20.92. From ; Molokai; Hawaiian Islands. Date: 7/ 9/1893. Source: R.C.L. Perkins. Material: adult M; whole/good skin.

Reg. no. 1895.7.20.93. From ; Molokai; Hawaiian Islands. Date: 24/ 8/1893. Source: R.C.L. Perkins. Material: adult F; whole/good skin.

Reg. no. 1939.12.9.51. From ; Molokai; Hawaiian Islands. Date: //1893. Source: H.C. Palmer; Rothschild coll.. Material: adult M; whole/good skin.

Moho apicalis

Oahu Oo

Formerly occurred on the island of Oahu in the Hawaiian Islands. It was last recorded in 1837 and the reason for its extinction is not known.

Reg. no. 1860.11.26.50. From ; [Oahu]; Hawaiian Islands. Date: //. Source: Gould coll.. Material: adult M; skin. Syntype. Not in the collection. Exchanged with Rothschild in 1895.

Reg. no. 1860.11.26.51. From ; [Oahu]; Hawaiian Islands. Date: //. Source: J. Gould. Material: adult F; whole/good skin. Syntype. Gould, *Proc. zool. Soc. Lond.* 1860: 381. Mount.

Reg. no. VEL.26.19a. From ; Oahu; Hawaiian Islands. Date: //. Source: Byron. Material: M; whole/good skin.

Moho nobilis

Hawaii Oo

Formerly in mountain forests of the island of Hawaii. Last seen in 1934 and presumed now extinct.

Reg. no. 1861.11.11.186. From ; Hawaii; Hawaiian Islands. Date: //. Source: Gould. Material: ; whole/good skin. Mount.

Reg. no. 1880.11.18.449. From Hilo; Hawaii; Hawaiian Islands. Date: //. Source: Challenger Expedition. Material: adult M; whole/good skin.

Reg. no. 1895.7.20.29. From Kona; Hawaii; Hawaiian Islands. Date: //1893. Source: R.C.L. Perkins. Material: M; whole/good skin.

Reg. no. 1895.7.20.30. From Mauna Loa; Hawaii; Hawaiian Islands. Date: / 6/1892. Source: R.C.L. Perkins. Material: imm ; whole/good skin. Date may not be correct.

Reg. no. 1897.10.28.3. From Amaulu; Hawaii; Hawaiian Islands. Date: /12/1895. Source: R.C.L. Perkins; Royal Society & British Association. Material: adult F; whole/good skin.

Reg. no. 1898.9.30.89. From ; Hawaii; Hawaiian Islands. Date: 26/ 5/1887. Source: S.B. Wilson; Seebohm coll.. Material: adult M; whole/good skin.

Reg. no. 1898.9.30.90. From ; Hawaii; Hawaiian Islands. Date: 26/ 5/1887. Source: S.B. Wilson; Seebohm coll.. Material: adult F; whole/good skin.

Reg. no. 1925.2.16.101. From ; Hawaii; Hawaiian Islands. Date: /11/1892. Source: S.B. Wilson. Material: M; whole/good skin.

Reg. no. 1925.2.16.102. From ; Hawaii; Hawaiian Islands. Date: /11/1892. Source: S.B. Wilson. Material: F; whole/good skin.

Reg. no. 1939.12.9.320. From ; Hawaii; Hawaiian Islands. Date: 31/10/1891. Source: H.C. Palmer; Rothschild coll.. Material: adult M; whole/good skin.

Reg. no. 1939.12.9.321. From ; Hawaii; Hawaiian Islands. Date: 31/10/1891. Source: H.C. Palmer; Rothschild coll.. Material: adult F; whole/good skin.

Reg. no. 1939.12.9.322. From ; Hawaii; Hawaiian Islands. Date: 7/11/1891. Source: H.C. Palmer; Rothschild coll.. Material: adult F; whole/good skin.

Reg. no. 1939.12.9.56. From ; Hawaii; Hawaiian Islands. Date: //. Source: Rothschild coll.. Material: adult M; whole/good skin.

Anthornis melanura melanocephala
Chatham Islands Bellbird

Formerly found in the Chatham Islands. It disappeared in the early years of the 20th century, probably as a result of forest destruction, although there was apparently still quite a lot of forest left in 1900 and the disturbance seemed to be minimal.

Reg. no. 1842.9.29.26. From ; ; Chatham Islands. Date: //. Source: Dieffenbach; New Zealand Company. Material: adult M; whole/good skin. Ex-mount. Holotype. G.R. Gray, *Dieffenbach's Travels in New Zealand*, 2 (1843): 188.

Reg. no. 1893.12.31.5. From ; ; Chatham Islands. Date: //. Source: H.O. Forbes; E. Gerrard. Material: M; whole/good skin.

Reg. no. 1893.12.31.6. From ; ; Chatham Islands. Date: //. Source: H.O. Forbes; E. Gerrard. Material: F; whole/good skin.

Reg. no. 1893.7.6.1. From Little Mangere; ; Chatham Islands. Date: //. Source: W. Hawkins. Material: M; whole/good skin. Eyes, blue-black with yellow-orange ring.

Reg. no. 1902.10.25.435,438. From ; ; Chatham Islands. Date: //. Source: Seymour; P. Crowley; Radcliffe-Saunders coll.. Material: 2 ; whole/good eggs (clutch).

Reg. no. 1902.10.25.437,439. From ; ; Chatham Islands. Date: //. Source: Seymour; P. Crowley; Radcliffe-Saunders coll.. Material: 2 ; whole/good eggs (clutch).

Reg. no. 1902.3.25.987. From ; ; Chatham Islands. Date: //. Source: Seymour; P. Crowley. Material: 1 ; whole/good egg.

Reg. no. E1941.9.4.909. From ; ; Chatham Islands. Date: //. Source: Rothschild. Material: 1 ; whole/good egg.

Ammodramus maritimus nigrescens
Dusky Seaside Sparrow

This form had a very restricted habitat in a few salt marshes on the eastern coast of Florida. By 1981 only 5 males (and no females) remained, and these were taken into captivity. The last bird died on 16 June 1987.

Reg. no. 1892.1.12.50. From ; Florida; United States. Date: 6/ 1/1886. Source: C.J. Maynard; C.B. Cory coll.; Salvin–Godman coll.. Material: adult M; whole/good skin.

Reg. no. 1892.1.12.51. From Indian River; Florida; United States. Date: 1/ 5/1886. Source: C.J. Maynard; C.B. Cory coll.; Salvin–Godman coll.. Material: adult F; whole/good skin.

Reg. no. 1906.12.7.2739. From Indian River; Florida; United States. Date: 26/ 2/1886. Source: C.J. Maynard; G.S. Miller coll.. Material: adult M; whole/good skin.

Reg. no. 1906.12.7.2740. From Indian River; Florida; United States. Date: 11/ 3/1886. Source: C.J. Maynard; G.S. Miller coll.. Material: adult F; whole/good skin. The day of collection is given as 16th in the register.

Reg. no. 1906.12.7.2741. From Merritts Island; Florida; United States. Date: 10/ 5/1886. Source: C.J. Maynard; G.S. Miller coll.. Material: adult M; whole/good skin.

Reg. no. 1988.21.1. From Indian River; Florida; United States. Date: //. Source: C.J. Maynard; H.W. Henshaw coll.. Material: adult F; whole/good skin.

Ammodramus baileyi

Sierra Madre Sparrow

Occurs in the highlands of parts of Mexico in areas of bunch–grass near pine trees. It is threatened by habitat destruction.

Reg. no. 1899.2.1.3261. From Sierra Bolanos; Jalisco; Mexico. Date: 6/ 3/1889. Source: W.B. Richardson; Salvin–Godman coll.. Material: adult F; whole/good skin. The day of collection is given as 5th in the register.

Reg. no. 1899.2.1.3262. From Sierra Bolanos; Jalisco; Mexico. Date: 9/ 3/1889. Source: W.B. Richardson; Salvin–Godman coll.. Material: adult M; whole/good skin. The day of collection is given as 10th in the register, where it is also stated that the specimen was exchanged with the US National Museum in 1939. The latter is either incorrect, or two specimens were registered with the same number.

Reg. no. 1899.2.1.3263. From Sierra Bolanos; Jalisco; Mexico. Date: 10/ 3/1889. Source: W.B. Richardson; Salvin–Godman coll.. Material: adult M; whole/good skin. The sex is given as female in the register.

Reg. no. 1899.2.1.3264. From Sierra Bolanos; Jalisco; Mexico. Date: 10/ 3/1889. Source: W.B. Richardson; Salvin–Godman coll.. Material: adult F; whole/good skin. The day of collection is given as 3rd in the register.

Reg. no. 1899.2.1.3265. From Sierra Bolanos; Jalisco; Mexico. Date: 3/ 3/1889. Source: W.B. Richardson; Salvin–Godman coll.. Material: adult M; whole/good skin. Two specimens of this species bear the same reg. no.. The register does not give the date of this one.

Reg. no. 1899.2.1.3265. From Sierra Bolanos; Jalisco; Mexico. Date: 9/ 3/1889. Source: W.B. Richardson; Salvin–Godman coll.. Material: adult M; whole/good skin. Two specimens of this species bear the same reg. no.. The register gives the date of this one.

Poospiza rubecula

Rufous–breasted Warbling–Finch

Restricted to mountain scrub on the western slopes of the Andes of northern Peru. It is rare and local, little–known, and its habitat is disappearing.

Reg. no. 1896.10.6.329. From 10,400 ft., Huamachuco; ; n. Peru. Date: / 2/1894. Source: O.T. Baron; Salvin–Godman coll.. Material: adult M; whole/good skin. Syntype. Salvin, *Novitates Zoologicae* 2 (1895): 8. There are two other syntypes in the collection.

Reg. no. 1896.10.6.330. From Cajabamba; ; Peru. Date: / 1/1894. Source: O.T. Baron; Salvin–Godman coll.. Material: adult M; whole/good skin. Syntype.

Reg. no. 1896.10.6.331. From Cajabamba; ; Peru. Date: / 1/1894. Source: O.T. Baron; Salvin–Godman coll.. Material: juv M; whole/good skin. Syntype.

Reg. no. 1902.3.13.441. From Surco; Lima; Peru. Date: 9/ 2/1900. Source: P.O. Simons. Material: adult M; whole/good skin.

Dacnis albiventris

White–bellied Dacnis

Occurs widely in the lowlands of the upper Amazon basin, but it is very rare and local within this area.

Reg. no. 1843.5.24.101. From [Bogota]; ; [Colombia]. Date: //. Source: Leadbeater. Material: adult M; whole/good skin. Locality taken from the *Cat. Bds.*; not given on the label or in the register.

Reg. no. 1854.4.6.20. From Bogota; ; Colombia. Date: //. Source: Argent. Material: adult M; whole/good skin.

Reg. no. 1884.7.31.74. From Bogota; ; Colombia. Date: //. Source: Bell; Sclater coll.. Material: juv M; whole/good skin.

Reg. no. 1884.7.31.75. From Xeberos, [Amazons]; ; [Peru]. Date: //1867. Source: E. Bartlett; Sclater coll.. Material: adult M; whole/good skin.

Reg. no. 1884.7.31.76. From Xeberos, [Amazons]; ; [Peru]. Date: //1867. Source: E. Bartlett; Sclater coll.. Material: adult F; whole/good skin.

Reg. no. 1885.4.1.108. From Chemicuros, Andes; ; Peru. Date: 30/ 4/1875. Source: E. Bartlett; Salvin–Godman coll.. Material: adult M; whole/good skin. Two individuals of this species bear the same reg. no..

Reg. no. 1885.4.1.108. From Xeberos, Amazons; ; Peru. Date: / 1/1879. Source: E. Bartlett; Salvin–Godman coll.. Material: juv M; whole/good skin. Two individuals of this species bear the same reg. no..

Reg. no. 1885.4.1.109. From Sarayacu; ; Ecuador. Date: / 2/1880. Source: C. Buckley; Salvin–Godman coll.. Material: adult M; whole/good skin.

Reg. no. 1885.4.1.110. From ; ; Ecuador. Date: //. Source: C. Buckley; Salvin–Godman coll.. Material: ; skin. Details taken from the register. Not found 1988.

Reg. no. 1898.12.14.190. From Villavicencio to Medina, Llanos of Rio Meta; ; Colombia. Date: //. Source: T.H. Wheeler; Salvin–Godman coll.. Material: adult M; whole/good skin.

Reg. no. 1898.12.14.191. From Villavicencio to Medina, Llanos of Rio Meta; ; Colombia. Date: //. Source: T.H. Wheeler; Salvin–Godman coll.. Material: adult M; whole/good skin.

Reg. no. 1898.12.14.192. From Villavicencio to Medina, Llanos of Rio Meta; ; Colombia. Date: //. Source: T.H. Wheeler; Salvin–Godman coll.. Material: adult M; whole/good skin.

Reg. no. 1898.12.14.193. From Villavicencio to Medina, Llanos of Rio Meta; ; Colombia. Date: //. Source: T.H. Wheeler; Salvin–Godman coll.. Material: adult M; whole/good skin.

Reg. no. 1898.12.14.194. From Villavicencio to Medina, Llanos of Rio Meta; ; Colombia. Date: //. Source: T.H. Wheeler; Salvin–Godman coll.. Material: adult M; whole/good skin.

Reg. no. 1898.12.14.195. From Villavicencio to Medina, Llanos of Rio Meta; ; Colombia. Date: //. Source: T.H. Wheeler; Salvin–Godman coll.. Material: adult F; whole/good skin.

Reg. no. 1898.12.14.196. From Villavicencio to Medina, Llanos of Rio Meta; ; Colombia. Date: //. Source: T.H. Wheeler; Salvin–Godman coll.. Material: adult F; whole/good skin.

Dacnis berlepschi

Scarlet–breasted Dacnis

Restricted to southwestern Colombia and northwestern Ecuador, where it is very rare and local, and occurs mainly in wet forest edge and secondary growth.

Reg. no. 1902.5.5.1. From Carondelet, Rio Bogota; ; . Date: 26/10/1900. Source: Miketta & Flemming; Rosenberg. Material: juv M; whole/good skin. Iris brown, feet grey, bill black.

Reg. no. 1902.5.5.2. From Carondelet, Rio Bogota; ; . Date: 3/11/1900. Source: Miketta & Flemming; Rosenberg. Material: F; whole/good skin. Iris yellow, feet dark grey, bill black.

Reg. no. 1902.5.5.24. From Carondelet, Rio Bogota; ; . Date: 31/10/1900. Source: Miketta & Flemming; Rosenberg. Material: adult M; whole/good skin. Iris yellow, feet dark grey, bill black.

Vermivora bachmanii

Bachman's Warbler

Formerly bred in southeastern USA and wintered in Cuba. It declined as a result of habitat destruction both in its breeding and wintering areas. It has not definitely been seen since 1988 and may now be extinct.

Reg. no. 1885.3.8.889. From ; Florida; United States. Date: //. Source: Salvin-Godman coll.. Material: ; whole/good skin.

Reg. no. 1885.3.8.890. From ; Florida; United States. Date: //. Source: Salvin-Godman coll.. Material: ; whole/good skin.

Reg. no. 1892.4.1.31. From Key West; Florida; United States. Date: 9/ 8/1888. Source: J.W. Atkins; W.E.D. Scott; Salvin-Godman coll.. Material: adult F; whole/good skin.

Reg. no. 1892.4.1.32. From Key West; Florida; United States. Date: 9/ 8/1888. Source: J.W. Atkins; W.E.D. Scott; Salvin-Godman coll.. Material: adult M; whole/good skin.

Reg. no. 1892.4.1.33. From Key West; Florida; United States. Date: 30/ 8/1887. Source: J.W. Atkins; W.E.D. Scott; Salvin-Godman coll.. Material: adult M; whole/good skin.

Reg. no. 1892.4.1.34. From Key West; Florida; United States. Date: 26/ 7/1888. Source: J.W. Atkins; W.E.D. Scott; Salvin-Godman coll.. Material: juv M; whole/good skin.

Reg. no. 1892.4.1.35. From Key West; Florida; United States. Date: 6/ 8/1888. Source: J.W. Atkins; W.E.D. Scott; Salvin-Godman coll.. Material: adult ; whole/good skin.

Reg. no. 1892.4.1.36. From Key West; Florida; United States. Date: 8/ 8/1888. Source: J.W. Atkins; W.E.D. Scott; Salvin-Godman coll.. Material: adult F; whole/good skin.

Reg. no. 1892.4.1.37. From Key West; Florida; United States. Date: 8/ 8/1888. Source: J.W. Atkins; W.E.D. Scott; Salvin-Godman coll.. Material: adult M; whole/good skin.

Reg. no. 1892.4.1.38. From Key West; Florida; United States. Date: 28/ 7/1888. Source: J.W. Atkins; W.E.D. Scott; Salvin-Godman coll.. Material: juv F; whole/good skin.

Reg. no. 1892.4.1.39. From Key West; Florida; United States. Date: 28/ 7/1888. Source: J.W. Atkins; W.E.D. Scott; Salvin-Godman coll.. Material: juv F; whole/good skin.

Reg. no. 1892.4.1.40. From Key West; Florida; United States. Date: 28/ 7/1888. Source: J.W. Atkins; W.E.D. Scott; Salvin-Godman coll.. Material: juv F; whole/good skin.

Reg. no. 1906.12.7.3006. From Branford; Florida; United States. Date: 23/ 3/1892. Source: G.S. Miller coll.. Material: adult M; whole/good skin.

Reg. no. 1906.12.7.3007. From Branford; Florida; United States. Date: 24/ 3/1892. Source: G.S. Miller coll.. Material: adult M; whole/good skin.

Dendroica petechia petechia
Yellow Warbler

Barbados, West Indies. It was formerly widespread on the island, but it has been declining this century as a result of habitat destruction and parasitism by the Shiny Cowbird *Molothrus bonariensis* which colonised the island in 1916.

Reg. no. 1884.5.15.347. From ; ; Barbados. Date: //1873. Source: Briggs; Sclater coll.. Material: M; whole/good skin.

Reg. no. 1884.5.15.348. From ; ; [Barbados], (Brazil). Date: //. Source: Argent; Sclater coll.. Material: adult M; whole/good skin.

Reg. no. 1885.3.8.404. From ; ; Barbados. Date: //. Source: Wildebor; Smithsonian; Salvin–Godman coll.. Material: adult M; whole/good skin.

Reg. no. 1904.8.17.401. From ; ; Barbados. Date: 16/ 1/1904. Source: M.J. Nicoll; Earl of Crawford, voyage of R.Y.S. Valhalla. Material: adult F; whole/good skin.

Reg. no. 1904.8.17.402. From ; ; Barbados. Date: 21/ 1/1904. Source: M.J. Nicoll; Earl of Crawford, voyage of R.Y.S. Valhalla. Material: adult F; whole/good skin.

Reg. no. 1904.8.17.403. From ; ; Barbados. Date: 20/ 1/1904. Source: M.J. Nicoll; Earl of Crawford, voyage of R.Y.S. Valhalla. Material: adult F; whole/good skin.

Reg. no. 1904.8.17.404. From ; ; Barbados. Date: 16/ 1/1904. Source: M.J. Nicoll; Earl of Crawford, voyage of R.Y.S. Valhalla. Material: adult M; whole/good skin. Iris black.

Reg. no. 1904.8.17.405. From ; ; Barbados. Date: 16/ 1/1904. Source: M.J. Nicoll; Earl of Crawford, voyage of R.Y.S. Valhalla. Material: adult M; whole/good skin.

Reg. no. 1914.12.1.687. From ; ; Barbados. Date: 25/ 1/1907. Source: P.R. Lowe. Material: adult F; whole/good skin.

Reg. no. 1914.12.1.688. From ; ; Barbados. Date: 25/ 1/1907. Source: P.R. Lowe. Material: adult ; whole/good skin.

Reg. no. 1914.12.1.689. From ; ; Barbados. Date: 25/ 1/1907. Source: P.R. Lowe. Material: adult M; whole/good skin.

Reg. no. 1914.12.1.690. From ; ; Barbados. Date: 28/ 1/1907. Source: P.R. Lowe. Material: adult F; whole/good skin.

Reg. no. 1914.12.1.691. From ; ; Barbados. Date: 25/ 1/1907. Source: P.R. Lowe. Material: adult F; whole/good skin.

Reg. no. 1914.12.1.692. From ; ; Barbados. Date: 28/ 1/1907. Source: P.R. Lowe. Material: adult M; whole/good skin.

Reg. no. 1914.12.1.693. From ; ; Barbados. Date: 28/ 1/1907. Source: P.R. Lowe. Material: adult F; whole/good skin.

Reg. no. 1914.12.1.694. From ; ; Barbados. Date: 28/ 1/1907. Source: P.R. Lowe. Material: adult M; whole/good skin.

Reg. no. 1914.12.1.695. From ; ; Barbados. Date: 25/ 1/1907. Source: P.R. Lowe. Material: adult F; whole/good skin.

Reg. no. 1914.12.1.696. From ; ; Barbados. Date: 25/ 1/1907. Source: P.R. Lowe. Material: adult M; whole/good skin.

Reg. no. 1914.12.1.697. From ; ; Barbados. Date: 28/ 1/1907. Source: P.R. Lowe. Material: adult M; skin. Details taken from the register. Not in the collection; exchanged with Zool. Mus., Amsterdam, in 1953.

Reg. no. 1914.12.1.698. From ; ; Barbados. Date: 29/ 1/1907. Source: P.R. Lowe. Material: adult M; whole/good skin.

Dendroica kirtlandii

Kirtland's Warbler

Breeds in jack–pine woods in parts of Michigan and, to a lesser extent, Wisconsin, USA, and winters in the Bahamas and adjacent islands. The population was stable at only about 200 pairs for many years, but there were nearly 400 singing males in spring 1992, the best year since 1961.

Reg. no. 1966.9.3. From Luzerne, Oscoda County; Michigan; United States. Date: 6/ 7/1914. Source: M.M. Peet; University of Michigan. Material: adult M; whole/good skin.

Reg. no. 1966.9.4. From Luzerne, Oscoda County; Michigan; United States. Date: 12/ 8/1914. Source: M.M. Peet; University of Michigan. Material: adult F; whole/good skin.

Catharopeza bishopi

Whistling Warbler

Restricted to St. Vincent, where it occurs in mountain forest and is rare and local.

Reg. no. 1885.3.8.229. From ; ; St. Vincent. Date: / 2/1879. Source: C.E. Lister; Salvin–Godman coll.. Material: adult F; whole/good skin.

Reg. no. 1898.2.8.49. From ; ; St. Vincent. Date: 23/11/1889. Source: D.W. Smith; Salvin–Godman coll.. Material: juv M; whole/good skin.

Reg. no. 1898.2.8.50. From ; ; St. Vincent. Date: 22/11/1889. Source: D.W. Smith; Salvin–Godman coll.. Material: adult ; whole/good skin.

Reg. no. 1898.2.8.51. From ; ; St. Vincent. Date: 17/12/1889. Source: D.W. Smith; Salvin–Godman coll.. Material: adult ; whole/good skin.

Reg. no. 1898.2.8.52. From ; ; St. Vincent. Date: 3/ 4/1889. Source: D.W. Smith; Salvin–Godman coll.. Material: adult ; whole/good skin.

Reg. no. 1898.2.8.53. From ; ; St. Vincent. Date: / 5/1889. Source: D.W. Smith; Salvin–Godman coll.. Material: juv ; whole/good skin.

Reg. no. 1898.2.8.54. From ; ; St. Vincent. Date: 9/ 9/1889. Source: D.W. Smith; Salvin–Godman coll.. Material: juv ; whole/good skin.

Reg. no. 1898.2.8.55. From ; ; St. Vincent. Date: 13/ 4/1889. Source: D.W. Smith;

Salvin–Godman coll.. Material: adult F; whole/good skin.

Leucopeza semperi

Semper's Warbler

Confined to St. Lucia. It inhabited forest but was severely depleted in numbers following the introduction of the mongoose. It was last seen in 1972 and may now be extinct.

Reg. no. 1884.5.15.287. From ; ; St. Lucia. Date: //1875. Source: J. Semper; Sclater coll.. Material: adult ; whole/good skin. Syntype.

Reg. no. 1884.5.15.288. From ; ; St. Lucia. Date: //1875. Source: J. Semper; Sclater coll.. Material: adult ; whole/good skin. Syntype. P.L. Sclater, *Proc. zool. Soc. Lond.* 1876: 14, pl. 2. The other syntype is also in the collection.

Reg. no. 1885.3.8.227. From ; ; St. Lucia. Date: //1882. Source: H. Whitely; Salvin–Godman coll.. Material: ; whole/good skin.

Reg. no. 1885.3.8.228. From ; ; St. Lucia. Date: //1882. Source: H. Whitely; Salvin–Godman coll.. Material: ; whole/good skin.

Reg. no. 1889.8.5.15. From ; ; St. Lucia. Date: 3/ 4/1889. Source: West Indian Committee of the Royal Society. Material: adult F; whole/good skin. The register gives the sex as male, but the collector's label says female.

Reg. no. 1889.8.5.16. From ; ; St. Lucia. Date: 3/ 4/1889. Source: West Indian Committee of the Royal Society. Material: adult F; whole/good skin.

Myioborus pariae

Yellow–faced Redstart

Endemic to high cloud forest on the Paria Peninsula, Venezuela. It is common within this habitat, but the human threat to the forests is intense.

Reg. no. 1963.36.3. From Cerro Humo, Irapa; Sucre; Venezuela. Date: 2/ 5/1948. Source: W.H. Phelps. Material: adult M; whole/good skin.

Loxops sagittirostris

Greater Amakihi

Formerly occurred in rain forest on the windward side of the Island of Hawaii at altitudes of between 1,200 and 4,000 feet. It was very rare when discovered in 1892 and was not seen after 1900.

Reg. no. 1897.10.28.26. From Hilo; Hawaii; Hawaiian Islands. Date: / 1/1896. Source: R.C.L. Perkins; Royal Society & British Association. Material: adult M; whole/good skin.

Reg. no. 1897.10.28.27. From Hilo; Hawaii; Hawaiian Islands. Date: / 1/1896. Source: R.C.L. Perkins; Royal Society & British Association. Material: adult F; whole/good skin.

Reg. no. 1939.12.9.144. From Kaiwiki; Hawaii; Hawaiian Islands. Date: 14/ 2/1900. Source: H.W. Henshaw; Rothschild beq.. Material: adult M; whole/good skin.

Loxops maculata maculata

Oahu Creeper

Endemic to Oahu, where it is extremely rare, and probably confined to the Koolau Mountains. This form should now be called *Paroreomyza maculata* (monotypic).

Reg. no. 1895.7.20.131. From ; Oahu; Hawaiian Islands. Date: 2/ 4/1893. Source: R.C.L. Perkins; Royal Society & British Association. Material: adult F; whole/good skin.

Reg. no. 1895.7.20.132. From ; Oahu; Hawaiian Islands. Date: / 4/1893. Source: R.C.L. Perkins; Royal Society & British Association. Material: adult M; whole/good skin.

Reg. no. 1939.12.9.146. From Wailua; Oahu; Hawaiian Islands. Date: 24/ 4/1893. Source: H.C. Palmer; Rothschild coll.. Material: adult M; whole/good skin.

Reg. no. 1939.12.9.147. From Kawailoa; Oahu; Hawaiian Islands. Date: 4/ 6/1893. Source: H.C. Palmer; Rothschild coll.. Material: adult M; whole/good skin.

Loxops maculata flammea

Molokai Creeper

Endemic to the forests of Molokai above 1,500 feet, where it was last seen in 1963, and is probably extinct. This form should now be called *Paroreomyza flammea* (monotypic).

Reg. no. 1895.7.20.113. From ; Molokai; Hawaiian Islands. Date: / 6/1893. Source: R.C.L. Perkins; Royal Society & British Association. Material: adult M; whole/good skin.

Reg. no. 1895.7.20.114. From ; Molokai; Hawaiian Islands. Date: 18/ 6/1893. Source: R.C.L. Perkins; Royal Society & British Association. Material: adult M; whole/good skin.

Reg. no. 1895.7.20.115. From ; Molokai; Hawaiian Islands. Date: 18/ 6/1893. Source: R.C.L. Perkins; Royal Society & British Association. Material: adult M; whole/good skin.

Reg. no. 1895.7.20.116. From ; Molokai; Hawaiian Islands. Date: / 6/1893. Source: R.C.L. Perkins; Royal Society & British Association. Material: imm M; whole/good skin.

Reg. no. 1895.7.20.117. From ; Molokai; Hawaiian Islands. Date: / 6/1893. Source: R.C.L. Perkins; Royal Society & British Association. Material: adult F; whole/good skin.

Reg. no. 1895.7.20.118. From ; Molokai; Hawaiian Islands. Date: / 5/1893. Source: R.C.L. Perkins; Royal Society & British Association. Material: imm F; whole/good skin.

Reg. no. 1895.7.20.119. From ; Molokai; Hawaiian Islands. Date: 18/ 6/1893. Source: R.C.L. Perkins; Royal Society & British Association. Material: imm F; whole/good skin.

Reg. no. 1895.7.20.120. From ; Molokai; Hawaiian Islands. Date: / 5/1893. Source: R.C.L. Perkins; Royal Society & British

Association. Material: adult F; whole/good skin.

Reg. no. 1939.12.9.306. From ; Molokai; Hawaiian Islands. Date: 20/12/1892. Source: H.C. Palmer; Rothschild coll.. Material: adult M; whole/good skin.

Reg. no. 1939.12.9.307. From Kalae; Molokai; Hawaiian Islands. Date: / 6/1888. Source: S.B. Wilson; Cambridge University Museum; Rothschild coll.. Material: adult M; whole/good skin. Syntype. Wilson, *Proc. zool. Soc. Lond.* 1889 (1890): 445. There are two other syntypes in the University Museum of Zoology, Cambridge.

Reg. no. 1939.12.9.308. From ; Molokai; Hawaiian Islands. Date: //. Source: H.C. Palmer; Rothschild coll.. Material: adult ; whole/good skin.

Reg. no. 1939.12.9.309. From ; Molokai; Hawaiian Islands. Date: 20/12/1892. Source: H.C. Palmer; Rothschild coll.. Material: adult F; whole/good skin.

Reg. no. 1939.12.9.310. From ; Molokai; Hawaiian Islands. Date: 20/12/1892. Source: H.C. Palmer; Rothschild coll.. Material: adult F; whole/good skin.

Reg. no. 1939.12.9.311. From ; Molokai; Hawaiian Islands. Date: 26/12/1892. Source: H.C. Palmer; Rothschild coll.. Material: adult F; whole/good skin.

Reg. no. 1939.12.9.72. From ; Molokai; Hawaiian Islands. Date: 14/ 1/1893. Source: H.C. Palmer; Rothschild coll.. Material: adult M; whole/good skin.

Reg. no. 1939.12.9.73. From ; Molokai; Hawaiian Islands. Date: 26/12/1892. Source: H.C. Palmer; Rothschild coll.. Material: F; whole/good skin.

Reg. no. 1939.12.9.74. From ; Molokai; Hawaiian Islands. Date: 26/12/1892. Source: H.C. Palmer; Rothschild coll.. Material: F; whole/good skin.

Reg. no. 1940.12.8.63. From ; ; Hawaiian Islands (Sandwich Islands). Date: //. Source: H.C. Palmer; Rothschild coll.. Material: adult; whole/good alc. Collected between 1890 and 1893.

Loxops maculata montana

Lanai Creeper

Endemic to Lanai, where it was last seen in 1937 and is now extinct. This form should now be called *Paroreomyza montana montana*.

Reg. no. 1895.7.20.175. From ; Lanai; Hawaiian Islands. Date: /12/1893. Source: R.C.L. Perkins; Royal Society & British Association. Material: adult F; whole/good skin.

Reg. no. 1895.7.20.78. From ; Lanai; Hawaiian Islands. Date: / 7/1894. Source: R.C.L. Perkins; Royal Society & British Association. Material: adult M; whole/good skin.

Reg. no. 1895.7.20.79. From ; Lanai; Hawaiian Islands. Date: / 1/1894. Source: R.C.L. Perkins; Royal Society & British Association. Material: adult F; whole/good skin.

Reg. no. 1895.7.20.80. From ; Lanai; Hawaiian Islands. Date: / 1/1894. Source: R.C.L. Perkins; Royal Society & British Association. Material: adult M; whole/good skin.

Reg. no. 1895.7.20.81. From ; Lanai; Hawaiian Islands. Date: / 1/1895. Source: R.C.L. Perkins; Royal Society & British Association. Material: adult M; whole/good skin.

Reg. no. 1895.7.20.82. From ; Lanai; Hawaiian Islands. Date: / 7/1894. Source: R.C.L. Perkins; Royal Society & British Association. Material: adult F; whole/good skin.

Reg. no. 1895.7.20.83. From ; Lanai; Hawaiian Islands. Date: 23/ 2/1894. Source: R.C.L. Perkins; Royal Society & British Association. Material: imm M; whole/good skin.

Reg. no. 1895.7.20.84. From ; Lanai; Hawaiian Islands. Date: / 7/1894. Source: R.C.L. Perkins; Royal Society & British Association. Material: adult F; whole/good skin.

***Loxops maculata* subspecies**

The species formerly known as *L. maculata* has been split into five species: *Oreomystis mana* (Hawaii), *O. bairdi* (Kauai), *Paroreomyza maculata* (Oahu), *P. flammea* (Molokai) and *P. montana* (Lanai, *P. m. montana*; Maui, *P. m. newtoni*)(*Condor* 94 (1992): 836–846). It is not known to which of these species the following specimens belong.

Reg. no. . From ; ; [Hawaiian Islands]. Date: //. Source: Palmer; Rothschild coll.. Material: ; whole/good alc. There are 24 (the label on the jar, and *Bull. Br. Mus. nat. Hist. (Zool.)* 34 (1979): 174), say 26) spirit specimens in the collection. Two were not found 1988 and 1993. None of the 24 specimens has been identified beyond "*L. maculata*".

Loxops coccinea ochracea

Maui Akepa

Endemic to Maui, where it is now near extinction and restricted to high mountain forest.

Reg. no. 1895.7.20.149. From ; Maui; Hawaiian Islands. Date: / 4/1894. Source: R.C.L. Perkins; Royal Society & British Association. Material: adult M; whole/good skin.

Reg. no. 1895.7.20.150. From Haleakala; Maui; Hawaiian Islands. Date: / 4/1894. Source: R.C.L. Perkins; Royal Society & British Association. Material: adult M; whole/good skin.

Reg. no. 1895.7.20.151. From Haleakala; Maui; Hawaiian Islands. Date: / 4/1894. Source: R.C.L. Perkins; Royal Society & British Association. Material: adult M; whole/good skin.

Reg. no. 1895.7.20.152. From Haleakala; Maui; Hawaiian Islands. Date: / 4/1894. Source: R.C.L. Perkins; Royal Society & British Association. Material: adult M; whole/good skin.

Reg. no. 1895.7.20.153. From Haleakala; Maui; Hawaiian Islands. Date: / 4/1894. Source: R.C.L. Perkins; Royal Society & British Association. Material: adult M; whole/good skin.

Reg. no. 1895.7.20.154. From Haleakala; Maui; Hawaiian Islands. Date: / 4/1894. Source: R.C.L. Perkins; Royal Society & British Association. Material: adult F; whole/good skin.

Reg. no. 1895.7.20.155. From Haleakala; Maui; Hawaiian Islands. Date: / 3/1894. Source: R.C.L. Perkins; Royal Society & British Association. Material: adult F; whole/good skin.

Reg. no. 1895.7.20.156. From Haleakala; Maui; Hawaiian Islands. Date: / 4/1894. Source: R.C.L. Perkins; Royal Society & British Association. Material: imm F; whole/good skin.

Reg. no. 1897.10.28.47. From Haleakala; Maui; Hawaiian Islands. Date: / 5/1896. Source: R.C.L. Perkins; Royal Society & British Association. Material: imm F; whole/good skin.

Reg. no. 1897.10.28.48. From Haleakala; Maui; Hawaiian Islands. Date: /10/1896. Source: R.C.L. Perkins; Royal Society & British Association. Material: adult M; whole/good skin.

Reg. no. 1897.10.28.49. From ; Maui; Hawaiian Islands. Date: /10/1896. Source: R.C.L. Perkins; Royal Society & British Association. Material: imm M; whole/good skin.

Reg. no. 1939.12.9.62. From ; Maui (Mauai); Hawaiian Islands. Date: 10/ 8/1892. Source: H.C. Palmer; Rothschild coll.. Material: adult M; whole/good skin. Syntype. Rothschild, *Ibis* 1893: 112 also *Bull. Brit. Orn. Club* 1 (1893): 16.

Reg. no. 1939.12.9.65. From ; Maui; Hawaiian Islands. Date: 30/10/1892. Source: H.C. Palmer; Rothschild coll.. Material: M; whole/good skin.

Reg. no. S1961.11.44. From ; [Maui]; [Hawaiian Islands]. Date: //. Source: Rothschild coll.. Material: ; whole/good skeleton.

Loxops coccinea rufa

Oahu Akepa

Endemic to Oahu. Formerly thought to be extinct, but now considered to be only nearly so.

Reg. no. 1843.12.30.20. From ; Oahu; Hawaiian Islands. Date: //. Source: J. Gould. Material: ; whole/good skin.

Reg. no. VEL.20.187a. From ; [Oahu]; Hawaiian Islands. Date: //. Source: Bloxam; Byron, voyage of H.M.S. Blonde. Material: adult ; whole/good skin. Syntype. Bloxham (sic), 1826, in Byron, *Voyage of H.M.S. Blonde*, Appendix: 250. A second syntype is also in the collection.

Reg. no. VEL.20.187b. From ; [Oahu]; Hawaiian Islands. Date: //. Source: Bloxam; Byron, voyage of H.M.S. Blonde. Material: imm M; whole/good skin. Syntype.

***Loxops coccinea* subspecies**

Reg. no. . From ; ; [Hawaiian Islands]. Date: //. Source: Palmer; Rothschild coll.. Material: ; whole/good alc. There are 15 (the label on the jar, and *Bull. Br. Mus. nat. Hist. (Zool.)* 34 (1979): 174), say 17) spirit specimens in the collection. Two were not found 1988 and 1993. None of the 15 specimens has been identified to subspecies.

Hemignathus obscurus lanaiensis
Lanai Akialoa

Mountain forests of Lanai. Not seen since 1894 and presumed extinct as a result of habitat destruction.

Reg. no. 1939.12.9.55. From ; Lanai; Hawaiian Islands. Date: 15/11/1892. Source: H.C. Palmer; Rothschild coll.. Material: adult F; whole/good skin. Syntype. Rothschild, *Bull. Brit. Orn. Club* 1 (1893): 24–5. A male syntype is in the collection of the American Museum of Natural History. See also *Novit. Zool.* 26 (1919): 170.

Hemignathus obscurus obscurus
Lesser Akialoa

Mountain forests of Hawaii. It was common in 1894–1895 but has not certainly been seen since. There was an unconfirmed sighting in 1940.

Reg. no. 1895.7.20.19. From Mauna Loa; Hawaii; Hawaiian Islands. Date: / 8/1893. Source: R.C.L. Perkins; Royal Society & British Association. Material: adult M; whole/good skin.

Reg. no. 1895.7.20.20. From Mauna Loa; Hawaii; Hawaiian Islands. Date: 22/ 7/1892. Source: R.C.L. Perkins; Royal Society & British Association. Material: adult M; whole/good skin.

Reg. no. 1895.7.20.21. From Mauna Loa; Hawaii; Hawaiian Islands. Date: / 8/1892. Source: R.C.L. Perkins; Royal Society & British Association. Material: adult F; whole/good skin.

Reg. no. 1895.7.20.22. From Mauna Loa; Hawaii; Hawaiian Islands. Date: / 8/1892. Source: R.C.L. Perkins; Royal Society & British Association. Material: adult F; whole/good skin.

Reg. no. 1895.7.20.23. From Mauna Loa; Hawaii; Hawaiian Islands. Date: / 9/1892. Source: R.C.L. Perkins; Royal Society & British Association. Material: adult F; whole/good skin.

Reg. no. 1895.7.20.24. From Mauna Loa; Hawaii; Hawaiian Islands. Date: / 9/1892. Source: R.C.L. Perkins; Royal Society & British Association. Material: adult F; whole/good skin.

Reg. no. 1897.10.28.10. From Kaholuamano; Kauai; Hawaiian Islands. Date: /10/1895. Source: [R.C.L. Perkins]; Royal Society & British Association. Material: adult M; skin. Details taken from the register. Not found 1988.

Reg. no. 1897.10.28.7. From Kona; Hawaii; Hawaiian Islands. Date: / 3/1896. Source: R.C.L. Perkins; Royal Society & British Association. Material: adult M; whole/good skin.

Reg. no. 1897.10.28.8. From Kaholuamano; Kauai; Hawaiian Islands. Date: /10/1895. Source: [R.C.L. Perkins]; Royal Society & British Association. Material: adult M; skin. Details taken from the register. Not found 1988.

Reg. no. 1939.12.9.143. From ; Hawaii; Hawaiian Islands. Date: 11/ 9/1891. Source: H.C. Palmer; Rothschild coll.. Material: adult M; whole/good skin.

Reg. no. 1939.12.9.333. From ; Hawaii; Hawaiian Islands. Date: 1/10/1891. Source: H.C. Palmer; Rothschild coll.. Material: adult M; whole/good skin.

Hemignathus procerus/H. (obscurus) ellisianus

Greater Akialoa

Mountain forests of Kauai. It probably survived until about 1967, but is certainly now extinct. *H. procerus* is properly part of *H.(o.) ellisianus* (Gray) (Pratt, Bruner & Berrett, 1987, *Birds of Hawaii and the Tropical Pacific*: p. 302).

Reg. no. 1895.7.20.61. From ; Kauai; Hawaiian Islands. Date: / 5/1894. Source: R.C.L. Perkins; Royal Society & British Association. Material: adult M; whole/good skin.

Reg. no. 1895.7.20.62. From ; Kauai; Hawaiian Islands. Date: / 5/1894. Source: R.C.L. Perkins; Royal Society & British Association. Material: adult F; whole/good skin.

Reg. no. 1895.7.20.63. From ; Kauai; Hawaiian Islands. Date: / 6/1894. Source: R.C.L. Perkins; Royal Society & British Association. Material: imm M; whole/good skin.

Reg. no. 1897.10.28.10. From Kaholuamano; Kauai; Hawaiian Islands. Date: /10/1895. Source: R.C.L. Perkins; Royal Society & British Association. Material: adult F; whole/good skin.

Reg. no. 1897.10.28.11. From Kaholuamano; Kauai; Hawaiian Islands. Date: / 4/1895. Source: R.C.L. Perkins; Royal Society & British Association. Material: adult F; whole/good skin. With large eggs.

Reg. no. 1897.10.28.9. From Kaholuamano; Kauai; Hawaiian Islands. Date: /10/1895. Source: R.C.L. Perkins; Royal Society & British Association. Material: F; whole/good skin.

Reg. no. 1925.2.16.71. From ; Kauai; Hawaiian Islands. Date: //1892. Source: S.B. Wilson. Material: F; whole/good skin.

Reg. no. 1925.2.16.72. From ; Kauai; Hawaiian Islands. Date: / 9/1888. Source: S.B. Wilson. Material: F; whole/good skin.

Reg. no. 1925.2.16.73. From ; Kauai; Hawaiian Islands. Date: //1891. Source: S.B. Wilson. Material: F; whole/poor skin. Damaged: poor skin, bill broken, leg broken.

Reg. no. 1939.12.9.332. From ; Kauai; Hawaiian Islands. Date: 1/ 3/1891. Source: H.C. Palmer; Rothschild coll.. Material: adult F; whole/good skin.

Reg. no. 1939.12.9.54. From ; Kauai; Hawaiian Islands. Date: 5/ 3/1891. Source: H.C. Palmer; Rothschild coll.. Material: adult M; whole/good skin.

Reg. no. 1940.12.8.19. From ; ; Hawaiian Islands. Date: //. Source: Palmer; Rothschild coll.. Material: adult ; whole/good alc. Collected between 1890–1893. The subspecific identity has been questioned.

Hemignathus lucidus hanapepe **Kauai Nukupuu**

Known only from the headwaters of the Hanapepe River in the Alakai Swamp on Kauai. It was very rare when first discovered and has not certainly been seen since 1898.

Reg. no. 1843.7.22.41. From ; [Kauai]; Hawaiian Islands (Sandwich Islands). Date: //. Source: Sir E. Belcher. Material: imm ; whole/good skin.

Reg. no. 1895.7.20.68. From Alaka'i Swamp; Kauai; Hawaiian Islands. Date: / 5/1894. Source: R.C.L. Perkins; Royal Society & British Association. Material: adult M; whole/good skin. The reg. no. was erroneously noted as 1895.7.20.66 on the label.

Reg. no. 1897.10.28.18. From Kaholuamano; Kauai; Hawaiian Islands. Date: /10/1895. Source: R.C.L. Perkins; Royal Society & British Association. Material: adult M; whole/good skin.

Reg. no. 1897.10.28.19. From Kaholuamano; Kauai; Hawaiian Islands. Date: /10/1895. Source: R.C.L. Perkins; Royal Society & British Association. Material: imm M; whole/good skin.

Reg. no. 1897.10.28.20. From Kaholuamano; Kauai; Hawaiian Islands. Date: /10/1895. Source: R.C.L. Perkins; Royal Society & British Association. Material: imm M; whole/good skin.

Reg. no. 1897.10.28.21. From Alaka'i Swamp; Kauai; Hawaiian Islands. Date: / 2/1897. Source: R.C.L. Perkins; Royal Society & British Association. Material: adult F; whole/good skin.

Reg. no. 1898.10.20.2. From 3,000 ft., Hanapepe; Kauai; Hawaiian Islands. Date: / 9/1888. Source: S.B. Wilson; Seebohm coll.. Material: adult M; whole/good skin. Syntype. Wilson, *Ann. Mag. nat. Hist.*, series 6, 4 (1889): 401. There are two other syntypes in the University Museum of Zoology, Cambridge. See also *Ibis* 1890: 192–3, pl. 6.

Reg. no. 1925.2.16.78. From Alaka'i Swamp; Kauai; Hawaiian Islands. Date: //1891. Source: S.B. Wilson. Material: F; whole/good skin.

Reg. no. 1939.12.9.140. From Alaka'i Swamp; Kauai; Hawaiian Islands. Date: 7/ 3/1891. Source: H.C. Palmer; Rothschild coll.. Material: adult M; whole/good skin.

Reg. no. 1939.12.9.141. From Alaka'i Swamp; Kauai; Hawaiian Islands. Date: //. Source: Rothschild coll.. Material: imm M; whole/good skin.

Reg. no. 1940.12.8.43. From ; ; Hawaiian Islands. Date: //. Source: Palmer; Rothschild coll.. Material: adult ; whole/good alc. Collected between 1890–1893.

Reg. no. 1940.12.8.44. From ; ; Hawaiian Islands. Date: //. Source: Palmer; Rothschild coll.. Material: adult ; whole/good alc. Collected between 1890–1893.

Hemignathus lucidus affinis Maui Nukupuu

Forested slopes of Haleakala Mountain between 4,000 and 4,500 feet on Maui. It was last seen in 1896 and presumed extinct, but a small population was discovered in 1967. The form may now be extinct.

Reg. no. 1895.7.20.166. From Haleakala; Maui; Hawaiian Islands. Date: 4/ 4/1894. Source: R.C.L. Perkins; Royal Society & British Association. Material: adult F; whole/good skin. After breeding, with ovaries still enlarged.

Reg. no. 1895.7.20.167. From Haleakala; Maui; Hawaiian Islands. Date: 16/ 3/1894. Source: R.C.L. Perkins; Royal Society & British Association. Material: adult M; whole/good skin.

Reg. no. 1895.7.20.168. From Haleakala; Maui; Hawaiian Islands. Date: 30/ 3/1894. Source: R.C.L. Perkins; Royal Society & British Association. Material: adult M; whole/good skin.

Reg. no. 1895.7.20.169. From Haleakala; Maui; Hawaiian Islands. Date: 30/ 4/1894. Source: R.C.L. Perkins; Royal Society & British Association. Material: adult M; whole/good skin.

Reg. no. 1897.10.28.12. From Haleakala; Maui; Hawaiian Islands. Date: /10/1896. Source: R.C.L. Perkins; Royal Society & British Association. Material: adult M; whole/good skin.

Reg. no. 1897.10.28.13. From Haleakala; Maui; Hawaiian Islands. Date: / 5/1896. Source: R.C.L. Perkins; Royal Society & British Association. Material: adult F; whole/good skin.

Reg. no. 1897.10.28.14. From Haleakala; Maui; Hawaiian Islands. Date: /10/1896. Source: R.C.L. Perkins; Royal Society & British Association. Material: adult M; whole/good skin.

Reg. no. 1939.12.9.142. From Haleakala; Maui (Mauai); Hawaiian Islands. Date: 9/ 8/1892. Source: H.C. Palmer; Rothschild coll.. Material: adult M; whole/good skin. Syntype. Rothschild, *Ibis* 1893: 112.

Reg. no. S1961.11.39. From ; ; [Hawaiian Islands]. Date: //. Source: Rothschild coll.. Material: ; whole/good skeleton.

Hemignathus wilsoni/H. munroi Akiapolaau

Formerly widespread on the island of Hawaii but now confined to small patches of forest. About 1,500 birds are thought to survive. (If *Viridonia* is merged in *Hemignathus*, as has been suggested, then *wilsoni* (Rothschild) is preoccupied by *H.*

virens wilsoni (Rothschild), and should be called *H. munroi* Pratt).

Reg. no. 1895.7.20.10. From ; Hawaii; Hawaiian Islands. Date: / 9/1892. Source: R.C.L. Perkins; Royal Society & British Association. Material: adult F; whole/good skin.

Reg. no. 1895.7.20.11. From ; Hawaii; Hawaiian Islands. Date: //1893. Source: R.C.L. Perkins; Royal Society & British Association. Material: adult F; whole/good skin.

Reg. no. 1895.7.20.12. From Kona; Hawaii; Hawaiian Islands. Date: / 8/1894. Source: R.C.L. Perkins; Royal Society & British Association. Material: adult M; whole/good skin.

Reg. no. 1895.7.20.13. From ; Hawaii; Hawaiian Islands. Date: / 8/1892. Source: R.C.L. Perkins; Royal Society & British Association. Material: adult M; whole/good skin.

Reg. no. 1895.7.20.7. From ; Hawaii; Hawaiian Islands. Date: / 9/1892. Source: R.C.L. Perkins; Royal Society & British Association. Material: adult M; whole/good skin.

Reg. no. 1895.7.20.8. From ; Hawaii; Hawaiian Islands. Date: / 8/1892. Source: R.C.L. Perkins; Royal Society & British Association. Material: adult F; whole/good skin.

Reg. no. 1895.7.20.9. From ; Hawaii; Hawaiian Islands. Date: / 8/1894. Source: R.C.L. Perkins; Royal Society & British Association. Material: adult F; whole/good skin.

Reg. no. 1897.10.28.15. From ; Hawaii; Hawaiian Islands. Date: / 7/1895. Source: R.C.L. Perkins; Royal Society & British Association. Material: adult M; whole/good skin.

Reg. no. 1897.10.28.16. From Kau; Hawaii; Hawaiian Islands. Date: / 6/1895. Source: R.C.L. Perkins; Royal Society & British Association. Material: imm M; whole/good skin.

Reg. no. 1897.10.28.17. From Kau; Hawaii; Hawaiian Islands. Date: / 6/1895. Source: R.C.L. Perkins; Royal Society & British Association. Material: F; whole/good skin.

Reg. no. 1898.10.20.3. From ; Hawaii; Hawaiian Islands. Date: 12/ 3/1888. Source: S.B. Wilson; Seebohm coll.. Material: imm ; whole/good skin.

Reg. no. 1898.10.20.4. From ; Hawaii; Hawaiian Islands. Date: / 3/1888. Source: S.B. Wilson; Seebohm coll.. Material: adult M; whole/good skin.

Reg. no. 1898.10.20.5. From ; Hawaii; Hawaiian Islands. Date: /12/1887. Source: S.B. Wilson; Seebohm coll.. Material: imm M; whole/good skin.

Reg. no. 1925.2.16.167. From Mana; Hawaii; Hawaiian Islands. Date: / 3/1888. Source: S.B. Wilson. Material: juv M; whole/good skin.

Reg. no. 1925.2.16.73a. From ; Hawaii; Hawaiian Islands. Date: / 3/1888. Source:

S.B. Wilson. Material: ; whole/good skin. A specimen of *H. procerus* was found to have the same reg. no.; the no. of this bird has been altered with a suffix.

Reg. no. 1925.2.16.74. From Mana Kauai; Hawaii; Hawaiian Islands. Date: / 3/1888. Source: S.B. Wilson. Material: ; whole/good skin. "Hawaii" was printed on the label of this specimen, but had been deleted and "Kauai" substituted, apparently in error, for this species does not occur on Kauai.

Reg. no. 1925.2.16.75. From ; Hawaii; Hawaiian Islands. Date: //1891. Source: S.B. Wilson. Material: M; whole/good skin.

Reg. no. 1925.2.16.76. From ; Hawaii; Hawaiian Islands. Date: / 3/1888. Source: S.B. Wilson. Material: imm ; whole/good skin.

Reg. no. 1925.2.16.77. From ; Hawaii; Hawaiian Islands. Date: //1891. Source: S.B. Wilson. Material: imm F; whole/good skin.

Reg. no. 1939.12.9.138. From ; Hawaii; Hawaiian Islands. Date: 29/ 9/1891. Source: H.C. Palmer; Rothschild coll.. Material: adult M; whole/good skin. Syntype. Rothschild, 1893, *The Avifauna of Laysan* ..., p. 97–9, pl. 50. A female syntype is also in the collection.

Reg. no. 1939.12.9.139. From ; Hawaii; Hawaiian Islands. Date: 26/ 9/1891. Source: H.C. Palmer; Rothschild coll.. Material: adult F; whole/good skin. Syntype.

Reg. no. 1939.12.9.338. From ; Hawaii; Hawaiian Islands. Date: //. Source: Rothschild coll.. Material: ; whole/good skin.

Reg. no. 1939.12.9.362. From ; Maui; Hawaiian Islands. Date: 16/ 7/1892. Source: Rothschild coll.. Material: adult F; skin. Details taken from the register. Not found 1988.

Reg. no. 1939.12.9.61. From ; Maui; Hawaiian Islands. Date: 6/10/1892. Source: Rothschild coll.. Material: adult F; skin. Details taken from the register. Not found 1988.

Reg. no. 1940.12.8.45. From ; ; Hawaiian Islands. Date: //. Source: Palmer; Rothschild coll.. Material: adult ; whole/good alc. Collected between 1890–1893. The identity has been questioned.

Psittirostra psittacea

Ou

Formerly widespread on all the main islands of Hawaii, and latterly confined to the Alakai Swamp on Kauai, and a restricted area on Hawaii. The species is declining and may be extinct.

Reg. no. 1860.1.16.19. From ; ; Hawaiian Islands. Date: //. Source: J. Gould. Material: ; whole/good skin.

Reg. no. 1860.1.16.20. From ; ; Hawaiian Islands. Date: //. Source: J. Gould. Material: ; whole/good skin.

Reg. no. 1880.11.18.450. From Hilo; Hawaii; Hawaiian Islands. Date: //. Source: Challenger Expedition. Material: M; whole/good skin.

Reg. no. 1880.11.18.451. From Hilo; Hawaii; Hawaiian Islands. Date: //. Source: Challenger Expedition. Material: M; whole/good skin.

Reg. no. 1890.10.3.19. From ; Kauai; Hawaiian Islands. Date: //. Source: Knudsen coll.; Collett. Material: ; whole/good skin.

Reg. no. 1895.7.20.108. From ; Molokai; Hawaiian Islands. Date: / 5/1893. Source: R.C.L. Perkins; Royal Society & British Association. Material: adult M; whole/good skin.

Reg. no. 1895.7.20.76. From ; Lanai; Hawaiian Islands. Date: / 1/1894. Source: R.C.L. Perkins; Royal Society & British Association. Material: adult M; whole/good skin.

Reg. no. 1895.7.20.77. From ; Lanai; Hawaiian Islands. Date: / 7/1894. Source: R.C.L. Perkins; Royal Society & British Association. Material: adult F; whole/good skin.

Reg. no. 1897.10.28.28. From ; Kauai; Hawaiian Islands. Date: / 7/1895. Source: R.C.L. Perkins; Royal Society & British Association. Material: juv M; whole/good skin.

Reg. no. 1898.11.1.3. From Kaawaloa; Hawaii; Hawaiian Islands. Date: 19/ 5/1887. Source: S.B. Wilson; Seebohm coll.. Material: adult F; whole/good skin.

Reg. no. 1898.11.1.4. From ; Hawaii; Hawaiian Islands. Date: 19/ 5/1887. Source: S.B. Wilson; Seebohm coll.. Material: M; whole/good skin.

Reg. no. 1898.11.1.5. From ; Hawaii; Hawaiian Islands. Date: / 5/1887. Source: S.B. Wilson; Seebohm coll.. Material: ; whole/good skin.

Reg. no. 1925.2.16.79. From ; Kauai; Hawaiian Islands. Date: /11/1892. Source: S.B. Wilson. Material: juv M; whole/good skin.

Reg. no. 1925.2.16.80. From ; Kauai; Hawaiian Islands. Date: /11/1892. Source: S.B. Wilson. Material: imm M; whole/good skin.

Reg. no. 1925.2.16.81. From ; Kauai; Hawaiian Islands. Date: /11/1892. Source: S.B. Wilson. Material: F; whole/good skin.

Reg. no. 1925.2.16.82. From ; Hawaii; Hawaiian Islands. Date: //1888. Source: S.B. Wilson. Material: M; whole/good skin.

Reg. no. 1925.2.16.83. From ; Hawaii; Hawaiian Islands. Date: / 5/1887. Source: S.B. Wilson. Material: M; whole/good skin.

Reg. no. 1925.2.16.84. From ; Hawaii; Hawaiian Islands. Date: / 6/1887. Source: S.B. Wilson. Material: M; whole/good skin.

Reg. no. 1925.2.16.85. From Olaa; Hawaii; Hawaiian Islands. Date: / 9/1887. Source:

S.B. Wilson. Material: juv M; whole/good skin.

Reg. no. 1925.2.16.86. From Paauhau; Hawaii; Hawaiian Islands. Date: /11/1887. Source: S.B. Wilson. Material: juv M; whole/good skin.

Reg. no. 1925.2.16.87. From ; Hawaii; Hawaiian Islands. Date: / 6/1887. Source: S.B. Wilson. Material: M; whole/good skin.

Reg. no. 1925.2.16.88. From ; Hawaii; Hawaiian Islands. Date: /11/1892. Source: S.B. Wilson. Material: M; whole/good skin.

Reg. no. 1925.2.16.89. From ; Hawaii; Hawaiian Islands. Date: /11/1892. Source: S.B. Wilson. Material: M; whole/good skin.

Reg. no. 1925.2.16.90. From ; Hawaii; Hawaiian Islands. Date: /11/1892. Source: S.B. Wilson. Material: M; whole/good skin.

Reg. no. 1925.2.16.91. From ; Hawaii; Hawaiian Islands. Date: /11/1892. Source: S.B. Wilson. Material: F; whole/good skin.

Reg. no. 1925.2.16.92. From ; Hawaii; Hawaiian Islands. Date: /11/1892. Source: S.B. Wilson. Material: F; whole/good skin.

Reg. no. 1925.2.16.93. From ; Hawaii; Hawaiian Islands. Date: / 6/1887. Source: S.B. Wilson. Material: M; whole/good skin.

Reg. no. 1925.2.16.94. From ; Kauai; Hawaiian Islands. Date: / 2/1892. Source: S.B. Wilson. Material: M; whole/good skin.

Reg. no. 1939.12.9.127. From ; Kauai; Hawaiian Islands. Date: 21/ 1/1891. Source: H.C. Palmer; Rothschild coll.. Material: adult M; whole/good skin.

Reg. no. 1939.12.9.128. From ; Kauai; Hawaiian Islands. Date: 18/ 2/1891. Source: H.C. Palmer; Rothschild coll.. Material: adult F; whole/good skin.

Reg. no. 1939.12.9.334. From ; Kauai; Hawaiian Islands. Date: 19/ 1/1891. Source: H.C. Palmer; Rothschild coll.. Material: adult M; whole/good skin.

Reg. no. 1939.12.9.335. From ; Kauai; Hawaiian Islands. Date: / 6/1890. Source: H.C. Palmer; Rothschild coll.. Material: adult M; whole/good skin.

Reg. no. 1939.12.9.336. From ; ; Hawaiian Islands. Date: //. Source: H.C. Palmer; Rothschild coll.. Material: adult F; whole/good skin.

Reg. no. VEL.19.87a. From ; ; Hawaiian Islands. Date: //. Source: Lord Byron. Material: M; whole/good skin.

Reg. no. VEL.19.87b. From ; ; [Hawaiian Islands]. Date: //. Source: Material: ; skin. Details taken from the register. Not found 1988.

Reg. no. VEL.19.87c. From ; ; Hawaiian Islands. Date: //. Source: Lord Byron. Material: F; whole/good skin.

Psittirostra bailleui

Palila

Endemic to the island of Hawaii where formerly common, but now confined to the upper slopes of Mauna Kea where the population fluctuates between 2,000 and 6,000 birds.

Reg. no. 1880.4.30.36. From ; Hawaii (Sandwich Island); [Hawaiian Islands]. Date: //. Source: Bailleu; Paris Museum. Material: adult ; whole/good skin.

Reg. no. 1895.7.20.31. From Kona; Hawaii; Hawaiian Islands. Date: / 8/1894. Source: R.C.L. Perkins; Royal Society & British Association. Material: adult M; whole/good skin.

Reg. no. 1897.10.28.61. From Kona; Hawaii; Hawaiian Islands. Date: / 3/1896. Source: R.C.L. Perkins; Royal Society & British Association. Material: adult M; whole/good skin.

Reg. no. 1898.11.1.1. From ; Hawaii; Hawaiian Islands. Date: 10/ 6/1887. Source: S.B. Wilson; Seebohm coll.. Material: M; whole/good skin.

Reg. no. 1898.11.1.2. From ; Hawaii; Hawaiian Islands. Date: 13/ 6/1887. Source: S.B. Wilson; Seebohm coll.. Material: F; whole/good skin.

Reg. no. 1925.2.16.100. From ; Hawaii; Hawaiian Islands. Date: / 6/1887. Source: S.B. Wilson. Material: F; whole/good skin.

Reg. no. 1925.2.16.95. From Kona; Hawaii; Hawaiian Islands. Date: / 6/1896. Source: S.B. Wilson. Material: ; whole/good skin.

Reg. no. 1925.2.16.96. From Kona; Hawaii; Hawaiian Islands. Date: / 6/1896. Source: S.B. Wilson. Material: M; whole/good skin.

Reg. no. 1925.2.16.97. From Kona; Hawaii; Hawaiian Islands. Date: / 6/1896. Source: S.B. Wilson. Material: ; whole/good skin.

Reg. no. 1925.2.16.98. From Kona; Hawaii; Hawaiian Islands. Date: / 6/1896. Source: S.B. Wilson. Material: M; whole/good skin.

Reg. no. 1925.2.16.99. From Kona; Hawaii; Hawaiian Islands. Date: / 6/1896. Source: S.B. Wilson. Material: ; whole/good skin.

Reg. no. 1939.12.9.131. From ; Hawaii; Hawaiian Islands. Date: 13/ 6/1887. Source: S.B. Wilson; Rothschild coll.. Material: adult F; whole/good skin.

Reg. no. S1961.11.43. From ; [Hawaii]; [Hawaiian Islands]. Date: //. Source: Rothschild coll.. Material: ; whole/good skeleton. Probably a female (Gronwald).

Reg. no. 1940.12.8.21. From ; ; Hawaiian Islands. Date: //. Source: Palmer; Rothschild coll.. Material: adult ; whole/good alc. Collected between 1890–1893. The identity has been questioned.

Psittirostra palmeri

Greater Koa-Finch

Mountain koa-tree forests on the island of Hawaii. It has not been seen for certain since 1896, though there was an unconfirmed sighting in 1937.

Reg. no. 1895.7.20.33. From Mauna Loa; Hawaii; Hawaiian Islands. Date: 1/ 7/1892. Source: R.C.L. Perkins; Royal Society & British Association. Material: adult M; whole/good skin.

Reg. no. 1895.7.20.34. From Mauna Loa; Hawaii; Hawaiian Islands. Date: //1893. Source: R.C.L. Perkins; Royal Society & British Association. Material: adult M; whole/good skin.

Reg. no. 1895.7.20.35. From Mauna Loa; Hawaii; Hawaiian Islands. Date: / 9/1892. Source: R.C.L. Perkins; Royal Society & British Association. Material: juv M; whole/good skin.

Reg. no. 1895.7.20.36. From Mauna Loa; Hawaii; Hawaiian Islands. Date: 9/ 7/1892. Source: R.C.L. Perkins; Royal Society & British Association. Material: adult F; whole/good skin.

Reg. no. 1895.7.20.37. From Mauna Loa; Hawaii; Hawaiian Islands. Date: / 9/1892. Source: R.C.L. Perkins; Royal Society & British Association. Material: imm M; whole/good skin.

Reg. no. 1895.7.20.38. From Mauna Loa; Hawaii; Hawaiian Islands. Date: / 9/1892. Source: R.C.L. Perkins; Royal Society & British Association. Material: imm M; whole/good skin.

Reg. no. 1897.10.28.62. From Kona; Hawaii; Hawaiian Islands. Date: / 3/1892. Source: R.C.L. Perkins; Royal Society & British Association. Material: adult M; whole/good skin.

Reg. no. 1939.12.9.132. From ; Hawaii; Hawaiian Islands. Date: 26/ 6/1892. Source: H.C. Palmer; Rothschild coll.. Material: adult M; whole/good skin.

Reg. no. 1939.12.9.135. From ; Hawaii; Hawaiian Islands. Date: 16/10/1891. Source: H.C. Palmer; Rothschild coll.. Material: adult F; whole/good skin.

Psittirostra flaviceps

Lesser Koa–Finch

Mountain koa–tree forests on the island of Hawaii. Not seen since 1891, the species was always rare.

Reg. no. 1939.12.9.133. From [Kona]; Hawaii; Hawaiian Islands. Date: 19/10/1891. Source: H.C. Palmer; Rothschild coll.. Material: adult M; whole/good skin. Syntype. Rothschild, *Ann. Mag. nat. Hist.*, series 6, 10 (1892): 111. There is a female syntype in the collection.

Reg. no. 1939.12.9.134. From ; Hawaii; Hawaiian Islands. Date: 6/10/1891. Source: H.C. Palmer; Rothschild coll.. Material: adult F; whole/good skin. Syntype.

Psittirostra kona

Kona Grosbeak

Mountain forests of the island of Hawaii, where it was known only from a restricted area of about 4 square miles on the slopes of Mauna Loa. It was last seen in 1894.

Reg. no. 1895.7.20.39. From Mauna Loa; Hawaii; Hawaiian Islands. Date: 2/ 8/1892. Source: R.C.L. Perkins; Royal Society & British Association. Material: adult M; whole/good skin.

Reg. no. 1895.7.20.40. From Mauna Loa; Hawaii; Hawaiian Islands. Date: 14/ 7/1892. Source: R.C.L. Perkins; Royal

Society & British Association. Material: adult F; whole/good skin.

Reg. no. 1895.7.20.41. From Mauna Loa; Hawaii; Hawaiian Islands. Date: / 9/1892. Source: R.C.L. Perkins; Royal Society & British Association. Material: adult F; whole/good skin.

Reg. no. 1895.7.20.42. From Mauna Loa; Hawaii; Hawaiian Islands. Date: 18/ 7/1892. Source: R.C.L. Perkins; Royal Society & British Association. Material: adult M; whole/good skin.

Reg. no. 1895.7.20.43. From Mauna Loa; Hawaii; Hawaiian Islands. Date: / 9/1892. Source: R.C.L. Perkins; Royal Society & British Association. Material: adult M; whole/good skin.

Reg. no. 1895.7.20.44. From Mauna Loa; Hawaii; Hawaiian Islands. Date: 14/ 7/1892. Source: R.C.L. Perkins; Royal Society & British Association. Material: adult M; whole/good skin.

Reg. no. 1939.12.9.135a. From ; Hawaii; Hawaiian Islands. Date: 20/10/1891. Source: H.C. Palmer; Rothschild coll.. Material: M; whole/good skin.

Reg. no. 1939.12.9.298. From ; Hawaii; Hawaiian Islands. Date: 11/10/1891. Source: H.C. Palmer; Rothschild coll.. Material: adult M; whole/good skin.

Reg. no. 1939.12.9.299. From ; Hawaii; Hawaiian Islands. Date: 11/10/1891. Source: H.C. Palmer; Rothschild coll.. Material: adult M; whole/good skin.

Reg. no. 1939.12.9.300. From ; Hawaii; Hawaiian Islands. Date: 13/10/1891. Source: H.C. Palmer; Rothschild coll.. Material: adult F; whole/good skin.

Reg. no. 1940.12.8.1. From ; ; Hawaiian Islands. Date: //. Source: Rothschild coll.. Material: adult ; whole/good alc. Collected between 1890–1893. Alleged to be from Laysan, but the species was endemic to the island of Hawaii.

Himatione sanguinea freethii

Laysan Honeycreeper

Known only from Laysan Island where it became extinct in 1923 as a result of habitat destruction by rabbits.

Reg. no. 1896.1.8.4. From ; Laysan Island; Hawaiian Islands. Date: 20/ 6/1891. Source: H.C. Palmer; W. Rothschild. Material: M; whole/good skin. Syntype. Rothschild, *Ann. Mag. nat. Hist.*, series 6, 10 (1892): 109–10.

Reg. no. 1939.12.9.68. From ; Laysan Island; Hawaiian Islands. Date: 27/ 8/1896. Source: Prof. Schauinsland; Rothschild coll.. Material: adult F; whole/good skin.

Reg. no. 1939.12.9.69. From ; Laysan Island; Hawaiian Islands. Date: //. Source: Rothschild coll.. Material: adult M; whole/good skin.

Reg. no. N193.63. From ; Laysan Island; Hawaiian Islands. Date: //. Source: Prof. Schauinsland; Rothschild. Material: 1 ; whole/good nest.

Reg. no. N193.64. From ; Laysan Island; Hawaiian Islands. Date: //. Source: Prof. Schauinsland; Rothschild. Material: 1 ; whole/good nest.

Reg. no. E1941.9.4.856. From ; [Laysan Island]; [Hawaiian Islands]. Date: //. Source: Berne Museum; Rothschild. Material: 1 ; whole/good egg.

Ciridops anna

Ula-ai-hawane

Mountain forests of the island of Hawaii. Known only from 5 specimens, the last collected in 1892. There was an unconfirmed sight record in 1937.

Reg. no. 1939.12.9.58. From Kohala Mountains; Hawaii; Hawaiian Islands. Date: 20/ 2/1892. Source: H.C. Palmer; Rothschild coll.. Material: adult ; whole/good skin. Shot by a native. There is a trunk from the same bird in the spirit collection with the untraced reg. no. 1939.12.9.8. The skin was apparently prepared from a specimen originally in spirit and is thus bleached, see *Pacific Science* 46 (1992): 495–500.

Reg. no. 1939.12.9.8. From Mount Kohala; Hawaii; Hawaiian Islands. Date: 20/ 2/1892. Source: Rothschild coll.. Material: adult ; part/poor alc. Partly dissected trunk. The reg. no. is incorrect and untraced. This is the trunk from skin 1939.12.9.58.

Drepanis pacifica

Hawaii Mamo

Mountain forests of the island of Hawaii. Last seen in 1898 and now extinct.

Reg. no. 1939.12.9.52. From Oloa, Purea District; Hawaii; Hawaiian Islands. Date: //1859. Source: J. Mills; S.B. Wilson; Rothschild coll.. Material: M; whole/good skin.

Drepanis funerea

Black Mamo

Mountain forests of Molokai where it was last seen in 1907 and is presumed extinct.

Reg. no. 1895.7.20.94. From ; Molokai; Hawaiian Islands. Date: //. Source: [R.C.L. Perkins]; Royal Society & British Association. Material: ; skin. Syntype. Not in the collection; exchanged with Rothschild in 1897.

Reg. no. 1895.7.20.95. From 5,000 ft.; Molokai; Hawaiian Islands. Date: 18/ 6/1893. Source: R.C.L. Perkins; Royal Society & British Association. Material: adult M; whole/good skin. Syntype. Newton, *Proc. zool. Soc. Lond.* 1893 (publ. 1894): 690. One of the other syntypes is also in the collection.

Reg. no. 1895.7.20.96. From ; Molokai; Hawaiian Islands. Date: 13/ 9/1893. Source: R.C.L. Perkins; Royal Society & British Association. Material: adult M; whole/good skin. Syntype.

Reg. no. 1897.10.28.6. From 4,000 ft. in the mountains; Molokai; Hawaiian Islands. Date: / 6/1896. Source: R.C.L. Perkins; Royal Society & British Association. Material: adult F; whole/good skin. Ova enlarged.

Icterus leucopteryx bairdi

Grand Cayman Oriole

This poorly differentiated race was endemic to Grand Cayman Island, where it became extinct in 1938.

Reg. no. 1925.12.27.6. From ; ; Grand Cayman. Date: 1/ 5/1911. Source: W.W. Brown; Museum of Comparative Zoology. Material: adult M; whole/good skin.

Icterus bonana

Martinique Oriole

Endemic to Martinique where it now occurs mainly in the arid hills of the south. Its decline is due largely to parasitism by a recent coloniser, the Shiny Cowbird *Molothrus bonariensis*.

Reg. no. 1845.7.6.126. From ; ; [Martinique]. Date: //. Source: Sir A. Smith. Material: adult ; whole/good skin. Ex-mount.

Reg. no. 1851.1.24.18. From ; ; [Martinique]. Date: //. Source: Warwick. Material: adult ; whole/good skin. The register gives the locality as Rio Grande.

Reg. no. 1885.11.2.432. From ; ; Martinique. Date: //. Source: Salvin-Godman coll.. Material: adult M; whole/good skin.

Reg. no. 1885.7.12.232. From ; ; Martinique. Date: //1870. Source: Boucard; Sclater coll.. Material: adult ; whole/good skin.

Reg. no. 1885.7.12.233. From ; ; Martinique. Date: //. Source: Deyrolle; Sclater coll.. Material: adult ; whole/good skin.

Reg. no. 1893.12.16.9. From ; ; Martinique. Date: 11/11/1886. Source: W.B. Richardson; Sclater coll.. Material: adult F; whole/good skin.

Reg. no. 1905.6.28.192. From ; ; Martinique. Date: / 4/1863. Source: E. Cavendish Taylor coll.. Material: adult ; whole/good skin. The month of collection is given as May in the register.

Reg. no. 1937.1.2.2. From ; ; Captivity. Date: 8/ 1/1937. Source: A. Ezra. Material: ; whole/good alc.

Quiscalus palustris

Slender-billed Grackle

Formerly restricted to the marshes of the Rio Lerma in Mexico. The last known specimens were collected in 1910, since when the bird has not been seen.

Reg. no. 1885.11.2.639. From ; ; Mexico. Date: //. Source: Boucard; Salvin-Godman coll.. Material: adult F; whole/good skin. Said to be from the vicinity of Mexico City.

Reg. no. 1885.11.2.640. From ; ; Mexico. Date: / 1/1879. Source: E. Bartlett; Salvin-Godman coll.. Material: adult M; whole/good skin. Ex-mount.

Reg. no. 1885.7.12.353. From ; ; Mexico. Date: //. Source: Parzudaki; Sclater coll.. Material: adult M; whole/good skin.

Reg. no. 1885.7.12.354. From ; ; Mexico. Date: //. Source: Boucard; Sclater coll.. Material: adult M; whole/good skin.

Reg. no. 1885.7.12.355. From ; ; Mexico. Date: //. Source: Boucard; Sclater coll.. Material: adult F; whole/good skin.

Neospiza concolor

Sao Tome Canary

Endemic to Sao Tome Island, where it has not been seen since 1888. It is presumed to have become extinct some time between 1888 and 1909, when there was the first of a number of unsuccessful searches.

Reg. no. 1891.8.20.14. From ; ; Sao Tome (Sao Thome Island). Date: / 9/1890. Source: B. du Bocage. Material: M; whole/good skin.

Carduelis cucullata

Red Siskin

Formerly occurred in a semi–nomadic state over much of northern Venezuela and northeastern Colombia with populations on Gasparee, Monos and Trinidad. It was much sought–after as a cage bird and, although now legally protected, is still trapped. Numbers have recently declined in Venezuela to such an extent that few are now being caught.

Reg. no. 1847.7.16.15. From Caracas; ; Venezuela. Date: //. Source: M.A. Salle; C. Dyson; Cuming. Material: adult M; whole/good skin. Dyson is not mentioned in the register.

Reg. no. 1857.10.16.49. From ; ; Trinidad. Date: //. Source: J. Gould. Material: adult F; whole/good skin.

Reg. no. 1885.12.14.1181. From ; ; Venezuela. Date: //. Source: Salvin–Godman coll.. Material: adult M; whole/good skin.

Reg. no. 1885.12.14.1182. From Carupano; ; Venezuela. Date: / 2/1867. Source: A. Goering; Salvin–Godman coll.. Material: adult M; whole/good skin.

Reg. no. 1885.2.10.648. From ; ; Trinidad. Date: //. Source: Argent; P.L. Sclater coll.. Material: adult M; whole/good skin.

Reg. no. 1885.2.10.649. From ; ; Venezuela. Date: //1870. Source: Verdey; P.L. Sclater coll.. Material: juv M; whole/good skin.

Reg. no. 1885.2.10.650. From ; ; Venezuela. Date: //. Source: P.L. Sclater coll.. Material: ; skin. Details taken from the register. Not found 1988.

Reg. no. 1888.9.12.572. From Carupano; ; Venezuela. Date: //. Source: A. Goering; Tweeddale coll.. Material: adult M; whole/good skin.

Reg. no. 1925.2.20.1. From ; ; Captivity. Date: 22/ 2/1925. Source: H.L. Sich. Material: adult F; whole/good skin. Bred in captivity.

Reg. no. 1934.10.15.1. From ; ; Captivity. Date: 26/10/1934. Source: D.J. Cameron. Material: F; whole/good skin. Died in captivity. The day is erroneously given as 25th in the register.

Reg. no. 1988.21.17. From ; ; Trinidad. Date: //. Source: Material: ; whole/good skin.

Reg. no. 1988.21.18. From ; ; Cuba. Date: //. Source: Material: adult M; skin. Mount. Specimen g. in the *Cat. Bds.*. The label bears the number 286a, but this is untraced.

Reg. no. . From ; ; . Date: //. Source: ; ; Captivity. Material: adult ; whole/good alc. Died in captivity.

Carduelis siemiradzkii

Saffron Siskin

Known from a very restricted area in southwest Ecuador and northwest Peru, where it is at great risk, presumably due to habitat destruction.

Reg. no. 1885.12.14.1215. From Balzar Mountains; ; Ecuador. Date: / 3/1880. Source: Illingworth; Salvin–Godman coll.. Material: adult M; whole/good skin.

Reg. no. 1899.4.30.18. From Guayaquil; ; Ecuador. Date: 26/11/1898. Source: P.O. Simons. Material: adult M; whole/good skin.

Reg. no. 1899.4.30.19. From Guayaquil; ; Ecuador. Date: 26/11/1898. Source: P.O. Simons; Rosenberg. Material: adult M; whole/good skin.

Reg. no. 1899.4.30.20. From Guayaquil; ; Ecuador. Date: 29/11/1898. Source: P.O. Simons; Rosenberg. Material: adult F; whole/good skin.

Reg. no. 1899.4.30.21. From Guayaquil; ; Ecuador. Date: 26/11/1898. Source: P.O. Simons; Rosenberg. Material: juv M; whole/good skin.

Carpodacus mexicanus mcgregori

McGregor's House-Finch

Endemic to San Benito Island, off Baja California. It was very rare by 1905 and definitely extinct by 1938.

Reg. no. 1900.7.20.50. From San Benito Island; [Baja California]; [Mexico]. Date: 9/3/1899. Source: R.H. Beck; C.K. Worthen; Salvin–Godman coll.. Material: M; whole/good skin.

Reg. no. 1900.7.20.51. From San Benito Island; Baja California; [Mexico]. Date: 9/3/1899. Source: R.H. Beck; C.K. Worthen; Salvin–Godman coll.. Material: adult F; whole/good skin.

Reg. no. 1900.7.20.52. From San Benito Island; [Baja California]; [Mexico]. Date: 9/3/1899. Source: R.H. Beck; C.K. Worthen; Salvin–Godman coll.. Material: adult M; whole/good skin.

Chaunoproctus ferreorostris

Bonin Grosbeak

Bonin Islands. The only known specimens were collected on Peel Island, where it was extinct by 1854, though it was believed to have survived on the Bailley group until about 1890. The reasons for extinction are not known.

Reg. no. 1855.12.19.39. From Sima; Bonin Islands; Japan. Date: //. Source: Beechey, voyage of H.M.S. Blossom; Zoological Society. Material: adult M; whole/good skin. Holotype. Vigors, *Zool. J. Lond.* 4 (1828, publ. 1829): 354.

Reg. no. 1855.12.19.71. From Sima; Bonin Islands; Japan. Date: //. Source: Beechey, voyage of H.M.S. Blossom; Zoological Society. Material: adult F; whole/good skin.

Pyrrhula pyrrhula murina

Azores Bullfinch

Known only from the eastern end of the island of Sao Miguel in the Azores. In spite of persecution it was common until 1866 but was considered close to extinction in 1903. Following the rediscovery of 30–40 pairs in the late 1970s, a survey in September 1989 found about 370 birds in a 500 ha. fragment of native vegetation.

Reg. no. 1879.4.5.767. From ; Sao Miguel; Azores Islands. Date: / 3/1865. Source: F. Godman; Salvin–Godman coll.. Material: adult M; whole/good skin. Syntype.

Reg. no. 1879.4.5.768. From ; Sao Miguel; Azores Islands. Date: / 4/1865. Source: F. Godman; Salvin–Godman coll.. Material: adult M; whole/good skin. Syntype. Godman, *Ibis* 1866: 97, pl. 3. There are five other syntypes in the collection.

Reg. no. 1879.4.5.769. From ; Sao Miguel; Azores Islands. Date: / 4/1865. Source: F. Godman; Salvin–Godman coll.. Material: M; whole/good skin. Syntype.

Reg. no. 1879.4.5.770. From ; Sao Miguel; Azores Islands. Date: / 3/1865. Source: F. Godman; Salvin–Godman coll.. Material: adult F; whole/good skin. Syntype. The register erroneously gives the month of collection as April.

Reg. no. 1879.4.5.771. From ; Sao Miguel; Azores Islands. Date: / 4/1865. Source: F. Godman; Salvin–Godman coll.. Material: adult F; whole/good skin. Syntype.

Reg. no. 1879.4.5.772. From ; Sao Miguel; Azores Islands. Date: / 3/1865. Source: F. Godman; Salvin–Godman coll.. Material: adult F; whole/good skin. Syntype.

Reg. no. 1904.12.31.114. From Furnas; Sao Miguel; Azores Islands. Date: / 4/1903. Source: W.R. Ogilvie Grant; Rothschild coll.. Material: adult M; whole/good skin. Iris brown, bill black, legs very dark horn–brown.

Reg. no. 1904.12.31.115. From n. of Furnas; Sao Miguel; Azores Islands. Date: / 4/1903. Source: W.R. Ogilvie Grant; Rothschild coll.. Material: adult M; whole/good skin. Iris brown, bill black, legs very dark horn–brown.

Reg. no. 1904.12.31.116. From ; Sao Miguel; Azores Islands. Date: / 4/1903. Source: W.R. Ogilvie Grant; Rothschild coll.. Material: M; whole/good skin.

Reg. no. 1904.12.31.117. From n. of Furnas Valley; Sao Miguel; Azores Islands. Date: 25/ 3/1903. Source: W.R. Ogilvie Grant; Rothschild coll.. Material: F; whole/good skin. Iris brown, bill black, legs brown. The sex is given as male in the register, but female on the label.

Reg. no. 1904.12.31.118. From n. of Furnas Valley; Sao Miguel; Azores Islands. Date: 25/ 3/1903. Source: W.R. Ogilvie Grant; Rothschild coll.. Material: M; whole/good skin. Iris brown, bill black, legs very dark horn–brown. The sex is given as female in the register, but male on the label.

Reg. no. 1915.5.1.571. From Furnas; Sao Miguel; Azores Islands. Date: 26/ 5/1907. Source: P.R. Lowe; Cleaves coll.. Material: M; whole/good skin.

Reg. no. E1988.4.28. From ; San Miguel; Azores Islands. Date: 2/ 6/1900. Source: Stadtrat Ochs; Otto Bamberg; Jourdain coll.. Material: 2 ; whole/good eggs (clutch). Exhibited, see *Bull. Brit. Orn. Club* 25 (1910): 118–9. Bannerman & Bannerman, 1966, *Birds of the Atlantic Islands*, 3: 168–9, do not consider these eggs to be sufficiently authenticated.

Reg. no. 1903.6.10.1. From n. of Furnas; Sao Miguel; Azores Islands. Date: //. Source: W.R. Ogilvie Grant. Material: adult ; whole/good alc. There are five skins collected by Ogilvie Grant in 1903.

Reg. no. 1903.6.10.2. From n. of Furnas; Sao Miguel; Azores Islands. Date: //. Source: W.R. Ogilvie Grant. Material: adult ; whole/good alc. There are five skins collected by Ogilvie Grant in 1903.

Ploceus golandi

Clarke's Weaver

Known only from the Arabuko–Sokoke Forest in southeastern Kenya, where it is threatened by illegal but persistent logging.

Reg. no. 1913.1.17.1. From Mombassa; ; Kenya. Date: //. Source: S.R. Clarke. Material: adult M; whole/good skin. Holotype. Clarke, *Bull. Brit. Orn. Club* 31 (1913): 32.

Reg. no. 1959.36.1. From Sokoke Forest, Kilifi; ; Kenya. Date: 7/ 9/1959. Source: I. Parker; J.G. Williams; Coryndon Museum. Material: adult F; whole/good skin. Iris brown, feet pale flesh, bill horn–grey, base whitish.

Reg. no. 1959.37.2. From Sokoke Forest, Kilifi; ; Kenya. Date: 7/ 9/1959. Source: J.G. Williams; I. Parker; Coryndon Museum. Material: adult M; whole/good skin. Iris rich brown, feet pale flesh, bill black.

Reg. no. 1983.1.2. From Sokoke Forest; ; Kenya. Date: 6/ 7/1980. Source: H.A. Britton; P.L. Britton. Material: adult F; whole/good skin. Killed by a Lizard Buzzard in a mist net.

Malimbus ibadanensis

Ibadan Malimbe

Occurs in a very restricted area of southwestern Nigeria where it appears to have become very rare in recent years.

Reg. no. 1958.9.1. From Ibadan; ; Nigeria. Date: 31/ 1/1958. Source: J.H. Elgood.

Material: adult F; whole/good skin. Shot in a tree in a garden in University College.

Reg. no. 1958.9.2. From Ibadan; ; Nigeria. Date: 28/ 2/1958. Source: J.H. Elgood. Material: adult M; whole/good skin. Holotype. Elgood, *Ibis* 1958: 621–4.

Reg. no. 1958.9.3. From Ibadan; ; Nigeria. Date: 15/ 5/1958. Source: J.H. Elgood. Material: adult M; whole/good skin.

Reg. no. 1959.25.1. From Ibadan; ; Nigeria. Date: //1959. Source: J.H. Elgood. Material: M; whole/good skin. Collected in May or June. Possibly not fully adult.

Reg. no. 1959.25.2. From Ibadan; ; Nigeria. Date: //1959. Source: J.H. Elgood. Material: adult F; whole/good skin. Collected in May or June.

Reg. no. N253.1. From Egbado College, Ilaro; ; w. Nigeria. Date: 4/ 9/1965. Source: J.A. Button. Material: 1 ; whole/good nest. An egg is also in the collection, E1966.11.1. The nest was built in a dead tree 30 ft. from the ground, and near the top. The tree was one of several left standing in a clearing of 2/3 acre made for farming in well matured secondary bush. Maize growing at the time.

Reg. no. E1966.11.1. From Egbado College, Ilaro; ; w. Nigeria. Date: 4/ 9/1965. Source: J.A. Button. Material: 1 ; whole/good egg (clutch). This is the first egg taken of this species. The nest is also in the collection, N253.1. Second brood; egg probably laid about 28 August, based on state of embryo. First brood successfully reared (2 young ?); egg–laying date approximately 10 July. Nest completed by this time.

Foudia rubra

Mauritius Fody

Now restricted to the remaining native forest of Mauritius where it is constantly threatened by alien predators.

Reg. no. 1841.4.2.383. From ; ; Mauritius. Date: //. Source: Warwick. Material: adult M; whole/good skin.

Reg. no. 1845.7.6.18. From ; ; Mauritius. Date: //. Source: Sir A. Smith. Material: adult M; whole/good skin.

Reg. no. 1873.10.25.1. From ; ; Mauritius. Date: //. Source: R.B. Sharpe. Material: adult F; whole/good skin.

Reg. no. 1889.7.25.255. From ; ; Mauritius. Date: //. Source: Bewsher; Shelley coll.. Material: adult F; whole/good skin.

Reg. no. S1975.11.1. From Vacoas; ; Mauritius. Date: /10/1974. Source: A. Cheke. Material: M; whole/good skeleton.

Reg. no. N1990.2.2. From s. facing scarp above Bassin Blanc; ; Mauritius. Date: 23/12/1989. Source: R.J. Safford. Material: ; whole/good nest. Had been 8m up a *Eucalyptus robusta* in small grove. Taken by a predator during incubation and found empty on ground. Predator (*Rattus rattus*?) chewed hole c.5cm across near bottom of nest (hole has expanded as nest dried out).

Reg. no. N1992.3.1. From Pigeon Wood, Savanne Mountains; ; Mauritius. Date: 19/

9/1991. Source: R.J. Safford. Material: ; whole/good nest. The nest had been deserted.

Reg. no. E1992.5.1. From Savanne Mountains; ; Mauritius. Date: /12/1991. Source: R.J. Safford. Material: 2 ; whole/good eggs (clutch). The nest was in a *Cryptomeria japonica* grove on a south-facing scarp. Complete clutch of two eggs, died (? chilled) a few days before they were due to hatch. One egg cracked.

Reg. no. A1975.10.2. From Vacoas; ; Mauritius. Date: 30/ 9/1974. Source: A. Cheke. Material: ; whole/good alc.

Foudia flavicans

Rodriguez Fody

Restricted to a small area in the north of Rodriguez. In 1983 about 100 birds were thought to survive.

Reg. no. 1876.3.14.10. From ; ; Rodriguez. Date: //. Source: H.H. Slater, Transit of Venus Expedition; Royal Society. Material: adult F; whole/good skin. Presumably collected at the same time as the Gulliver specimens (October/November 1874).

Reg. no. 1876.3.14.11. From ; ; Rodriguez. Date: //. Source: H.H. Slater, Transit of Venus Expedition; Royal Society. Material: adult F; whole/good skin. Presumably collected at the same time as the Gulliver specimens (October/November 1874).

Reg. no. 1876.3.14.5. From ; ; Rodriguez. Date: //. Source: H.H. Slater, Transit of Venus Expedition; Royal Society. Material: adult M; whole/good skin. Presumably collected at the same time as the Gulliver specimens (October/November 1874).

Reg. no. 1876.3.14.6. From ; ; Rodriguez. Date: //. Source: H.H. Slater, Transit of Venus Expedition; Royal Society. Material: adult M; whole/good skin. Presumably collected at the same time as the Gulliver specimens (October/November 1874).

Reg. no. 1876.3.14.63. From ; ; Rodriguez. Date: //. Source: H.H. Slater, Transit of Venus Expedition; Royal Society. Material: adult M; whole/good skin. Presumably collected at the same time as the Gulliver specimens (October/November 1874).

Reg. no. 1876.3.14.7. From ; ; Rodriguez. Date: //. Source: H.H. Slater, Transit of Venus Expedition; Royal Society. Material: adult F; whole/good skin. Presumably collected at the same time as the Gulliver specimens (October/November 1874).

Reg. no. 1876.3.14.8. From ; ; Rodriguez. Date: //. Source: H.H. Slater, Transit of Venus Expedition; Royal Society. Material: adult M; whole/good skin. Presumably collected at the same time as the Gulliver specimens (October/November 1874).

Reg. no. 1876.3.14.9. From ; ; Rodriguez. Date: //. Source: H.H. Slater, Transit of Venus Expedition; Royal Society. Material: adult F; whole/good skin. Presumably collected at the same time as the Gulliver specimens (October/November 1874).

Reg. no. 1876.3.15.10. From ; ; Rodriguez. Date: /11/1874. Source: H. Gulliver, Transit of Venus Expedition; Royal Society. Material: adult F; whole/good skin.

Reg. no. 1876.3.15.11. From ; ; Rodriguez. Date: /10/1874. Source: H. Gulliver, Transit of Venus Expedition; Royal Society. Material: adult M; whole/good skin.

Reg. no. 1876.3.15.12. From ; ; Rodriguez. Date: /10/1874. Source: H. Gulliver, Transit of Venus Expedition; Royal Society. Material: adult F; whole/good skin.

Reg. no. 1876.3.15.9. From ; ; Rodriguez. Date: /11/1874. Source: H. Gulliver, Transit of Venus Expedition; Royal Society. Material: adult M; whole/good skin.

Reg. no. 1876.3.14.66. From ; ; Rodriguez. Date: //. Source: Slater. Material: 1 ; whole/good nest. Subsequently re-registered as N137.1.

Reg. no. 1876.3.15.1. From ; ; Rodriguez. Date: //. Source: G. Gulliver. Material: ; whole/good nest. Not found 1988.

Reg. no. 1876.3.15.2. From ; ; Rodriguez. Date: //. Source: G. Gulliver. Material: ; whole/good nest. Not found 1988.

Reg. no. 1876.3.15.5–6. From ; ; Rodriguez. Date: /12/1874. Source: G. Gulliver. Material: 2 ; whole/good eggs.

Reg. no. 1876.3.15.19–22. From ; ; Rodriguez. Date: //. Source: G. Gulliver, Transit of Venus Expedition; Royal Society. Material: 4; alc. Listed as *F. flavicans* in the register, but subsequently reidentified as *F. madagascariensis*.

Aplonis corvina

Kittlitz's or Kosrae Starling

Known only from skins collected by Kittlitz on Kosrae Island in the Carolines in 1828 and now in St. Petersburg. When next sought, in 1880, the bird was no longer extant.

Two incomplete skeletons and an eviscerated spirit specimen were said to be in the collection (*Bull. Br. Mus. nat. Hist. (Zool.)* 34 (1979): 178). The skeletal specimens were mis-labelled. The spirit specimen was not found 1988 but was doubtfully of this species.

Aplonis mavornata

Mysterious Starling

Known only from the type, now considered to have been collected by Andrew Bloxam on Mauke Island in 1825.

Reg. no. VEL.12.192a. From ; [Mauke]; [Cook Islands]. Date: 9/ 8/1825. Source: [A. Bloxam, voyage of H.M.S. Blonde]. Material: adult ; whole/good skin. Holotype. Ex-mount. Buller, 1888, *A History of the Birds of New Zealand*, 2nd ed., 1: 25. See *Notornis* 33 (1986): 197–208. The specimen was previously thought to have come from Raiatea, Society Islands. Killed hopping about a tree, in the afternoon, before 4 pm.

Aplonis fusca fusca

Norfolk Island Starling

Endemic to Norfolk Island. Extinct since 1925.

Reg. no. 1839.8.2.102. From ; ; [Norfolk Island] (Australia). Date: //. Source: Sir T. Mitchell. Material: adult M; whole/good skin. The label bears the reg. no. 1839.8.12.102 and 'Gould', but this appears to be erroneous. That reg. no. is untraced, but 1839.8.2.102 is listed in the register as *Aplonis*, without further data, and would seem to refer to this specimen.

Reg. no. 1858.4.3.114. From ; ; [Norfolk Island] (Australia). Date: //. Source: Zoological Society. Material: adult F; whole/good skin. The label claims the specimen is from the Gould coll., but this would appear to be erroneous.

Reg. no. 1870.3.31.129. From ; ; Norfolk Island. Date: / 6/1865. Source: J. Brenchley. Material: adult M; whole/good skin. Iris reddish–brown. Food, guava seeds and small brown seeds.

Reg. no. 1870.3.31.130. From ; ; Norfolk Island. Date: / 6/1865. Source: J. Brenchley. Material: adult F; whole/good skin. Iris reddish–brown. Food, guava seeds and small brown seeds.

Reg. no. 1881.5.1.3160. From ; ; [Norfolk Island] (India). Date: //. Source: Gould coll.. Material: adult M; whole/good skin. This may be the missing type, but there is no confirmatory evidence.

Reg. no. 1898.11.10.169. From ; ; Norfolk Island. Date: //. Source: Seebohm coll.. Material: adult ; skin. Details taken from the register. Not found 1988.

Reg. no. 1919.7.15.79. From nr. Cascades; ; Norfolk Island. Date: 1/ 4/1913. Source: G.M. Mathews. Material: adult M; whole/good skin. Iris bright orange, feet greyish–black, bill black.

Reg. no. 1919.7.15.80. From nr. Cascades; ; Norfolk Island. Date: 1/ 4/1913. Source: G.M. Mathews. Material: adult F; whole/good skin. Iris bright orange–yellow, feet greyish–black, bill black.

Reg. no. 1919.7.15.81. From nr. Cascades; ; Norfolk Island. Date: 3/ 4/1913. Source: G.M. Mathews. Material: adult F; whole/good skin. Iris bright yellow, feet greyish–black, bill black.

Reg. no. 1919.7.15.82. From nr. Cascades; ; Norfolk Island. Date: 5/ 4/1913. Source: G.M. Mathews. Material: adult F; whole/good skin. Iris bright orange, feet greyish–black, bill black. The day of collection is given as 15th in the register.

Reg. no. 1919.7.15.83. From Dixey; ; Norfolk Island. Date: 25/ 4/1913. Source: G.M. Mathews. Material: adult M; whole/good skin. Iris bright orange–red, feet black, bill black.

Reg. no. 1919.7.15.84. From nr. Cascades; ; Norfolk Island. Date: 24/ 4/1913. Source: G.M. Mathews. Material: adult M; whole/good skin. Iris bright orange–red, feet greyish, bill black. The day of collection is given as 28th in the register.

Reg. no. 1919.7.15.85. From nr. Cascades; ; Norfolk Island. Date: 29/ 4/1913. Source: G.M. Mathews. Material: adult M; whole/good skin. Iris bright orange, feet greyish–black, bill black.

Reg. no. 1919.7.15.86. From nr. Cascades; ; Norfolk Island. Date: 5/ 5/1913. Source: G.M. Mathews. Material: adult M; whole/good skin. Iris bright orange, feet greyish–black, bill black.

Reg. no. 1919.7.15.87. From nr. Kingston; ; Norfolk Island. Date: 8/ 5/1913. Source: G.M. Mathews. Material: adult M; whole/good skin. Iris bright orange, feet greyish–black, bill black.

Reg. no. 1919.7.15.88. From nr. Cascades; ; Norfolk Island. Date: 22/ 5/1913. Source: G.M. Mathews. Material: adult M; whole/good skin. Iris bright orange–red, feet black tinged greyish, bill black.

Reg. no. 1919.7.15.89. From nr. Cascades; ; Norfolk Island. Date: 22/ 5/1913. Source: G.M. Mathews. Material: adult F; whole/good skin. Iris bright reddish–orange, feet greyish–black, bill black.

Reg. no. 1919.7.15.90. From nr. Cascades; ; Norfolk Island. Date: 2/ 6/1913. Source: G.M. Mathews. Material: adult F; whole/good skin. Iris bright yellow, feet greyish–black, bill black.

Reg. no. 1919.7.15.91. From Steel's Point; ; Norfolk Island. Date: 21/ 8/1912. Source: G.M. Mathews. Material: adult M; whole/good skin. Iris orange–yellow, feet black tinged brown, bill black.

Reg. no. 1939.12.9.2511. From nr. Middlegate; ; Norfolk Island. Date: 28/ 4/1913. Source: G.M. Mathews; Rothschild coll.. Material: adult M; whole/good skin. Iris bright orange, feet greyish–black, bill black.

Reg. no. 1939.12.9.2512. From nr. Cascades; ; Norfolk Island. Date: 29/ 4/1913. Source: G.M. Mathews; Rothschild coll.. Material: adult M; whole/good skin. Iris bright orange–red, feet greyish–black, bill black.

Reg. no. 1939.12.9.2513. From nr. Cascades; ; Norfolk Island. Date: 3/ 3/1913. Source: G.M. Mathews; Rothschild coll.. Material: adult M; whole/good skin. Iris bright orange, feet greyish–black, bill black.

Reg. no. 1939.12.9.2514. From nr. Cascades; ; Norfolk Island. Date: 15/ 4/1913. Source: G.M. Mathews; Rothschild coll.. Material: adult M; whole/good skin. Iris bright orange, feet greyish–black, bill black.

Reg. no. 1939.12.9.2515. From nr. Cascades; ; Norfolk Island. Date: 29/ 4/1913. Source: G.M. Mathews; Rothschild coll.. Material: adult M; whole/good skin. Iris bright orange–red, feet greyish–black, bill black.

Reg. no. 1939.12.9.2516. From nr. Cascades; ; Norfolk Island. Date: 12/ 6/1913. Source: G.M. Mathews; Rothschild coll.. Material: adult M; whole/good skin. Iris bright yellow, feet greyish–black, bill black.

Reg. no. 1939.12.9.2517. From nr. Cascades; ; Norfolk Island. Date: 29/ 4/1913. Source: G.M. Mathews; Rothschild coll.. Material: adult F; whole/good skin. Iris bright orange, feet black tinged grey, bill black.

Reg. no. 1939.12.9.2518. From nr. Cascades; ; Norfolk Island. Date: 29/ 4/1913. Source: G.M. Mathews; Rothschild coll.. Material: adult F; whole/good skin. Iris bright orange, feet greyish–black, bill black.

Reg. no. 1988.25.2. From ; ; [Norfolk Island] (Australia). Date: //. Source: Dr G. Bennett, Australian Museum. Material: adult M; whole/good skin.

Reg. no. 1988.25.3. From ; ; [Norfolk Island]. Date: //. Source: Material: adult F; whole/good skin. Ex–mount. From the Old Collection.

Reg. no. N193.511. From ; ; Norfolk Island. Date: //. Source: B. Grinnell; Rothschild. Material: 1 ; whole/good nest.

Reg. no. E1941.2.1.197. From ; ; Norfolk Island. Date: 8/12/1912. Source: Mathews coll.; Rothschild. Material: 2 ; whole/good eggs.

Reg. no. E1941.2.1.198. From ; ; Norfolk Island. Date: 25/11/1912. Source: Mathews coll.; Rothschild. Material: 2 ; whole/good eggs.

Reg. no. E1941.2.1.199. From ; ; Norfolk Island. Date: 23/11/1912. Source: Mathews coll.; Rothschild. Material: 3 ; whole/good eggs.

Reg. no. E1941.2.1.200. From ; ; Norfolk Island. Date: 21/12/1912. Source: Mathews coll.; Rothschild. Material: 3 ; whole/good eggs.

Aplonis fusca hulliana

Lord Howe Island Starling

Endemic to Lord Howe Island. It disappeared soon after rats escaped from a wreck in 1918, before which it had been very common.

Reg. no. 1881.5.1.532. From ; ; Lord Howe Island. Date: 8/ 9/1853. Source: J. MacGillivray; J. Gould. Material: F; whole/good skin.

Reg. no. 1881.5.1.533. From ; ; Lord Howe Island. Date: 8/ 9/1853. Source: J. MacGillivray; J. Gould. Material: adult M; whole/good skin.

Reg. no. 1888.10.5.13. From ; ; Lord Howe Island. Date: //. Source: E. Saunders. Material: ; whole/good skin. The label had 'Norfolk Island' written on it, but it had been crossed out and 'Lord Howe Island' substituted. Listed in the register as *Aplonis*, Norfolk Island.

Reg. no. 1888.10.5.14. From ; ; [Lord Howe Island]. Date: //. Source: E. Saunders. Material: ; skin. Not found 1988. Listed in the register as *Aplonis*, Norfolk Island, but see the comments under 1888.10.5.13,15, which were similarly registered. According to the *Cat. Bds.*, there were three Saunders specimens from Lord Howe Island; this was undoubtedly the third.

Reg. no. 1888.10.5.15. From ; ; Lord Howe Island. Date: //. Source: E. Saunders. Material: ; whole/good skin. The label had 'Norfolk Island' written on it, but it had been crossed out and 'Lord Howe Island' substituted. Listed in the register as *Aplonis*, Norfolk Island.

Reg. no. 1898.11.10.166. From ; ; Lord Howe Island. Date: / 9/1887. Source: Sydney Museum; Seebohm coll.. Material: adult F; whole/good skin.

Reg. no. 1898.11.10.167. From ; ; Lord Howe Island. Date: / 9/1887. Source: Sydney Museum; Seebohm coll.. Material: adult M; whole/good skin.

Reg. no. 1898.11.10.168. From ; ; Lord Howe Island. Date: /10/1897. Source: Seebohm coll.. Material: adult ; skin. Details taken from the register. Not found 1988.

Reg. no. 1919.7.15.144. From the Pines; ; Lord Howe Island. Date: 5/11/1913. Source: G.M. Mathews. Material: adult M; whole/good skin. Bill black, eye orange, feet black.

Reg. no. 1919.7.15.145. From ; ; Lord Howe Island. Date: 13/ 1/1914. Source: G.M. Mathews. Material: adult M; whole/good skin. Bill black, eye bright orange, feet black.

Reg. no. 1919.7.15.146. From the Pines; ; Lord Howe Island. Date: 27/ 2/1914. Source: G.M. Mathews. Material: adult M; whole/good skin. Bill black, eye orange, feet greyish–black.

Reg. no. 1919.7.15.147. From the Pines; ; Lord Howe Island. Date: 12/ 5/1914. Source: G.M. Mathews. Material: adult F; whole/good skin. Bill black, eye orange, feet black.

Reg. no. 1919.7.15.148. From the Pines; ; Lord Howe Island. Date: 11/ 5/1914. Source: G.M. Mathews. Material: adult F; whole/good skin. Bill black, feet black, eye orange.

Reg. no. 1919.7.15.149. From the Pines; ; Lord Howe Island. Date: 13/ 1/1914. Source: G.M. Mathews. Material: adult F; whole/good skin. Bill black, eye orange, feet black.

Reg. no. 1919.7.15.150. From Mrs Nichol's; ; Lord Howe Island. Date: 24/ 7/1913. Source: G.M. Mathews. Material: adult F; whole/good skin. Iris bright orange, bill black, feet black.

Reg. no. E1941.2.1.201. From ; ; Lord Howe Island. Date: 10/10/1913. Source: Mathews coll.; Rothschild. Material: 3 ; whole/good eggs.

Reg. no. E1941.2.1.202. From ; ; Lord Howe Island. Date: 3/11/1913. Source: Mathews coll.; Rothschild. Material: 3 ; whole/good eggs.

Fregilupus varius

Reunion Starling

Formerly occurred on Reunion where it disappeared during the 19th century. Various possible dates have been given for its extinction. This species may not have been a starling.

Reg. no. 1889.5.30.15. From ; ; Reunion. Date: //. Source: Riocour; A. Boucard. Material: ; whole/good skin.

Leucopsar rothschildi

Rothschild's or Bali Myna

Endemic to Bali and now confined to a small national park on the extreme northwestern tip of the island, where it continues to be captured for the cage bird trade. By 1992, there were 55 birds in the wild, up from 18 in 1990. There is a considerable captive population, and some birds have recently been reintroduced to the wild in what is intended to be an ongoing programme.

Reg. no. 1925.11.23.1. From ; n.w. Bali; Indonesia. Date: 29/ 2/1925. Source: Baron von Plessen; Dr E. Stresemann. Material: adult M; whole/good skin.

Reg. no. 1934.12.3.2. From ; ; Captivity. Date: 14/12/1934. Source: A. Ezra. Material: adult F; whole/good skin. Bred in captivity.

Reg. no. 1936.1.1.6. From ; ; Captivity. Date: 20/ 1/1936. Source: A. Ezra. Material: M; whole/good skin. Bred in captivity.

Reg. no. 1936.8.1.3. From ; ; Captivity. Date: 20/ 8/1936. Source: A. Ezra. Material: adult F; whole/good skin. Died in captivity.

Reg. no. 1969.41.1058. From ; Bali; Indonesia. Date: //. Source: Hewitt coll.. Material: M; whole/good skin.

Reg. no. 1969.41.820. From ; Bali; Indonesia. Date: //. Source: Hewitt coll.. Material: F; whole/good skin.

Reg. no. 1969.41.821. From ; Bali; Indonesia. Date: //. Source: Hewitt coll.. Material: M; whole/good skin.

Reg. no. 1931.7.17.1. From ; [Bali]; [Indonesia], Captivity. Date: 24/ 7/1931. Source: A. Ezra. Material: ; whole/good skeleton. Died in captivity. Rough skeleton only.

Reg. no. S1982.13.2. From ; [Bali]; [Indonesia], Captivity. Date: 21/ 1/1982. Source: Waddesdon Manor. Material: ; whole/good skeleton. Died in captivity.

Reg. no. S1982.150.2. From ; Bali; Indonesia, Captivity. Date: 2/ 4/1982. Source: Jersey Wildlife Preservation Trust. Material: F; whole/good skeleton. Died in captivity, caught in Bali.

Reg. no. 1931.7.1.2. From ; ; Captivity. Date: 5/ 7/1931. Source: A. Ezra. Material: juv ; whole/good alc. Died in captivity.

Reg. no. A1982.51.8. From ; ; Captivity. Date: 1/12/1981. Source: Jersey Wildlife Preservation Trust. Material: F; part/good alc. Bred in captivity; hatched in 1980. Not in breeding condition. Eviscerated specimen.

Oriolus xanthonotus assimilis

Cebu Oriole

Endemic to Cebu in the Philippines, where it became extinct in 1906 as a result of the destruction of the native forest.

Reg. no. 1887.11.20.525. From ; Cebu; Philippines. Date: / 3/1877. Source: A.H. Everett; Tweeddale coll.. Material: adult M;

whole/good skin. Holotype. Tweeddale, *Proc. zool. Soc. Lond.* 1877: 760–1, pl. 76.

Callaeas cinerea cinerea

South Island Kokako

Found on the southern islands of New Zealand where it was thought to be extinct until rediscovered on Stewart Island. The population was estimated at about 1000 birds in 1989.

Reg. no. 1849.12.12.25. From Port Cooper; South Island; New Zealand. Date: 4/ 5/1849. Source: F. Strange; Stokes. Material: adult M; whole/good skin. The sex is given as female in the register.

Reg. no. 1849.12.12.26. From Pigeon Bay; South Island; New Zealand. Date: //. Source: Stokes. Material: juv ; whole/good skin.

Reg. no. 1849.12.12.27. From Port Cooper; South Island; New Zealand. Date: 4/ 5/1849. Source: F. Strange; Stokes. Material: adult F; whole/good skin.

Reg. no. 1875.7.2.1. From ; ; New Zealand. Date: //. Source: Colonial Museum. Material: juv ; whole/good skin.

Reg. no. 1875.7.2.2. From ; ; New Zealand. Date: //. Source: Colonial Museum. Material: adult ; whole/good skin.

Reg. no. 1875.7.2.3. From ; ; New Zealand. Date: //. Source: Colonial Museum. Material: adult ; whole/good skin.

Reg. no. 1897.12.6.27. From Otago; South Island; New Zealand. Date: //. Source: Otago University Museum. Material: M; whole/good skin.

Reg. no. 1897.12.6.28. From Otago; South Island; New Zealand. Date: //. Source: Otago University Museum. Material: F; whole/good skin.

Reg. no. 1903.7.17.1. From Dusky Bay; South Island; New Zealand. Date: / 6/1901. Source: Earl of Ranfurly. Material: M; whole/good skin.

Reg. no. 1926.12.22.29. From ; South Island; New Zealand. Date: //. Source: Imperial Institute. Material: ; whole/good skin.

Reg. no. 1926.12.22.31. From ; South Island; New Zealand. Date: //. Source: Imperial Institute. Material: ; whole/good skin.

Reg. no. 1927.12.18.167. From Milford Sound; South Island; New Zealand. Date: / 1/1889. Source: J.J. Lister coll.. Material: F; whole/good skin.

Reg. no. 1927.12.18.4. From Dusky Sound; South Island; New Zealand. Date: //1884. Source: Miss Lister. Material: F; whole/good skin.

Reg. no. 1927.6.14.4. From ; ; New Zealand. Date: //. Source: S. Jackson. Material: ; whole/good skin. Apart from one specimen (1927.6.14.3), the collection from which this bird came was apparently never entered in the registers.

Reg. no. 1949.64.202. From ; South Island; New Zealand. Date: / 8/1887. Source: A.

Reischek; G.M. Mathews. Material: M; whole/good skin.

Reg. no. 1955.6.N.4.9. From ; South Island; New Zealand. Date: //. Source: F.G.M. Boileau; J.H. Gurney coll. Material: adult ; whole/good skin. Presented in August 1895; date of death unknown.

Reg. no. 1958.3.27. From ; South Island; New Zealand. Date: //. Source: Dundee Museum. Material: ; whole/good skin.

Reg. no. 1988.21.19. From ; [South Island]; [New Zealand]. Date: //. Source: Material: ; whole/good skin. Mount.

Reg. no. 1988.21.20. From ; [South Island]; [New Zealand]. Date: //. Source: Material: ; whole/good skin.

***Callaeas cinerea* subspecies**

Reg. no. . From ; ; New Zealand. Date: //1894. Source: Caley Webster coll.; Rothschild coll.. Material: ; part skin. Flat skin from spirit specimen 1940.12.8.109. Details taken from the register. Not found 1988.

Reg. no. 1927.6.14.3. From ; ; New Zealand. Date: //. Source: Mrs Jackson. Material: ; skin. Presumed to be a skin. Subspecies not determined. Not in the collection; exchanged with the University of Michigan in 1966. This is the only bird from this collection which was entered in the registers, although numbers were allocated to specimens; see 1927.6.14.4 above.

Reg. no. 1902.10.25.422–3. From ; ; New Zealand. Date: //. Source: Leopold Field coll.; W. Radcliffe-Saunders coll.. Material: 2 ; whole/good eggs (clutch). Collected during the 19th century. These are probably *C.c. cinerea* as the two races would have been considered separate species at the time they were collected.

Reg. no. 1940.12.8.109. From ; ; New Zealand. Date: //1894. Source: Caley Webster coll.; Rothschild coll.. Material: ; part/good alc. Partially dissected. Head detached.

Reg. no. A1973.1.5. From ; ; New Zealand. Date: //. Source: Bennett. Material: ; whole/good alc. Head partially dissected. See *Bull. Br. Mus. nat. Hist. (Zool.)* 27 (1974): 1–48.

Reg. no. A1973.1.6. From ; ; [New Zealand]. Date: //. Source: Material: adult ; part/good alc. Dissected specimen in several pieces. See *Bull. Br. Mus. nat. Hist. (Zool.)* 27 (1974): 1–48.

Reg. no. A1973.1.7. From ; ; [New Zealand]. Date: //. Source: Material: adult ; whole/good alc. Dissected specimen, apparently complete.

Heterolocha acutirostris

Huia

Formerly confined to North Island, New Zealand. It became extinct in 1907 for reasons which are unclear.

Reg. no. . From ; ; [New Zealand]. Date: //. Source: Material: adult F; part/good skin. Stuffed head only, in Victorian display of bird anatomy in gallery at South Kensington.

Reg. no. . From ; ; [New Zealand]. Date: //. Source: Material: adult M; part/good skin. Stuffed head only, in Victorian display of bird anatomy in gallery at South Kensington.

Reg. no. . From ; ; New Zealand. Date: //1894. Source: Rothschild coll.. Material: adult M; part skin. Flat skin from spirit specimen 1940.12.8.107. Details taken from the register. Not found 1988.

Reg. no. . From ; ; New Zealand. Date: //. Source: Sir G. Grey. Material: adult F; skin. Mount. Details taken from *Cat. Bds.*. Not found 1988.

Reg. no. . From ; ; New Zealand. Date: //. Source: Percy Earl. Material: adult M; skin. Mount. Details taken from *Cat. Bds.*. Not found 1988.

Reg. no. 1847.1.8.23. From ; [North Island]; New Zealand. Date: //. Source: Sir G. Grey. Material: ; whole/good skin.

Reg. no. 1875.7.2.10. From Wainuiomata Valley, Rimutaka Range, Wellington; [North Island]; New Zealand. Date: //. Source: Colonial Museum, Wellington. Material: adult F; whole/good skin.

Reg. no. 1875.7.2.11. From Wainuiomata Valley, Rimutaka Range, Wellington; [North Island]; [New Zealand]. Date: //. Source: Dr Hector; Colonial Museum, Wellington. Material: adult M; whole/good skin.

Reg. no. 1875.7.2.12. From Wainuiomata Valley, Rimutaka Range, Wellington; [North Island]; New Zealand. Date: //. Source: Colonial Museum, Wellington. Material: adult F; whole/good skin.

Reg. no. 1875.7.2.13. From Wainuiomata Valley, Rimutaka Range, Wellington; [North Island]; New Zealand. Date: //. Source: Colonial Museum, Wellington. Material: imm F; whole/good skin.

Reg. no. 1875.7.2.14. From Wainuiomata Valley, Rimutaka Range, Wellington; ; New Zealand. Date: //. Source: Colonial Museum, Wellington. Material: adult F; whole/good skin.

Reg. no. 1875.7.2.81. From Wainuiomata Valley, Rimutaka Range, Wellington; North Island; New Zealand. Date: //. Source: [Dr Hector]; Colonial Museum, Wellington. Material: adult M; whole/good skin.

Reg. no. 1875.7.2.9. From Wainuiomata Valley, Rimutaka Range, Wellington; [North Island]; New Zealand. Date: //. Source: Dr Hector; Colonial Museum, Wellington. Material: imm M; whole/good skin. Not fully adult.

Reg. no. 1897.6.1.394. From ; [North Island]; New Zealand. Date: //. Source: Seebohm coll.. Material: imm F; whole/good skin. Almost adult.

Reg. no. 1897.6.1.395. From ; [North Island]; New Zealand. Date: //. Source: Seebohm coll.. Material: adult F; whole/good skin.

Reg. no. 1926.10.10.1. From Tararua Mountains; North Island; New Zealand. Date: / 9/1878. Source: S.R. Wason. Material: adult M; whole/good skin. A specimen of *Apteryx* from Wason was registered with the same number by mistake.

Reg. no. 1926.10.10.2. From Tararua Mountains; North Island; New Zealand. Date: / 9/1878. Source: S.R. Wason. Material: adult F; whole/good skin. A specimen of *Apteryx* from Wason was registered with the same number by mistake.

Reg. no. 1939.12.9.14. From Makuri Ranges; [North Island]; New Zealand. Date: //. Source: Rothschild coll.. Material: adult M; whole/good skin.

Reg. no. 1939.12.9.15a. From ; [North Island]; [New Zealand]. Date: //. Source: Rothschild coll.. Material: adult F; whole/good skin.

Reg. no. 1945.58.10. From ; North Island; New Zealand. Date: //. Source: Larkworthy. Material: ; whole/good skin. Mount. Not found 1988.

Reg. no. 1945.58.11. From ; [North Island]; [New Zealand]. Date: //. Source: Larkworthy. Material: adult M; whole/good skin. Ex-mount.

Reg. no. 1949.64.214. From ; South Island; New Zealand. Date: //. Source: G.M. Mathews. Material: adult M; whole/good skin.

Reg. no. 1949.64.215. From ; North Island; New Zealand. Date: //. Source: G.M. Mathews. Material: adult M; whole/good skin.

Reg. no. 1949.64.216. From ; [North Island]; [New Zealand]. Date: //. Source: G.M. Mathews. Material: adult M; whole/good skin.

Reg. no. 1949.64.217. From ; North Island; New Zealand. Date: //. Source: G.M. Mathews. Material: adult F; whole/good skin.

Reg. no. 1953.33.14. From ; North Island; New Zealand. Date: //. Source: C.F. Tebbutt. Material: F; whole/good skin.

Reg. no. 1953.33.15. From ; [North Island]; [New Zealand]. Date: //. Source: C.F. Tebbutt. Material: imm F; whole/good skin. A note in the register states that this bird was one of '49 skins sent to duplicates'.

Reg. no. 1953.33.16. From ; [North Island]; [New Zealand]. Date: //. Source: C.F. Tebbutt. Material: adult M; whole/good skin. A note in the register states that this bird was one of '49 skins sent to duplicates'.

Reg. no. 1955.6.N.4.10. From ; [North Island]; New Zealand. Date: 24/ 8/1895. Source: F.G.M. Boileau; J.H. Gurney coll.. Material: adult M; whole/good skin. The registed implies the locality was South Island.

Reg. no. 1881.1.17.91. From ; ; New Zealand. Date: //. Source: T.C. Eyton; E. Gerrard. Material: adult M; whole/good skeleton. Previously registered as Skeleton Vellum Cat. 1429a. Figured in Eyton, 1867, *Osteologia Avium*, pl. 9f, as *Neomorpha gouldi*. See *Bull. Br. Mus. nat. Hist. (Zool.)* 27 (1974): 1–48.

Reg. no. S1994.10.1. From ; ; [New Zealand]. Date: //. Source: Material: adult F; part/good skeleton. Skull only. This may be the skull from spirit specimen A1973.1.1. See *Bull. Br. Mus. nat. Hist. (Zool.)* 27 (1974): 1–48, especially p. 6.

Reg. no. 1940.12.8.107. From ; ; New Zealand. Date: //1894. Source: Caley Webster coll.; Rothschild coll.. Material: adult M; part/good alc. Dissected specimen. See *Bull. Br. Mus. nat. Hist. (Zool.)* 27 (1974): 1–48. Also made into a flat skin.

Reg. no. 1940.12.8.108. From ; ; New Zealand. Date: //1894. Source: Caley Webster coll.; Rothschild coll.. Material: adult F; whole/good alc. See *Bull. Br. Mus. nat. Hist. (Zool.)* 27 (1974): 1–48.

Reg. no. A1973.1.1. From ; ; New Zealand. Date: //. Source: Hector. Material: adult F; part/good alc. 'Spirit specimen with cranium and upper jaw completely skeletonized, removed and maintained as dried specimen'; see *Bull. Br. Mus. nat. Hist. (Zool.)* 27 (1974): 1–48, especially p. 6. See S1994.10.1 above.

Reg. no. A1973.1.2. From ; ; [New Zealand]. Date: //. Source: Material: adult M; whole/good alc. See *Bull. Br. Mus. nat. Hist. (Zool.)* 27 (1974): 1–48.

Reg. no. A1973.1.3. From ; ; [New Zealand]. Date: //. Source: Material: adult M; whole/good alc. See *Bull. Br. Mus. nat. Hist. (Zool.)* 27 (1974): 1–48.

Reg. no. A1973.1.4. From ; ; [New Zealand]. Date: //. Source: Dr Hector. Material: adult F; part/good alc. The main part of the specimen was not found 1987, although some of the internal organs and the bill were located. See *Bull. Br. Mus. nat. Hist. (Zool.)* 27 (1974): 1–48.

Strepera graculina crissalis

Lord Howe Island Currawong

Endemic to Lord Howe Island, where it is now very rare and declining, for reasons unknown.

Reg. no. 1875.11.8.18. From ; ; Lord Howe Island. Date: 12/ 9/1853. Source: J. MacGillivray; J. Gould. Material: adult M; whole/good skin. Holotype. Sharpe, *Cat. Bds.* 3 (1877): 58–9, pl. 2.

Reg. no. 1919.7.15.286. From Red Point; ; Lord Howe Island. Date: 13/ 3/1914. Source: G.M. Mathews. Material: adult M; whole/good skin.

Reg. no. 1919.7.15.287. From ; ; Lord Howe Island. Date: 13/ 3/1914. Source: G.M. Mathews. Material: adult F; whole/good skin.

Reg. no. 1919.7.15.288. From Little Slope; ; Lord Howe Island. Date: 24/ 3/1914. Source: G.M. Mathews. Material: adult M; whole/good skin.

Reg. no. 1919.7.15.289. From Erskine Valley; ; Lord Howe Island. Date: 25/ 3/1914. Source: G.M. Mathews. Material: adult F; whole/good skin.

Reg. no. 1919.7.15.290. From Mount Lidgbird; ; Lord Howe Island. Date: 1/ 4/1914. Source: G.M. Mathews. Material: adult F; whole/good skin.

Reg. no. 1919.7.15.291. From Mount Lidgbird; ; Lord Howe Island. Date: 1/4/1914. Source: G.M. Mathews. Material: adult M; whole/good skin.

Reg. no. E1941.2.1.151. From ; ; Lord Howe Island. Date: 22/10/1913. Source: R. Bell coll.; Mathews coll.. Material: 3 ; whole/good eggs.

Reg. no. E1941.2.1.152. From ; ; Lord Howe Island. Date: 28/10/1913. Source: R. Bell coll.; Mathews coll.. Material: 2 ; whole/good eggs.

Reg. no. E1941.2.1.153. From ; ; Lord Howe Island. Date: 25/10/1913. Source: R. Bell coll.; Mathews coll.. Material: 3 ; whole/good eggs.

Reg. no. E1941.2.1.155. From ; ; Lord Howe Island. Date: //. Source: R. Bell coll.; Mathews coll.. Material: 3 ; whole/good eggs.

Reg. no. E1941.2.1.154. From ; ; Lord Howe Island. Date: 24/10/1913. Source: R. Bell coll.; Mathews coll.. Material: 1 ; whole/good egg.

Cyanolyca nana

Dwarf Jay

Occurs in humid montane forest in southern Mexico, where it is severely threatened by habitat destruction.

Reg. no. 1857.6.10.13. From Cordova; ; Mexico. Date: //. Source: Verreaux. Material: adult ; whole/good skin.

Reg. no. 1885.2.21.5. From ; ; Mexico. Date: //. Source: de Oca; Sclater coll.. Material: adult ; whole/good skin.

Reg. no. 1887.3.9.66. From Jalapa; ; Mexico. Date: //. Source: de Oca; Salvin–Godman coll.. Material: adult ; whole/good skin.

Reg. no. 1887.3.9.67. From nr. Mexico City; ; Mexico. Date: //1867. Source: Salvin–Godman coll.. Material: adult ; whole/good skin.

Reg. no. 1890.5.30.295. From Cape de Perote; ; Mexico. Date: /11/1889. Source: M. Trujillo; Salvin–Godman coll.. Material: adult M; whole/good skin.

Reg. no. 1890.5.30.296. From Cape de Perote, Vera Cruz; ; Mexico. Date: / 7/1888. Source: Salvin–Godman coll.. Material: adult M; whole/good skin.

Reg. no. 1890.5.30.297. From Cape de Perote, Vera Cruz; ; Mexico. Date: / 7/1888. Source: Salvin–Godman coll.. Material: adult M; whole/good skin.

Reg. no. 1890.5.30.298. From Las Vigas, Canton de Jalapa; ; Mexico. Date: //1888. Source: F. Ferrari Perez; Salvin–Godman coll.. Material: adult M; whole/good skin.

Reg. no. 1890.5.30.299. From San Juan a Calton; ; Mexico. Date: / 6/1888. Source: Salvin–Godman coll.. Material: adult M; whole/good skin.

Cyanolyca mirabilis

White-throated Jay

Occurs in humid mountain forests in Guerrero and Oaxaca, Mexico, where it is threatened by habitat destruction.

Reg. no. 1971.7.2. From Omilteme; Guerrero; Mexico. Date: 29/ 5/1949. Source: W.W. Brown; Delaware Museum. Material: adult F; whole/good skin. Iris brown. Ovaries minute.

Corvus kubaryi

Mariana Crow

Confined to woodland on the islands of Guam and Rota in the Marianas Islands, where it is believed to be declining.

Reg. no. 1896.7.16.21. From ; Guam; Marianas Islands. Date: 21/12/1894. Source: Rothschild coll.. Material: ; whole/good skin.

Reg. no. 1898.4.29.25. From ; Guam; Marianas Islands. Date: 19/ 7/1895. Source: Rothschild coll.. Material: ; whole/good skin.

Reg. no. 1898.4.29.26. From ; Guam; Marianas Islands. Date: 10/ 9/1895. Source: Rothschild coll.. Material: F; whole/good skin.

Corvus tropicus/C. hawaiiensis

Hawaiian Crow

Endemic to the island of Hawaii, where the present wild population is believed to be 12 birds. A further 11 are held in captivity, but these are now suffering from inbreeding. Captive breeding is taking place with the intention of releasing birds back into the wild. This species should be called *C. hawaiiensis* Peale, as *C. tropicus* properly refers to a drongo.

Reg. no. 1897.10.28.1. From Kona; Hawaii; Hawaiian Islands. Date: / 3/1896. Source: R.C.L. Perkins; Royal Society & British Association. Material: adult F; whole/good skin.

Reg. no. 1897.10.28.2. From Kona; Hawaii; Hawaiian Islands. Date: / 3/1896. Source: R.C.L. Perkins; Royal Society & British Association. Material: adult M; whole/good skin.

Reg. no. 1925.2.16.176. From ; Hawaii; Hawaiian Islands. Date: / 6/1888. Source: S.B. Wilson. Material: M; whole/good skin.

Reg. no. 1925.2.16.177. From ; Hawaii; Hawaiian Islands. Date: //. Source: S.B. Wilson. Material: ; whole/good skin.

Reg. no. 1925.2.16.178. From ; Hawaii; Hawaiian Islands. Date: //. Source: S.B. Wilson. Material: ; whole/good skin.

SPECIES PRESUMED EXTINCT PRIOR TO 1600, OF WHICH THERE ARE SPECIMENS AT TRING.

Ratite species

There are a number of ratite bones in the public galleries at Tring and some feathers from moas *Dinornis* in the bird collections. Most of the Museum's ratite remains are held in the Palaeontology Department.

Aepyornis maximus

Reg. no. . From ; ; [Madagascar]. Date: //. Source: Material: ; part/good skeleton. Casts of leg bones in Tring gallery.

Aepyornis hildebrandti

Reg. no. . From ; ; [Madagascar]. Date: //. Source: Material: ; whole/good skeleton. In Tring gallery.

Megalapteryx didinus

Reg. no. . From ; ; [New Zealand]. Date: //. Source: Material: ; whole/good skeleton. In Tring gallery.

Reg. no. . From ; ; [New Zealand]. Date: //. Source: Material: ; whole/good skeleton. In Tring gallery.

Emeus huttonii

Reg. no. . From ; ; [New Zealand]. Date: //. Source: Material: ; whole/good skeleton. In Tring gallery.

Euryapteryx gravis

Reg. no. . From ; ; [New Zealand]. Date: //. Source: Material: ; whole/good skeleton. In Tring gallery.

***Dinornis* sp.**

Reg. no. 1896.5.28.1–60. From Lake Wakatipu, nr. Queenstown, Otago; [South Island]; New Zealand. Date: /11/1874. Source: Taylor White coll.. Material: ; part/good skin. Feathers only. From a cave on the shores of the lake. There are 60 feathers under this entry; each one is registered separately from 1896.5.28.1 to 1896.5.28.60. Further specimens are held in the Palaeontology Department. See *Trans. Proc. N.Z. Inst.* 8 (1876): 97–9, 101–2; 18 (1886): 83–84, pls. 2–3.

Reg. no. 1988.16.1. From nr. Queenstown, Otago; [South Island]; New Zealand. Date: /11/1874. Source: Taylor White coll.. Material: ; part/good skin. Feathers only. Further specimens are held in the Palaeontology Department.

Dinornis maximus

Reg. no. . From ; ; [New Zealand]. Date: //. Source: Material: ; whole/good skeleton. In Tring gallery.

Reg. no. . From ; ; [New Zealand]. Date: //. Source: Material: ; part/good skeleton. Cast of incomplete skeleton in Tring gallery. Skull missing.

Coracopsis nigra barklyi
Seychelles Black Parrot

Further specimens, continued from page 150.

Reg. no. 1880.6.3.5. From ; Praslin Island; Seychelles. Date: //1879. Source: H.L. Warry; Tristram coll.. Material: adult ; whole/good skin.

Reg. no. 1890.10.10.36–37. From ; ; . Date: //. Source: Kirk; Shelley coll.. Material: 2; skins. Listed in the register as *barklyi* under reg. nos. 1890.10.10.36–37 and on the labels as 1890.10.10.37–38. The specimens come from Anjouan and Grand Comoro. They do not belong to this race.

Reg. no. 1889.1.20.640. From ; ; Seychelles. Date: //1880. Source: Tristram coll.; Tweeddale coll.. Material: adult ; whole/good skin.

Reg. no. 1906.12.21.189. From ; Praslin Island; Seychelles. Date: 3/ 4/1906. Source: M.J. Nicoll; Earl of Crawford, voyage of R.Y.S. Valhalla. Material: imm F; whole/good skin. Iris black. Bill brown. Tarsi and toes black, soles dull yellow.

Reg. no. 1912.1.12.1. From ; Praslin Island; Seychelles. Date: 12/ 1/1912. Source: H.P. Thomasset. Material: adult ; whole/good skin.

Reg. no. 1912.1.12.2. From ; Praslin Island; Seychelles. Date: 12/ 1/1912. Source: H.P. Thomasset. Material: adult ; whole/good skin.

Reg. no. 1927.12.18.129. From ; Praslin Island; Seychelles. Date: 8/ 2/1888. Source: J.J. Lister. Material: adult ; whole/good skin. Possible female.

Reg. no. 1946.75.18. From ; Praslin Island; Seychelles. Date: / 4/1940. Source: Sapsworth & Goodfellow; A.D. Sapsworth. Material: adult F; whole/good skin. Iris brown. Feet black. Bill black at base, greyish at tip.

Reg. no. 1946.75.19. From ; Praslin Island; Seychelles. Date: / 4/1940. Source: Sapsworth & Goodfellow; A.D. Sapsworth. Material: adult F; whole/good skin. Iris brown. Feet black. Bill black at base, greyish at tip. Cere yellowish.

REFERENCES

Blandamer, J.S. & Burton, P.J.K. 1979. Anatomical specimens of birds in the collections of the British Museum (Natural History). *Bull. Br. Mus. nat. Hist. (Zool.)*, 34: 125–180.

Brodkorb, P. 1963–78. Catalogue of fossil birds. *Bull. Fla St. Mus. biol. Sci.* 7 (1963): 179–293; 8 (1964): 195–335; 11 (1967): 99–220; 15 (1971): 163–266; 23 (1978): 139–228.

Collar, N.J. & Andrew, P. 1988a. *Birds to Watch.* ICBP technical publication no. 8. Cambridge.

Collar, N. & Andrew, P. 1988b. A systematic list of the world's threatened bird species. Pp. 211–40 in Mountfort (1988).

Collar, N.J., Gonzaga, L.P., Krabbe, N., Madroño Nieto, A., Naranjo, L.G., Parker, T.A. III & Wege, D.C. 1992. *Threatened Birds of the Americas*. Cambridge.

Collar, N.J. & Stuart, S.N. 1985. *Threatened Birds of Africa and Related Islands.* Cambridge.

Diamond, A.W. (ed.) 1987. *Studies of Mascarene Island Birds.* Cambridge.

Fuller, E. 1987. *Extinct Birds.* London.

Greenway, J.C. 1967. *Extinct and Vanishing Birds of the World.* 2nd edition. New York.

King, W.B. 1978–1979. *Red Data Book. Vol. 2, Aves.* 2 vols. Morges, Switzerland.

Knox, A.G. 1993. Richard Meinertzhagen – a case of fraud examined. *Ibis* 135: 320–325.

Knox, A.G. & Walters, M. 1992. Under the skin: the bird collections of the Natural History Museum. *Bull. Brit. Orn. Club Centenary Suppl.* 112A: 169–190.

Morony, J.J., Bock, W.J. & Farrand, J. 1975. *Reference List of the Birds of the World.* New York.

Mountfort, G. 1988. *Rare Birds of the World.* London.

Oates, E.W. (& Ogilvie-Grant, W.R.) 1901–1912. *Catalogue of the Collection of Birds' Eggs in the British Museum (Natural History).* 5 vols. London.

Orenstein, R. 1985. Extinct birds. Pp. 196–9 in Campbell, B. & Lack, E. (eds.), *A Dictionary of Birds.* Calton.

Peters, J.L. & others. 1931–1987. *Check-list of Birds of the World.* 16 vols. Cambridge, Mass. Reference to Peters (1934–1987) excludes the first edition of vol. 1 (1931).

Rothschild, W. 1907. *Extinct Birds.* London.

Sharpe, R.B. 1906. Birds. Pp. 79–515 in, *The History of the Collections Contained in the Natural History Departments of the British Museum, volume 2.* London.

Sharpe, R.B. & others. 1874–1898. *Catalogue of Birds in the British Museum.* 27 vols. London.

Sibley, C.G. & Monroe, B.L., Jr. 1990. *Distribution and Taxonomy of Birds of the World*. New Haven.

Stattersfield, A. 1988. A systematic list of birds presumed to have become extinct since 1600. Pp. 241–4 in Mountfort (1988).

Warren, R.L.M. & Harrison, C.J.O. 1966–1973. *Type–specimens of Birds in the British Museum (Natural History)*. 3 vols. London.

INDEX

of genera, species and subspecies

abbotti, Dryolimnas, 12, 106
abbotti, Sula, 9, 41
Aburria, 10, 75
Accipiter, 10, 57
Acestrura, 17, 177
Acrocephalus, 5, 20, 214
acutirostris, Heteralocha, 5, 24, 271
adalberti, Aquila, 10, 60
aedon, Troglodytes, 19, 195
Aegotheles, 16
Aepyornis, 24, 277
aequatorialis, Penelope, 75
affinis, Hemignathus, 22, 248
afra, Nectarinia, 21
Agelastes, 11, 99
Alauda, 190
albaria, Ninox, 16, 170
albatrus, Diomedea, 8, 35
albicaudatus, Buteo, 67
albifacies, Sceloglaux, 16, 170
albifrons, Sterna, 13
albipectus, Pyrrhura, 15
albipennis, Penelope, 10, 74
albiventris, Dacnis, 22, 235
albogularis, Zosterops, 21, 226
albus, Porphyrio, 12
Alca, 134
aldabranus, Nesillas, 20, 216
Alectroenas, 14
Alethe, 19, 198
alleni, Grallaria, 18
Alopochen, 9
alphonsi, Circus, 10
alpinus, Anairetes, 18
altera, Coracina, 19, 192
altirostris, Moupinia, 20, 213
amabilis, Lorius, 14
amaka, Zonotrichia, 22
amaura, Speotyto, 16, 171
Amaurocichla, 20, 217
Amaurolimnas, 12, 107
Amazona, 15, 157
americana, Grus, 11, 100
americanus, Siphonorhis, 17, 172
amethystina, Phapitreron, 14
Ammodramus, 22, 233
Amphispiza, 22
amsterdamensis, Diomedea, 8
Anairetes, 18
Anas, 9,50
anatum, Falco, 10, 65
andersoni, Megapodius, 10, 73
andinus, Podiceps, 8
angustipluma, Chaetoptila, 21
anna, Ciridops, 5, 23, 256
Anodorhynchus, 15, 152
anomalus, Eleothreptus, 17, 172
antarcticus, Rallus, 11, 103
Anthornis, 21, 233
Anthus, 18
Apalopteron, 21, 230
Aphanapteryx, 12
Aphanolimnas, 12
apicalis, Loriculus, 151
apicalis, Moho, 21, 232
Aplonis, 5, 23, 264
Apteryx, 273
Aquila, 10, 60
aquilonis, Acrocephalus, 20
Ara, 15, 153
Aramides, 12, 107
ararauna, Ara, 153
Aratinga, 15
arausiaca, Amazona, 16, 161

arfakiana, Melanocharis, 21
Argusianus, 11, 99
aricomae, Geositta, 6, 18, 184
assimilis, Oriolus, 24, 269
astrolabii, Acrocephalus, 20
astur, Eutriorchis, 10, 57
aterrima, Pterodroma, 8
Athene, 16, 171
Atlantisia, 11, 105
Atlapetes, 22
atra, Chalcopsitta, 14
atra, Pyriglena, 18, 185
Atrichornis, 18, 189
atrifrons, Odontophorus, 80
aucklandica, Coenocorypha, 13, 132
auratus, Colaptes, 17, 181
aureonucha, Ploceus, 23
auriceps, Cyanoramphus, 14, 148
australis, Mergus, 5, 10, 50
axillaris, Pterodroma, 9, 39
Aythya, 10, 49

bachmanii, Vermivora, 22, 236
baileyi, Ammodramus, 22, 234
bailleui, Psittirostra, 23, 252
bairdi, Icterus, 23, 257
bairdi, Oreomystis, 243
bairdii, Campephilus, 17, 182
bannermani, Tauraco, 16, 167
barbatus, Dendrortyx, 11, 78
barklyi, Coracopsis, 15, 150, 278
barrierensis, Coenocorypha, 13
baudinianus, Dromaius, 32
Bebrornis, 20, 215
beli, Gennaeus, 91
beli, Lophura, 91
belli, Amphispiza, 22
bensoni, Lophopsittacus, 15
berlepschi, Acestrura, 17
berlepschi, Dacnis, 22, 236
bewickii, Thryomanes, 19, 195
bipunctatus, Argusianus, 11, 99
bishopi, Catharopeza, 22, 239
bishopi, Moho, 21, 231
bitorquatus, Rhinoptilus, 13, 126
blewitti, Athene, 16, 171
blumenbachii, Crax, 11, 78
bocagei, Lampribis, 9
bocagii, Amaurocichla, 20, 217
Bolborhynchus, 15
bombus, Acestrura, 17, 177
bonana, Icterus, 23, 257
bonariensis, Molothrus, 238, 257
bonasia, Aphanapteryx, 12
Borbonibis, 9
borbonica, Palaeornis, 152
borealis, Numenius, 5, 13, 25, 128
botelensis, Otus, 16
Bowdleria, 20, 217
boyciana, Ciconia, 9, 42
braccatus, Moho, 21, 230
bracei, Chlorostilbon, 17
brachyurus, Ramphocinclus, 19, 197
Branta, 47
brasiliensis, Amazona, 16, 158
brevicauda, Thryomanes, 19, 195
brevipes, Poliolimnas, 12, 111
broadbenti, Dasyornis, 20, 217
browni, Sterna, 13
burnabyi, Megapodius, 10, 74
Buteo, 67

Cabalus, 11, 104
caboti, Tragopan, 11, 86
caffer, Acrocephalus, 20, 214
cahow, Pterodroma, 8, 37
Calandrella, 18, 190
caledonicus, Nycticorax, 9, 41
Calidris, 13
californianus, Gymnogyps, 10, 51
Callaeas, 24, 270
Caloenas, 14, 143
camelus, Struthio, 8, 27, 29
Campephilus, 17, 182

Camptorhynchus, 10, 50
canadensis, Grus, 100
caninde, Ara, 15, 153
capensis, Turnagra, 21, 223
capnodes, Otus, 16, 169
Carduelis, 23, 258
caribbaea, Pterodroma, 8, 37
carinatum, Electron, 17, 179
carolinensis, Conuropsis, 5, 15, 154
Carpodacus, 23, 259
Carpophaga, 145
caryophyllacea, Rhodonessa, 9, 47
cassidix, Meliphaga, 21, 229
cassini, Falco, 68
Catharopeza, 22, 239
Catreus, 11, 92
cebuensis, Copsychus, 19, 200
cebuensis, Coracina, 18, 191
celaenops, Turdus, 20
Ceratotriccus, 18
cerverai, Ferminia, 19, 195
cervina, Rhipidura, 21, 222
Cettia, 20
chacoensis, Anthus, 18
Chaetoptila, 21
Chalcopsitta, 14
Charadrius, 13, 127
Charmosyna, 14
chathamensis, Cyanoramphus, 148
chathamensis, Haematopus, 12, 123
chathamensis, Hemiphaga, 14, 145
chathamica, Carpophaga, 145
Chaunoproctus, 23, 259
cherriei, Synallaxis, 18
chilensis, Colymbus, 34
chloroptera, Aratinga, 15
chloropus, Gallinula, 12, 112
Chlorostilbon, 17
choloensis, Alethe, 19, 198
Chondrohierax, 10, 53
Choriotis, 12, 115
chrysonotus, Loriculus, 15, 150
chrysopterygius, Psephotus, 146
chthonia, Grallaria, 18
Cichlherminia, 20, 201
ciconia, Ciconia, 9, 42
Ciconia, 9, 42
Cinclocerthia, 19, 196
Cinclodes, 185
cinclus, Cinclus, 19, 194
Cinclus, 19, 194
cinerea, Callaeas, 24, 270
cinereus, Poliolimnas, 12, 111
cinnamomina, Halcyon, 17, 179
Circus, 10
Ciridops, 5, 23, 256
clamosus, Atrichornis, 18, 189
clementeae, Amphispiza, 22
coccinea, Loxops, 22, 243
Coenocorypha, 13, 132
coerulescens, Coracina, 19, 192
Colaptes, 17, 181
colchicus, Phasianus, 93
Colinus, 11, 79
Columba, 13, 135
Colymbus, 34
comeri, Porphyriornis, 112
commersoni, Otus, 16
concolor, Amaurolimnas, 12, 107
concolor, Neospiza, 23, 258
consobrinus, Pipilo, 22
Conuropsis, 5, 15, 154
cookii, Cyanoramphus, 14, 147
cookii, Pterodroma, 9, 39
cooperi, Calidris, 13
coppingeri, Streptopelia, 13, 137
Copsychus, 19, 199
Coracina, 18, 191
Coracopsis, 15, 150, 278
cornutus, Eunymphicus, 15, 149
corvina, Aplonis, 5, 23, 264
Corvus, 24, 276
Coturnix, 11, 80
Coua, 16, 168

couesi, Anas, 9
courseni, Synallaxis, 18
crassirostris, Nycticorax, 9, 41
crassirostris, Turnagra, 224
Crax, 11, 76
crissalis, Strepera, 24, 274
cristata, Tadorna, 9, 46
crossoptilon, Crossoptilon, 92
Crossoptilon, 11, 90
Crypturellus, 8
cubensis, Ara, 154
cucullata, Carduelis, 23, 258
cucullatus, Raphus, 13, 134
cuculoides, Falco, 57
cunicularia, Speotyto, 16, 171
cupido, Tympanuchus, 11, 78
cuvieri, Dryolimnas, 12, 106
Cyanolyca, 24, 275
cyanophrys, Eupherusa, 17
Cyanopsitta, 15, 153
Cyanoramphus, 14, 147
Cygnus, 9

Dacnis, 22, 235
dacotiae, Saxicola, 19, 201
Dasyornis, 20, 217
debooyi, Nesotrochis, 12
delalandei, Coua, 16, 168
Dendroica, 22, 238
Dendrortyx, 11, 78
derbianus, Oreophasis, 11, 75
derogata, Diomedea, 35
diadema, Charmosyna, 14
Dicaeum, 21, 224
didinus, Megalapteryx, 24, 277
dieffenbachii, Nesolimnas, 11, 104
diemenensis, Dromaius, 8, 32
diemenianus, Dromaius, 8, 32
Dinornis, 24, 277
Diomedea, 8, 35
diphone, Cettia, 20
dohrnii, Glaucis, 17, 173
domicellus, Lorius, 145
Drepanis, 23, 256
Dromaius, 8, 32
Dryocopus, 17, 181
Dryolimnas, 12, 106
Ducula, 14, 144
dufresniana, Amazona, 16, 159
dukhunensis, Motacilla, 102
dulcis, Neomorphus, 16, 168
Dysmorodrepanis, 23

echo, Psittacula, 15, 152
Ectopistes, 5, 13, 138
Electron, 17, 179
elegans, Platycercus, 147
elegans, Rallus,11, 102
Eleothreptus, 17, 172
elisabeth, Myadestes, 19
ellioti, Syrmaticus, 11, 97
ellisi, Prosobonia, 132
ellisianus, Hemignathus, 22, 246
elpenor, Atlantisia, 11
Emeus, 24, 277
Empidonax, 18
Eos, 14
eques, Psittacula, 15, 152
eremita, Geronticus, 9, 42
Eriocnemis, 17, 174
Erithacus, 19
erythronotos, Myrmotherula, 18, 185
erythrophthalmus, Pipilo, 22
erythroptera, Ortalis, 10, 74
erythrotis, Cyanoramphus, 14, 148
euleri, Empidonax, 18
Eulidia, 6, 177
Eunymphicus, 15, 149
Eupherusa, 17, 173
Eurapteryx, 24, 277
Eutrichomyias, 21
Eutriorchis, 10, 57
everetti, Zosterops, 21, 224
excelsior, Geositta, 6, 18, 184

exsul, Psittacula, 15
exsul, Salpinctes, 19

Falco, 10, 57, 64
familiare, Apalopteron, 21, 230
familiaris, Acrocephalus, 5, 20, 214
fanovanae, Newtonia, 20
Ferminia, 19, 195
ferreorostris, Chaunoproctus, 23, 259
ferruginea, Gallicolumba, 14, 143
ferrugineifrons, Bolborhynchus, 15
finschi, Haematopus, 124
flammea, Loxops, 22, 241
flammea, Paroreomyza, 241, 243
flavicans, Foudia, 23, 263
flaviceps, Atlapetes, 22
flaviceps, Psittirostra, 23, 254
flavigula, Serinus, 23
flaviventris, Pezoporus, 15, 149
flosculus, Loriculus, 15, 151
forbesi, Cyanoramphus, 14, 148
Foudia, 23, 262
francesii, Accipiter, 10, 57
Francolinus, 11
freethii, Himatione, 23, 255
Fregilupus, 24, 268
frontalis, Phapitreron, 14
Fulica, 12
fuliginosa, Rhipidura, 21, 222
funerea, Drepanis, 23, 256
furcatus, Ceratotriccus, 18
fusca, Aplonis, 23, 264

galeata, Ducula, 14, 144
galeatus, Dryocopus, 17, 181
Gallicolumba, 14, 143
Gallinula, 12, 112
gambieri, Halcyon, 17
garleppi, Pterocnemia, 32
garretti, Acrocephalus, 20, 214
Gennaeus, 91
Geocichla, 224
geoffroyi, Neomorphus, 16, 168
georgica, Anas, 9
Geositta, 6, 18, 184
Geronticus, 9, 42
Gerygone, 20, 218
gibberifrons, Anas, 9
gigantea, Grallaria, 18, 185
gigas, Podilymbus, 8, 33
Glaucis, 17, 173
glaucogularis, Ara, 153
glaucus, Anodorhynchus, 15, 152
globulosa, Crax, 11, 77
godini, Eriocnemis, 17, 176
godmanae, Columba, 13, 135
golandi, Ploceus, 23, 261
goodfellowi, Eos, 14
gouldi, Neomorpha, 274
gracilipes, Amazona, 15, 157
graculina, Strepera, 24, 274
Grallaria, 18, 185
graminea, Zonotrichia, 22
grandis, Loxigilla, 22
gravis, Eurapteryx, 24, 277
graysoni, Mimodes, 19, 196
graysoni, Zenaida, 13, 142
greyii, Strigops, 164
Grus, 11, 100
Gryphus, 173
guadeloupensis, Speotyto, 16
guadeloupensis, Troglodytes, 19, 195
guildingii, Amazona, 16, 161
gundlachii, Accipiter, 10
gutturalis, Aramides, 12, 107
gutturalis, Cinclocerthia, 19, 196
Gymnocrex, 12, 108
Gymnogyps, 10, 51

habroptilus, Strigops, 16, 164
haematonotus, Psephotus, 147
Haematopus, 12, 123
hahashima, Apalopteron, 230
Halcyon, 17, 179

Haliaeetus, 10, 53
hamiltoni, Gennaeus, 91
hamiltoni, Lophura, 91
hanapepe, Hemignathus, 22, 247
hasitata, Pterodroma, 8, 37
hawaiiensis, Corvus, 276
heliaca, Aquila, 10, 60
heliodor, Acestrura, 179
helleri, Turdus, 20, 204
Hemignathus, 22, 245
Hemiphaga, 14, 144
Heteralocha, 5, 24, 271
hildebrandti, Aepyornis, 24, 277
himantopus, Himantopus, 124
Himantopus, 12, 124
Himatione, 23, 255
hulliana, Aplonis, 23, 267
huttonii, Emeus, 24, 277
hylodroma, Grallaria, 185
Hylonympha, 17, 173
Hylorchilus, 19, 195
hyperythrus, Odontophorus, 80
Hypsipetes, 19

ibadanensis, Malimbus, 23, 261
icterotis, Ognorhynchus, 15, 154
Icterus, 23, 257
impennis, Alca, 134
impennis, Pinguinus, 13, 132
imperialis, Amazona, 16, 163
imperialis, Campephilus, 17, 183
indicus, Struthio, 8, 31
innominatus, Strigops, 165
innotata, Aythya, 10, 49
inornata, Columba, 13
insulans, Colinus, 11
insularis, Gerygone, 20, 218
intactus, Gymnocrex, 12, 108
intermedia, Psittacula, 15, 152
Iodopleura, 18, 186
iredalei, Coenocorypha, 13, 132
ireneae, Otus, 16, 169
italica, Perdix, 11, 80

Jacamaralcyon, 17, 180
jacutinga, Aburria, 10, 75
javensis, Dryocopus, 17, 182
jefferyi, Pithecophaga, 10, 59
johnstonei, Empidonax, 18
jouyi, Columba, 13, 135
jubatus, Rhynochetos, 12, 113

kirtlandii, Dendroica, 22, 239
komadori, Erithacus, 19
kona, Psittirostra, 5, 23, 254
kubaryi, Corvus, 24, 276

labradorius, Camptorhynchus, 10, 50
lafresnayanus, Tricholimnas, 11, 105
Lalage, 19, 192
Lampribis, 9
lanaiensis, Hemignathus, 22, 245
lanaiensis, Phaeornis, 20, 203
Lanius, 19, 194
laticincta, Platysteira, 20, 219
latipes, Borbonibis, 9
leari, Anodorhynchus, 15, 153
leguati, Aphanapteryx, 12
leguati, Necropsar, 23
Lepidopyga, 17
Leptasthenura, 18
Leptotila, 13, 142
leucocephalus, Haliaeetus, 10, 54
leucocephalus, Himantopus, 124
leucogeranus, Grus, 11, 101
leucomelana, Lophura, 91
Leucopeza, 22, 240
leucophrys, Thryomanes, 19, 195
Leucopsar, 24, 269
leucoptera, Prosobonia, 13, 132
leucopteryx, Icterus, 23, 257
leucopyga, Lalage, 19, 192
leucopygia, Iodopleura, 186
levipes, Rallus, 11, 102

lherminieri, Cichlherminia, 20, 201
lhuysii, Lophophorus, 11, 89
lilliae, Lepidopyga, 17
lineata, Lophura, 91
lineatus, Gennaeus, 91
litoralis, Dasyornis, 20, 217
litsitsirupa, Geocichla, 224
longipes, Xenicus, 18, 187
longirostris, Acrocephalus, 20
longirostris, Haematopus, 123
Lophophorus, 11, 89
Lophopsittacus, 15, 150
Lophura, 91
Loriculus, 15, 150
Lorius, 14, 145
Loxigilla, 22
Loxops, 22, 240
lucidus, Hemignathus, 22, 247
ludovicianus, Conuropsis, 15, 157
ludovicianus, Lanius, 19
luscinia, Acrocephalus, 20
luteirostris, Zosterops, 21, 225
lutosus, Polyborus, 10, 64
lyalli, Xenicus, 18, 189

macgillivrayi, Rhipidura, 222
macquariensis, Rallus, 11, 104
macrocephala, Petroica, 218
macrocerca, Hylonympha, 17, 173
macrodactyla, Oceanodroma, 9, 40
macropterus, Vanellus, 13, 126
macroura, Zenaida, 13, 142
maculata, Loxops, 22, 241
maculata, Paroreomyza, 241
madagascariensis, Foudia, 264
madeira, Pterodroma, 9, 38
maderensis, Columba, 13
magentae, Pterodroma, 8, 37
malherbi, Cyanoramphus, 14, 148
Malimbus, 23, 261
mana, Oreomystis, 243
mantchuricum, Crossoptilon, 11, 90
mantelli, Notornis, 12, 113
mareensis, Turdus, 20, 205
margarettae, Phaethornis, 17
maritimus, Ammodramus, 22, 233
martinicensis, Troglodytes, 19, 196
mascarinus, Mascarinus, 15
Mascarinus, 15
maugei, Aratinga, 15
mauritianus, Alopochen, 9
mauritianus, Lophopsittacus, 15, 150
mauritianus, Nycticorax, 9
mavornata, Aplonis, 23, 264
maximus, Aepyornis, 24, 277
maximus, Dinornis, 24, 277
mayeri, Columba, 13, 135
mayottensis, Zosterops, 21, 227
mcgregori, Carpodacus, 23, 259
meadewaldoi, Haematopus, 12, 124
mearnsi, Lanius, 19
meeki, Microgoura, 14, 143
megacephalus, Nycticorax, 9
Megalapteryx, 24, 277
Megapodius, 10, 73
Megaxenops, 18
melanocephala, Anthornis, 21, 233
melanocephala, Sylvia, 20, 216
melanocephalus, Tragopan, 11, 81
Melanocharis, 21
melanops, Meliphaga, 229
melanura, Anthornis, 21, 233
melas, Himantopus, 126
meleagrides, Agelastes, 11, 99
Meliphaga, 21, 229
melodia, Zonotrichia, 22
mendozae, Pomarea, 21, 221
mercierii, Ptilinopus, 14, 143
Mergus, 5, 10, 50
merula, Turdus, 206
mesoleucus, Troglodytes, 19, 196
mexicanus, Carpodacus, 23, 259
Microgoura, 14, 143
migratorius, Ectopistes, 5, 13, 138

milleri, Grallaria, 18, 186
millsi, Pennula, 5, 12, 110
Mimodes, 19, 196
minor, Dromaius, 8
mirabilis, Cyanolyca, 24, 276
mirus, Chondrohierax, 10, 53
mitu, Crax, 11, 76
miyakoensis, Halcyon, 17
modesta, Zosterops, 21, 228
modestus, Cabalus, 11, 104
Moho, 21, 230
mollis, Pterodroma, 9, 38
Molothrus, 238, 257
monasa, Aphanolimnas, 12
montana, Loxops, 22, 242
montana, Paroreomyza, 242
monticola, Hypsipetes, 19
moquini, Haematopus, 12, 124
Motacilla, 102
Moupinia, 20, 213
munroi, Dysmorodrepanis, 23
munroi, Hemignathus, 248
murielae, Saxicola, 19, 201
murina, Pyrrhula, 23, 260
murivora, Ninox, 16
murivora, Tyto, 16
musae, Acrocephalus, 20, 215
Myadestes, 19
myadestinus, Phaeornis, 20, 202
Myioborus, 22, 240
Myrmotherula, 18, 185
Myrtis, 177

naevioides, Aquila, 63
namuli, Alethe, 199
nana, Cyanolyca, 24, 275
nana, Rhea, 8, 32
Necropsar, 23
Necropsittacus, 15
Nectarinia, 21
Nemosia, 22
Neomorpha, 274
Neomorphus, 16, 168
Neospiza, 23, 258
Nesillas, 20, 216
nesiotis, Porphyriornis, 12, 112
Nesoclopeus, 5, 12, 107
Nesolimnas, 11, 104
Nesophylax, 12, 110
Nesotrochis, 12
Nestor, 14, 146
newtoni, Fulica, 12
newtoni, Lanius, 19, 194
newtoni, Paroreomyza, 243
Newtonia, 20
nicefori, Thryothorus, 19
niceforoi, Anas, 9
nicobarica, Caloenas, 14, 143
niger, Copsychus, 19, 200
niger, Haliaeetus, 10, 56
niger, Nesophylax, 12, 110
nigra, Coracopsis, 15, 150, 278
nigra, Pomarea, 21, 221
nigrescens, Ammodramus, 22, 233
nigriceps, Choriotis, 12, 115
nigricollis, Podiceps, 8
nigrivestis, Eriocnemis,17, 174
nigrofumosus, Cinclodes, 185
Ninox, 16, 170
nippon, Nipponia, 9, 45
Nipponia, 9, 45
nitidissima, Alectroenas, 14
nobilis, Moho, 21, 232
noguchii, Sapheopipo, 17, 184
norfolciensis, Gallicolumba, 14
norissae, Sylvia, 20, 216
Notornis, 12, 113
novaehollandiae, Dromaius, 8, 32
novaeseelandiae, Charadrius, 13, 127
novaeseelandiae, Hemiphaga, 14, 144
novaeseelandiae, Ninox, 16, 170
novaezelandiae, Coturnix, 11, 80
novaezelandiae, Cyanoramphus, 14, 147
novaezelandiae, Himantopus, 12, 124

nukuhivae, Pomarea, 21, 221
Numenius, 5, 13, 25, 128
nycthemera, Lophura, 91
Nycticorax, 9, 41

oahensis, Phaeornis, 20, 203
obscurus, Hemignathus, 22, 245
obscurus, Phaeornis, 20, 202
obsoletus, Salpinctes, 19
occipitalis, Podiceps, 34
oceanica, Ducula, 14
Oceanodroma, 9, 40
ochracea, Loxops, 22, 243
ochraceiventris, Leptotila, 13, 143
ochropectus, Francolinus, 11
Odontophorus, 11, 79
Ognorhynchus, 15, 154
olivacea, Lampribis, 9
olympicus, Cinclus, 19, 194
Ophrysia, 11, 81
Oreomystis, 243
Oreophasis, 11, 75
orii, Parus, 21
orii, Troglodytes, 19
Oriolus, 24, 269
Ortalis, 10, 74
ostralegus, Haematopus, 12, 123
Otus, 16, 169

pachyrhyncha, Rhynchopsitta, 15
pacifica, Drepanis, 23, 256
pacificus, Pareudiastes, 5, 12, 112
pacificus, Rallus, 11, 104
Palaeornis, 152
pallidus, Dicaeum, 21, 224
palmeri, Phaeornis, 20, 203
palmeri, Porzanula, 5, 12, 108
palmeri, Psittirostra, 23, 253
palumbus, Columba, 13
palustris, Quiscalus, 23, 257
Pareudiastes, 5, 12, 112
pariae, Myioborus, 22, 240
parkinsoni, Procellaria, 9, 40
parnaguae, Megaxenops, 18
Paroreomyza, 241
Parus, 21
pelagicus, Haliaeetus, 10, 56
pelewensis, Caloenas, 14, 143
pelzelni, Aplonis, 23
pelzelni, Tachybaptus, 33
Penelope, 10, 74
pennata, Pterocnemia, 8, 32
Pennula, 5, 12, 110
perdix, Perdix, 11, 80
Perdix, 11, 80
peregrinus, Falco, 10, 65
pernambucensis, Tinamus, 8
perspicax, Penelope, 10, 74
perspicillatus, Phalacrocorax, 9, 41
petechia, Dendroica, 22, 238
Petroica, 20, 218
Pezophaps, 13, 134
Pezoporus, 15, 149
phaeopygia, Pterodroma, 9, 38
Phaeornis, 20, 202
Phaethornis, 17
Phalacrocorax, 9, 41
Phapitreron, 14
Phasianus, 93
philippensis, Loriculus, 15, 150
philippensis, Rallus, 11, 104
picturata, Streptopelia, 13, 137
pileata, Leptasthenura, 18
Pinguinus, 13, 132
pipile, Aburria, 10
Pipilo, 22
pipra, Iodopleura, 18, 186
Pithecophaga, 10, 59
Platycercus, 147
Platysteira, 20, 219
Ploceus, 23, 261
plumbeiventris, Gymnocrex, 108
podarces, Atlantisia, 11, 105
podiceps, Podilymbus, 33

Podiceps, 8, 34
Podilymbus, 8, 33
poeciloptera, Nesoclopeus, 5, 12, 107
poliocephalus, Turdus, 20, 204
poliocerca, Eupherusa, 17, 173
Poliolimnas, 12, 111
Polyborus, 10, 64
pomarea, Pomarea, 21, 221
Pomarea, 21, 221
Poospiza, 22, 235
Porphyrio, 12
Porphyriornis, 12, 112
portoricensis, Loxigilla, 22
Porzanula, 5, 12, 108
pretrei, Amazona, 15, 158
prigoginei, Nectarinia, 21
princeps, Accipiter, 10
principalis, Campephilus, 17, 182
pritchardi, Megapodius, 74
pritzbueri, Turdus, 20, 204
Procellaria, 9, 40
procerus, Hemignathus, 22, 246, 250
productus, Nestor, 14, 146
Prosobonia, 13, 132
Psammornis, 8, 31
Psephotus, 14, 146
psittacea, Psittirostra, 23, 250
Psittacula, 15, 152
Psittacus, 154
Psittirostra, 5, 23, 253
Pterocnemia, 8, 32
Pterodroma, 8, 37
Ptilinopus, 14, 143
pulcherrimus, Psephotus, 14, 146
punctata, Bowdleria, 20, 217
punctatus, Falco, 10, 64
purpurascens, Penelope, 75
pusillus, Accipiter, 10, 57
Pyriglena, 18, 185
pyrrhula, Pyrrhula, 23, 260
Pyrrhula, 23, 260
Pyrrhura, 15

quadricolor, Dicaeum, 21, 224
Quiscalus, 23, 257

radiolosus, Neomorphus, 16, 168
Rallus, 11, 102
Ramphocinclus, 19, 197
rapax, Aquila, 63
Raphus, 13, 134
ravidus, Turdus, 20, 213
rayneri, Platycercus, 147
razae, Calandrella, 18, 190
Razocorys, 190
remissa, Anas, 9
restricta, Cettia, 20
retrusus, Myadestes, 19
Rhea, 8, 32
Rhinoptilus, 13, 126
Rhipidura, 21, 222
rhodocorytha, Amazona, 16, 159
Rhodonessa, 9, 47
Rhynchopsitta, 15
Rhynochetos, 12, 113
richardsi, Dryocopus, 17, 182
ricordii, Chlorostilbon, 17
ridgwayi, Colinus, 11, 79
rodericana, Alectroenas, 14
rodericanus, Bebrornis, 20, 215
rodericanus, Necropsar, 24
rodericanus, Necropsittacus, 15
Rogibyx, 126
rolland, Rollandia, 34
Rollandia, 34
rossi, Thinornis, 127
rostrata, Streptopelia, 13, 137
rothschildi, Lampribis, 9
rothschildi, Leucopsar, 24, 269
rothschildi, Psammornis, 8, 31
rourei, Nemosia, 22
rowleyi, Eutrichomyias, 21
rubecula, Poospiza, 22, 235
rubra, Foudia, 23, 262
rubripes, Anas, 50

rufa, Loxops, 22, 244
rufa, Trichocichla, 20
rufescens, Bowdleria, 20, 217
ruficauda, Cinclocerthia, 19, 196
ruficollis, Tachybaptus, 33
rufifacies, Sceloglaux, 16, 170
rufipileus, Colaptes, 17, 181
rufocinerea, Grallaria, 18, 186
rufolavatus, Tachybaptus, 8, 33
ruki, Rukia, 21, 228
Rukia, 21, 228
rutha, Phaeornis, 20, 203
rutilus, Otus, 16, 169

sacer, Sarcoramphus, 10
sagittirostris, Loxops, 22, 240
Salpinctes, 19
Saltuarius, 8
salvadorii, Loriculus, 15, 151
sanctaeluciae, Cichlherminia, 20, 201
sanctaeluciae, Ramphocinclus, 197
sandvicensis, Gallinula, 12, 112
sandwichensis, Pennula, 5, 12, 110
sandwichensis, Pterodroma, 39
sanguinea, Himatione, 23, 255
Sapheopipo, 17, 184
Sarcoramphus, 10
Sarothrura, 12, 110
savesi, Aegotheles, 16
Saxicola, 19, 201
Sceloglaux, 16, 170
scops, Otus, 16
sechellarum, Copsychus, 19, 199
semiflava, Zosterops, 21, 227
semperi, Leucopeza, 22, 240
Serinus, 23
sharpii, Stictolimnas, 12
siemiradzkii, Carduelis, 23, 259
Siphonorhis, 17, 172
siquijorensis, Hypsipetes, 19
siquijorensis, Loriculus, 15, 151
solitaria, Pezophaps, 13, 134
solitarius, Raphus, 13
solitarius, Tinamus, 8
soumagnei, Tyto, 16, 168
spadicea, Hemiphaga, 14, 144
spectabilis, Chalcopsitta, 14
Speotyto, 16, 171
spixi, Gryphus, 173
spixii, Cyanopsitta, 15, 153
splendida, Zosterops, 225
stairi, Megapodius, 10, 74
Sterna, 13
Stictolimnas, 12
stokesii, Xenicus, 18, 187
strenua, Zosterops, 21, 225
strepera, Anas, 9
Strepera, 24, 274
Streptopelia, 13, 137
striata, Coracina, 18, 191
Strigops, 16, 164
strophium, Odontophorus, 11, 79
Struthio, 8, 27, 29
subflavescens, Cyanoramphus, 14, 147
subrufus, Erithacus, 19
Sula, 9, 41
sumichrasti, Hylorchilus, 19, 195
sumnerensis, Cygnus, 9
superciliosa, Ophrysia, 11, 81
sylvestris, Tricholimnas, 11, 105
Sylvia, 20, 216
Synallaxis, 18
syriacus, Struthio, 8, 27, 29
Syrmaticus, 11, 97

Tachybaptus, 8, 33
taczanowskii, Podiceps, 8, 34
Tadorna, 9, 46
tanagra, Turnagra, 21, 223
tarapacensis, Pterocnemia, 8, 32
Tauraco, 16, 167
temmincki, Tragopan, 93
teraokai, Ducula, 14
terrestris, Zoothera, 19

terrisi, Rhynchopsitta, 15
theodori, Anas, 9
Thinornis, 127
Thryomanes, 19, 195
Thryothorus, 19
tibetanum, Crossoptilon, 92
tibialis, Lorius, 14, 145
Tinamus, 8
toxopei, Charmosyna, 14
Tragopan, 11, 81, 93
traversi, Petroica, 20, 218
Trichocichla, 20
Tricholimnas, 11, 105
tricolor, Ara, 15, 154
tricolor, Rogibyx, 126
tricolor, Vanellus, 126
tridactyla, Jacamaralcyon, 17, 180
trigonostigma, Dicaeum, 21, 224
tristrami, Ptilinopus, 14, 143
troglodytes, Troglodytes, 19
Troglodytes, 19, 195
tropicus, Corvus, 24, 276
tuberosa, Crax, 77
tundrius, Falco, 10, 65
Turdus, 20, 204
Turnagra, 21, 223
Tympanuchus, 11, 78
Tyto, 16, 168

ulietanus, Cyanoramphus, 15, 149
ulietensis, Turdus, 20, 213
uncinatus, Chondrohierax, 10, 53
uvaeensis, Eunymphicus, 15, 149

Vanellus, 13, 126
variabilis, Xenicus, 18, 189
variegatus, Odontophorus, 80
varius, Fregilupus, 24, 268
varius, Parus, 21
varius, Psephotus, 146
Vermivora, 22, 236
versicolor, Amazona, 16, 160
versicolor, Columba, 13, 135
vestitus, Eriocnemis, 176
vinacea, Amazona, 16, 159
vinitinctus, Turdus, 20, 211
virens, Hemignathus, 249
virginianus, Colinus, 11, 79
Viridonia, 248
vitiensis, Columba, 13, 135
vittata, Amazona, 15, 157
vociferoides, Haliaeetus, 10, 53

wakensis, Rallus, 11, 104
wallichii, Catreus, 11, 92
wallicus, Pezoporus, 15, 149
wardi, Psittacula, 15, 152
watersi, Sarothrura, 12, 110
wellsi, Leptotila, 13, 142
wetmorei, Columba, 13
wilsoni, Bowdleria, 20
wilsoni, Hemignathus, 23, 248
wilsonii, Chondrohierax, 10, 53

xanthonotus, Oriolus, 24, 269
Xenicus, 18, 187
xenothorax, Leptasthenura, 18

yakushimensis, Turdus, 20
yarrellii, Eulidia, 6, 177
yarrellii, Myrtis, 177

zealandicus, Cyanoramphus, 15, 149
Zenaida, 13, 142
Zonotrichia, 22
Zoothera, 19
Zosterops, 21, 224